CHEMICAL ENGINEERING METHODS AND TECHNOLOGY

PHTHALATES

CHEMICAL PROPERTIES, IMPACTS ON HEALTH AND THE ENVIRONMENT

Chemical Engineering Methods and Technology

Additional books in this series can be found on Nova's website
under the Series tab.

Additional e-books in this series can be found on Nova's website
under the e-book tab.

Environmental Health - Physical, Chemical and Biological Factors

Additional books in this series can be found on Nova's website
under the Series tab.

Additional e-books in this series can be found on Nova's website
under the e-book tab.

CHEMICAL ENGINEERING METHODS AND TECHNOLOGY

PHTHALATES

CHEMICAL PROPERTIES, IMPACTS ON HEALTH AND THE ENVIRONMENT

GERARDO L. MORETTI
AND
DRAGO ROMANO
EDITORS

nova publishers

Nova Science Publishers, Inc.
New York

For permission to use material from this book please contact us:
Telephone 631-231-7269; Fax 631-231-8175
Web Site: http://www.novapublishers.com

LIBRARY OF CONGRESS CATALOGING-IN-PUBLICATION DATA

Phthalates : chemical properties, impacts on health, and the environment / editors, Gerardo L. Moretti and Drago Romano.
p. cm.
Includes index.
ISBN 978-1-62081-994-4 (hardcover)
1. Phthalate esters--Environmental aspects. 2. Phthalate esters--Health aspects. I. Moretti, Gerardo L. II. Romano, Drago.
TD196.P47P48 2012
363.738'4--dc23

2012018473

Published by Nova Science Publishers, Inc. † New York

CONTENTS

PREFACE

Phthalates are esters of phthalic acid and are a group of chemicals of high environmental relevance due to their production rates and ecotoxicological potential. Phthalates are mainly used as plasticizers in PVC plastics, wherein many consumer products contain specific members of this family of chemicals. In this book, the authors present current research on the chemical properties and impacts on health and the environment of phthalates. Topics include phthalate contamination in two estuaries from the Basque Coast; bioaccumulation and intracellular signal modifications of in vivo and in vitro models of phthalate esters; DEHP behavior in contact with liquid food simulants, blood, saliva and soil; removal of phthalates from the environment; gas phase fragmentation reactions of phthalate ions; the effects of DEHP exposure on brain development; and phthalate toxicogenomics and health effects.

Chapter 1 - Coordination polymers represent an area of growing interest in chemistry and material sciences due their unique properties and potential applications in fields such as catalysis, gas storage and separation, magnetism, photonics, and drug delivery. *ortho*-Phthalic acid is a versatile dicarboxylic ligand which has been widely used in the design of numerous coordination polymers. This chapter covers the progress made by this ligand in the formation of such inorganic-organic hybrid materials.

Chapter 2 - Phthalates are esters of the phthalic acid and are a group of chemicals of high environmental relevance due to their production rates and ecotoxicological potential. Phthalates are mainly used as plasticizers in PVC plastics. Therefore, many consumer products contain specific members of this family of chemicals, including building materials, household furnishings, clothing, cosmetics, pharmaceuticals, nutritional supplements, medical devices, dentures, children's toys, glow sticks, modelling clay, food packaging, automobiles, lubricants, waxes, cleaning materials and insecticides. Di-(2-ethylhexyl) phthalate (DEHP) is one of the most widespread phthalate plasticizer, used in numerous consumer products, commodities, and building materials. Due to its classification as toxic to reproduction, the corresponding labelling (skull and crossbones) and several restrictions on marketing and use (2003/36/EC) DEHP is replaced with other phthalate plasticizers as di-isodecyl phthalate (DIDP) and di-isononyl phthalate (DINP). With the total amount of plasticizer consumption remaining unchanged in Europe, the consumption of DEHP dropped from 42% in 1999 to 21% in 2006, whereas the share of DIDP and DINP increased from 35% to 60% within the same time period (Cardogan, 2007). Also other phthalates such as di-ethyl-phthalate (DEP), di-n-butyl phthalate (DBP), butyl benzyl phthalate (BBP), and di-n-octyl phthalate (DnOP) are widely used. As the phthalate plasticizers are not chemically bound to PVC, they can leach, migrate or evaporate into indoor air and atmosphere, foodstuff, other materials, etc. Once released to the environment they tend to adsorb on particles and, in this way, have been detected in different estuarine compartments as sediment, water or biota. The effects of phthalates on aquatic wildlife have been extensively studied. Phthalate compounds are weak xenoestrogenic compounds to aquatic organisms and can alter the reproduction of exposed organism. Furthermore, they can also modulate cellular pathways such as lipid metabolism leading to oxidative stress processes. Proliferation of peroxisomes and induction of peroxisomal enzymes such as catalase and acyl-CoA oxidase have been reported on fish and molluscs. Our work was developed in two selected estuaries from Bay of Biscay, the estuary of Bilbao and Urdaibai, showing different pollution loads. The estuary of Bilbao is one of the most industrialised and populated areas of the Bay of Biscay. Due to its geographic location, the area has been subjected to intense industrial pressure. For many years, urban and industrial effluents were discharged into the estuary without further treatment. In 1979, an Integral Plan for the Drainage of the estuary was approved and new environmental protection policies resulted in improved wastewater treatment systems. This, together with the closure of several industries, has produced a gradual recovery of the estuarine environment, with a continuing decrease in heavy metal concentrations and an increase in oxygenation. The tidal zone of the Ibaizabal River forms the estuary, although four tributaries (Kadagua, Asua, Galindo and Gobela) also discharge in the estuary.The docks and secondary channels are the most affected by urban and industrial wastes. The estuary of Urdaibai and its surroundings have been a Biosphere Reserve by UNESCO since 1984, and are located in a rural environment about 50 km east from the Ibaizabal estuary. The Urdaibai region is characterised by the basin of the Oka River, and by some shorter rivers, including the Golako, Mape, Artike and Laga. The estuary, some 13 km long and 500 m wide, is located in relatively close proximity to several industries (metallurgic, shipyards, dyes, cutlery manufacture) located in the surroundings of Gernika (population ~20,000) upstream the Oka River.

Chapter 3 - Phthalates, a class of chemicals used in a whole range of industrial applications, are considered to be ubiquitous global contaminants and endocrine disruptors (EDs). These hormone-mimicking compounds bioaccumulate within organ tissues in several animals, including fishes, amphibians and mammals with different and selective patterns of distribution. The endocrine disrupting effects are probably related to the absorption of these chemicals via the alimentary canal. In particular, phthalates have been seen to be stored in the oxyntic cells of the gastric tubular glands in fishes and amphibians. A relation between bioconcentration and organ functions was also observed. Moreover, *in vitro* studies on rat osteoblastic cultures demonstrated that benzyl butyl phthalate and di-n-butyl phthalate bioaccumulate, modify actin cytoarchitecture and exert mitogenic effects involving microfilament disruption and nuclear actin/lamin A regulation. On the other hand, primary mouse calvarial osteoblasts, treated with the above chemicals showed DNA base lesions with an increase of apoptotic markers induced by the p53-related pathway. These data strongly suggest that chronic exposure to phthalates could probably affect new bone formation and matrix deposition with clinical implications on bone homeostasis and mineral density. Understanding the way of action of phthalates through the study of autochthon or laboratory animals could reveal the mechanism by which these compounds, as EDs, regulate the fate of the organism.

Chapter 4 - Polyvinyl chloride (PVC) is a versatile material that can be formulated for various applications, ranging from every day products to highly specialized ones. In addition to the polymeric component, low molecular compounds such as stabilizers, lubricants, plasticizers... are added for processing and achieving the desired chemical and mechanical properties. The one used in the highest proportion is the plasticizer. Among the most commonly used plasticizers are phthalates which have been used in the production of flexible PVC for over 50 years (cable and wire covers, foodstuffs packaging- fresh meat, vegetables, cheese...-, blood bags, toys and children's items, ...). However, for the past 20 years, phthalates have come under considerable attention from media, legislative and environmental concerns.

Chapter 5 - Phthalates are synthesized in massive amounts, and widely used in building materials, home furnishings, transportation, clothing, paints, lacquers, cosmetics, food and medical products packaging. Release of phthalates into the environment during the manufacturing, use and disposal has caused serious concerns, since some of them are suspected to be mutagenic, hepatotoxic and carcinogenic. Especially, phthalates have been considered as endocrine-disrupting chemicals, and could interfere with reproductive system and behavior of humans and wild life through disturbance of the endocrine system even at a very low concentration level. The removal of phthalates from environment has attracted increasing attention in recent years. In this chapter, removal of phthalates from environment by various physical, chemical and biological techniques are reviewed, and special attention is paid to adsorption, advanced oxidation process and biodegradation of phthalates. The review provides an account on our research work combined with a reference of the literature published in the past ten years.

Chapter 6 - The widespread industrial use of phthalates as plasticizers led to major concern in exposure to phthalates; therefore, many analytical approaches based on mass spectrometry (MS) have been developed to analyze phthalates in various matrices. The MS of phthalates have been reported under various ionization modes including electron ionization (EI), positive chemical ionization (PCI), and negative chemical ionization (NCI). Understanding the gas phase chemical reactions under the various ionization conditions provides useful structural information. Under EI, PCI, and NCI modes, fragmentation mechanisms of selected phthalates were studied by collision induced dissociations (CID). Relatively low energy CID spectra with collision energy range of 5-10 eV provided useful dissociations for phthalates. The CID of the molecular ion or the protonated molecular ion peak of phthalates depending on the ionization mode showed multipathway fragmentations. Under EI and PCI conditions, two major pathways were identified as loss of alkoxy group and McLafferty rearrangement pathways. Ion structures and reactions along these pathways were studied by density functional theoretical (DFT) methods and showed important features of the ion structures. In many cases, both cyclic and open ion structures had to be considered because the benzene ring framework of phthalates allows the design of energetically stable cyclic ions. The DFT studies led to a better understanding of the mechanisms in terms of energetics and transition states involved along the fragmentation pathways. Under the EI mode, new strategies using computational modeling of cation-molecule complexes were developed for the well-known McLafferty rearrangement of phthalates leading to protonated phthalic anhydride at m/z 149. The DFT results allowed the identification of two types of complexes as: hyperconjugated carbocation-molecule and radical carbocation-molecule complexes. The EI ionization made possible to study McLafferty rearrangement of an odd-electron M+•; on the other hand, the PCI spectra showed McLafferty rearrangement product from an even electron $[M+H]^+$. The DFT results confirmed that McLafferty rearrangement is energetically feasible from either $M^{+\bullet}$ or $[M+H]^+$. Under methane PCI conditions, CID spectra of $[M+H]^+$, $[M+C_2H_5]^+$ and $[M+C_3H_5]^+$ of phthalates were used to design fragmentation mechanisms and to provide useful structural information. The PCI spectra showed the protonated phthalic anhydride as well as adducts of phthalic anhydride with $C_2H_5^+$ and $C_3H_5^+$. These adducts are very useful for structural identification purposes.

Chapter 7 - Phthalates are synthetic chemicals used primarily to increase the flexibility and the durability of plastic products, such as food storage containers, children's toys, and personal care products. Given the widespread use of these products, it is likely that exposure to phthalates is unavoidable in our daily lives. There has been increasing concern over the dangers of exposure to these compounds, especially in developing organisms as key organ systems, such as the brain, are undergoing rapid development at this time. Phthalates have anti-androgenic properties, meaning these chemicals interfere with the biological role of androgens (i.e., hormones critical for the masculinization of the brain) as well as suppress enzyme activity, mRNA expression, and protein levels important for brain masculinization and steroid synthesis. Interfering with androgens may directly or indirectly alter or prevent brain masculinization resulting in a more feminized version of the brain. A recent study has shown that phthalate exposure also affects estrogen synthesis in both males and females in addition to interfering with normal steroid homeostasis. Research investigating the effects of phthalates on brain development is limited. In a recent study, male rats were more vulnerable to the effects of phthalate exposure during development showing obvious structural deficits in the hippocampus (important for learning and memory), whereas brains of phthalate-exposed females were similar to control females. Researchers have also linked phthalate exposure to alterations in the dopaminergic pathway (important for movement and reward). The mechanisms of action underlying phthalate-induced changes in neurodevelopment have not been fully elucidated; however, it is likely that altered steroidogenesis may play a vital role. While research on neurodevelopmental effects of phthalate exposure is increasing, information is incomplete on the underlying neural and molecular mechanisms of phthalate toxicity in the brain. We recommend more extensive animal studies on phthalate exposure during vulnerable developmental periods when rapid structural and functional changes in the brain are taking place. Animal models will be crucial in establishing a causal link between phthalate exposure and changes in neuro-circuitry and in understanding the neural and molecular mechanisms which contribute to phthalate-induced changes in the developing brain.

Chapter 8 - Phthalate esters comprise a family of common industrial compounds, dialkyl- or alkyl/aryl esters of phthalic acid, which are used as plasticizers to impart flexibility to polyvinylchloride (PVC) plastics, as well as additives in many chemical and pharmaceutical formulations. They are present in a wide range of consumer products including building materials, food packaging, clothing, toys, medical devices, cosmetics, lubricant oils and solvents. Phthalates are not chemically bound to PVC and can leach out of plastics, which contributes to their ubiquitous presence in the environment. The industrial application as well as toxicity profile of phthalates is largely dependent upon the characteristics of the alkyl/aryl side chains of the ester molecule. Although phthalates usually display low acute toxicity, they have been focus of significant concern because of the evidence of reproductive adverse effects in experimental animals. In general, phthalates with medium- (e.g. dibutyl phthalate

[DBP]) or branched long-side chains (e.g. di(2-ehtylhexyl phthalate [DEHP]) are considered reproductive toxicants while those with short- (diethyl phthalate) or linear long-side chains (di-n-octyl phthalate) are less active. Results of transgeneration studies in rats indicate that the reproductive system of developing animals is particularly vulnerable to active phthalates, which are able to reduce endogenous hormonal production by the fetal testis. *In utero* exposure of rats to certain phthalates results in a spectrum of effects that are remarkably similar to reproductive disorders that comprise the human testicular dysgenesis syndrome, including cryptorchidism, hypospadias and low sperm counts. In addition, there are also evidences of endocrine disruption and reproductive effects in wildlife, including alterations in aquatic organisms. In the present review, we describe the reproductive effects and possible mode of action of major phthalates on laboratory animals, findings in wildlife and humans, as well as the relevance of these data to understanding the risks for human health and the environment.

Chapter 9 - Di-(2-ethylhexyl) phthalate (DEHP) is an ester of phthalic acid, which is mainly used as plasticizer in polyvinyl chloride (PVC) to increase its flexibility, elongation and processability. The main drawback, related to its low molecular weight, is the leaching out from flexible PVC (pPVC), which affects the polymer properties limiting its employ. Moreover, the use of phthalate-based plasticizers is being questioned worldwide because of their potential toxicity to humans and environment. Migration of such harmful substances from PVC can be avoided by substituting the unsafe plasticizer or by preventing its mass transfer from polymer matrix. In this communication, promising alternatives to reduce phthalate mass transfer are discussed in detail: the chemical cross-linking of flexible PVC and the use of chlorinated polyethylene (CPE) as physical barrier against DEHP leaching. Cross-linked flexible PVC samples were obtained using a difunctional amine, namely isophoron diamine (IPDA), as cross-linking agent while two different CPE resins were employed for the preparation of new CPE/pPVC compounds. The effect of cross-linking and the efficiency of CPE in preventing plasticizer leaching were evaluated through migration tests and supported by tribological analysis. Thermal, mechanical and dynamic-mechanical properties were also studied.

Chapter 10 - Phthalates are a group of similar phthalic acid esters and are widely used as plasticizers to soften and increase the flexibility in polyvinyl chloride plastics. Since phthalates are not covalently bound to the plastic matrix, they can easily leach into the surrounding environment. There is sufficient evidence in rodents that phthalate exposure causes developmental and reproductive toxicities. We have recently analyzed the interactions between 16 phthalates and genes/proteins in the Comparative Toxicogenomic Database (CTD; http://ctd.mdibl.org), and a total of 445 interactions between the five most frequently curated phthalates (DEHP/MEHP and DBP/BBP/MBP) and 249 unique genes/proteins were found. The GeneOntology, pathways and networks of these 249 unique genes/proteins were fully analyzed using the computer software programs from MetaCore analytical Suite, Ingenuity Pathways Analysis and Pathway Studio. The pathways and networks of the top 34 genes/proteins were found to be very similar to those of the 249 unique genes/proteins. Thus, the top 34 genes/proteins may serve as molecular biomarkers to assay toxicities of phthalate exposure. The top three phthalate toxicity categories were found to be cardiotoxicity, hepatotoxicity and nephrotoxicity, and the top 20 diseases included cardiovascular, liver, urologic, endocrine and genital diseases. Recently, rat neonatal cardiomyocytes treated with clinically relevant DEHP concentrations were independently shown to exhibit global changes in expression of genes associated with cell electrical activity, calcium handling, adhesion and microtubular transport, and these changes may explain the arrythmogenic effects of phthalates on these cardiomyocytes.

In: Phthalates
Editors: Gerardo L. Moretti and Drago Romano
ISBN: 978-1-62081-994-4

Chapter 1

PHTHALATE-BASED COORDINATION POLYMERS

***Svetlana G. Baca*[1] *and Silvio Decurtins*[2]**
[1]Institute of Applied Physics and Institute of Chemistry,
Academy of Sciences of Moldova, Chisinau, Moldova
[2]Department of Chemistry and Biochemistry,
University of Bern, Bern, Switzerland

ABSTRACT

Coordination polymers represent an area of growing interest in chemistry and material sciences due their unique properties and potential applications in fields such as catalysis, gas storage and separation, magnetism, photonics, and drug delivery. *ortho*-Phthalic acid is a versatile dicarboxylic ligand which has been widely used in the design of numerous coordination polymers. This chapter covers the progress made by this ligand in the formation of such inorganic-organic hybrid materials.

INTRODUCTION

Coordination polymers (CPs) or hybrid organic-inorganic polymers have attracted a tremendous attention in recent decades [see for example 1-8]. Built from metal ions, functioning as nodes and organic ligands, acting as bridges, the coordination polymers demonstrate a wide variety of unique properties. Possessing the unprecedented electrical, optical, magnetic, biological and catalytic properties, the CPs represent strategic and revolutionary materials that find an extensive use.

The fields of possible applications range from adsorption, storage, separation, and catalysis to luminescence, magnetism and conductivity. Moreover, with an increasing demand for alternative sources of energy, the fabrication of new CPs has been driven by the rapid growth of their practical use as materials for energy storage and conversion, storage and transport of hydrogen, methane and oxygen.

As mentioned by Batten *et al.* in their book [1] that provides a wide overview of major aspects of the research in the field of CPs "they have gone from polymeric rubbish of unknown composition to materials of the future". Thus, sustainable development and rational design and synthesis of such materials present an exciting challenge in chemistry.

To date, several approaches for the design and fabrication of such extended architectures have been introduced. One of them, so-named "serendipitous assembly" [9], relies on bridging metal ions by organic ligands with versatile coordination modes usually in "one-pot" synthesis involving mixing the metal salts [nitrates, sulfates, acetates, etc.] with organic ligands [carboxylates, N-containing ligands, etc.] in different solvents and CPs are formed by self-assembly of starting materials [from room temperatures up to hydro(solvo)thermal conditions]. This way has already led to a large diversity of coordination polymers. The other approach, the "bottom-up assembly" – firstly proposed by Hoskins and Robson [10] and further developed by the groups of Yaghi [11], Zaworotko [12], and others – is based on the assembling of molecular building blocks or pre-designed, pre-synthesized metal-containing building units and multidentate organic ligands into coordination networks. In such cases, varying the size, connectivity, charge and functionality of complex "building blocks" and/or ligands allows to achieve the targeted properties of the final product. In some cases, it may be accomplished through the combination of the properties of separate inorganic and organic components.

In both approaches the nature of bridging organic ligands plays a crucial role. For many years dicarboxylic acids that display several different coordination modes are intensively used as linkers in the construction of novel CPs. Among them, *ortho*-phthalic acid (H_2pht, 1,2-benzenedicarboxylic acid) is a versatile dicarboxylic ligand and has been widely employed for the preparation of coordination polymers. There are several reasons for the sustained interest in this ligand. Firstly, *o*-phthalic acid and its derivatives [e.g. phthalate ethers such as di-2-ethylhexyl phthalate (DEHP), diisodecyl phthalate (DIDP), diisononyl phthalate (DINP), and benzylbutylphthalate (BBzP)) and others] are utilized in a large variety of substances including perfumes, pharmaceuticals, dyes, synthetic fibers, plasticizers, gelling agent, lubricants, corrosion inhibitors. They also found extensive practical applications in building materials, personal care products, detergents, glues and even food products, toys, textiles, and medical devices. Secondly, the position of two carboxylic groups in *ortho*-position (ca. 60°) promotes a variety of bonding to metal centers and as a result the formation of diverse topological structures ranging from discrete mononuclear complexes to one-dimensional (1D) chains, two-dimensional (2D) layers and three-dimensional (3D) networks. In addition, the transition metal complexes of phthalic acid display promising biological activity and catalytic properties. For example, bis-phenanthroline copper(II) [13] and platinum(IV) [14] phthalate complexes were reported to have anticancer activity, zinc(II) cyclam phthalate [15] can be used as a potent anti-HIV agent, and a Mn(II)bis-phenanthroline complex of the phthalate exhibits fungicidal properties [16]. Copper(II) and cobalt(II) phthalate-based CPs with imidazole [17] were shown to effectively catalyze the oxidation of alkenes.

This chapter provides highlights of recent advances made by the phthalate ligand in the formation of one-, two- and three-dimensional CPs. In most cases ancillary bridging or terminal ligands with N- and/or O-donor atoms have been used for the construction of such CPs. The coordination versatility displayed by this ligand in the metal complexes is also discussed.

1. Coordination Modes of the Phthalate Ligand

As highlighted in the review by Porai-Koshits [18] on crystallochemistry and stereochemistry of monobasic carboxylate of transition metals, a carboxylic group has an extraordinary coordination ability and can coordinate to four metal centers displaying 19 different coordination modes, including *anti-anti*, *anti-syn* and *syn-syn* conformations. The presence of two carboxylic groups in *o*-phthalic acid contributes additional possibilities to bind metal atoms.

A detailed analysis of coordination modes of the phthalate ligand in the existing examples of its metal complexes retrieved from the Cambridge Structural Database (CSD version 5.32 + 4 updates) [19] provides 54 different coordination types [20, 21]. All coordination features of *o*-phthalic acid are shown in Scheme 1.

The phthalate ligand can coordinate up to seven metal centers at once showing a variety of coordination modes that the carboxylic groups are able to adopt: from monodentate, 1,3-chelating, and 1,6-chelating linking modes, and various bridging modes to the combinations of mono-, bi- and tridentate and chelating coordinations as shown in Scheme 1.

A large number of different factors such as metal identity, the number and nature of ancillary ligands involved, the presence of charge-balancing outer-sphere cations and anions in the reactions, different reaction conditions and others have influence on the coordination features displayed by this ligand.

However, most frequently the phthalate ligand displays the 1,6-bridging mode (type 5 in Scheme 1) which promotes the formation of CPs [20]. Other dominant modes of the coordination for the carboxylate groups of the phthalate ligand which appear in its complexes are the monodentate mode where only one carboxylic group is coordinated to a metal ion adopting a μ_1-η^1:η^0-monodentate mode (type 1 in Scheme 1), as well as the bidentate chelating, type 3, with carboxylate groups adopting μ_1-η^1:η^1-chelating modes to ligate to one metal atom, and finally the tridentate type (type 6 in Scheme 1) in which carboxylate groups adopt μ_1-η^1:η^0-monodentate and μ_1-η^1:η^1-chelating modes. The simultaneous coordination of the phthalate ligands to seven metal centers is quite rare but it was found in several two-dimensional coordination polymers: in homonuclear $[Tl_2(pht)]_n$ [22] and $\{[Li_4(pht)_2(EtOH)(H_2O)]\}_n$ [23] with the phthalate ligands adopting μ_7-η^4:η^2:η^2:η^3 and μ_7-η^2:η^2:η^2:η^1 coordination modes, respectively, and in heteronuclear $[M_2Co_2(pht)_2(OH)_2]_n$ [24] were the phthalate ligand exhibits a μ_7-η^2:η^3:η^3:η^1 coordination mode (for M = K) and a μ_7-η^2:η^2:η^3:η^2 coordination mode (for M = Cs) (types 51 and 53, Scheme 1).

2. Phthalate-Based Coordination Polymers

In 1967, Lindh described the synthesis and crystal structure of a mercury(I) phthalate [25] and it was the first reported phthalate-based coordination polymer in which the *o*-phthalate ligand executing a bridging function coordinated to two Hg(I) atoms, and further each of these units joined to two others by a short Hg...Hg bond with the formation of linear chains.

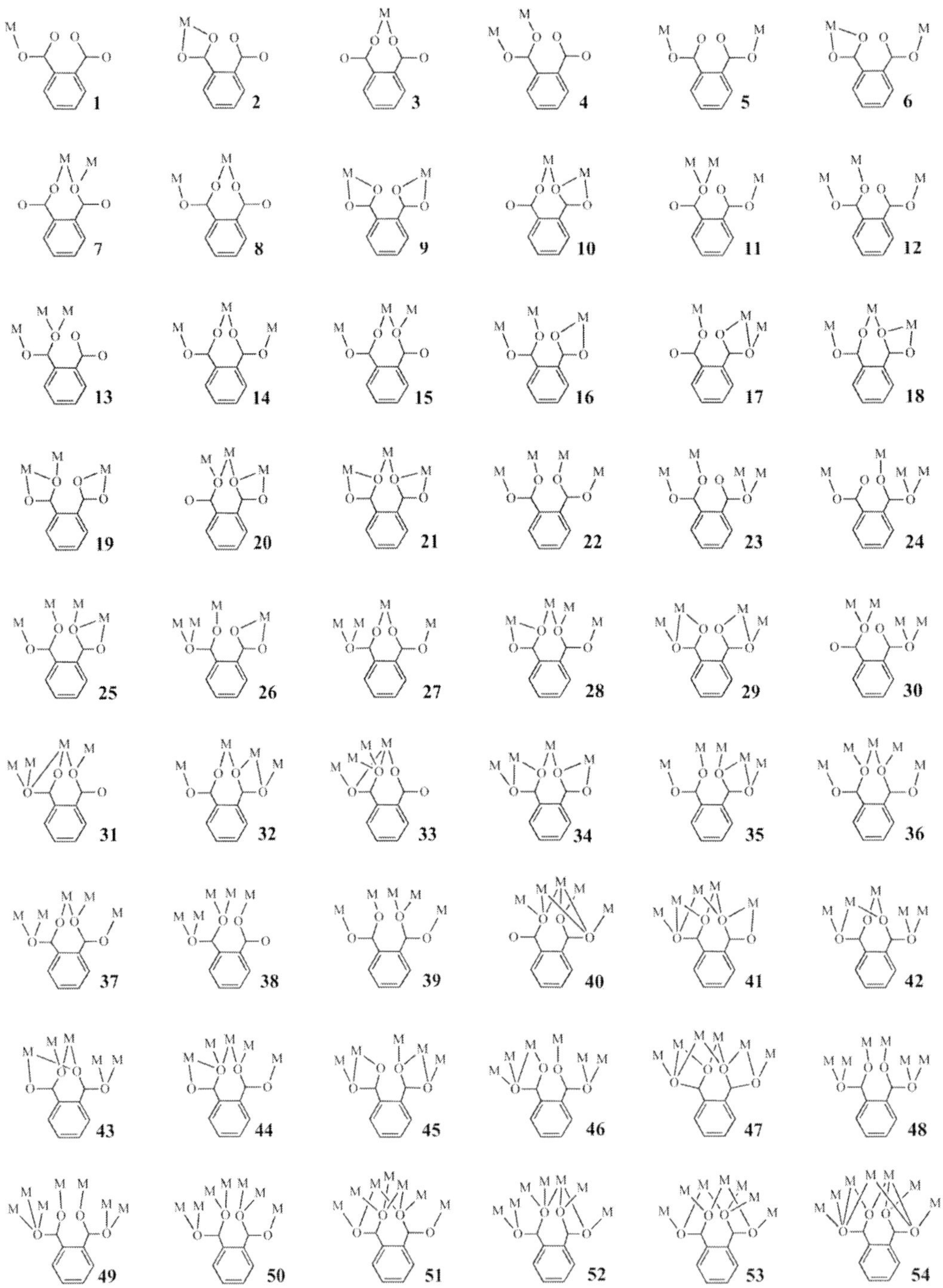

Scheme 1. Coordination modes displayed by the phthalate ligand in the metal complexes. The data were retrieved from the Cambridge Structural Database (CSD version 5.32 + 4 updates).

Several years later, Cingi and co-workers reported on a 1D diamine-copper(II) coordination polymer $[Cu(NH_3)_2(pht)]_n$ [26], followed by the synthesis and characterization of an extensive series of copper(II) phthalate coordination polymers containing different alkaline cations with general formula $\{M_2[Cu(pht)_2] \bullet nH_2O\}_n$ [n = 4, M = Li^+ [27]; n = 2, M

= Rb^+ [27], Cs^+ [27], Na^+ [28], K^+ [28]]. Here again, the phthalate ligand connects Cu(II) atoms into one-dimensional chains. Since the early reports, the bridging function of the title ligand has been extensively explored and numerous papers devoted to phthalate-based CPs have been published.

For developing multidimensional structures the phthalate ligand was often used in conjunction with other bridging and terminal polydentate ligands (Scheme 2). When the terminal *N,N'*-donor ligands stabilize low-dimensional coordination polymers by blocking the available metal coordination sites, for instance, *N,N'*-chelating 2,2'-bipyridine or 1,10'-phenanthroline with *endo*-topic coordination sites, the *N,N'*- or *N,O*-donor systems with an *exo*-topic functionality such as pyrazine, 4,4'-bipyridine, and others, it promotes the formation of both 2D and 3D networks.

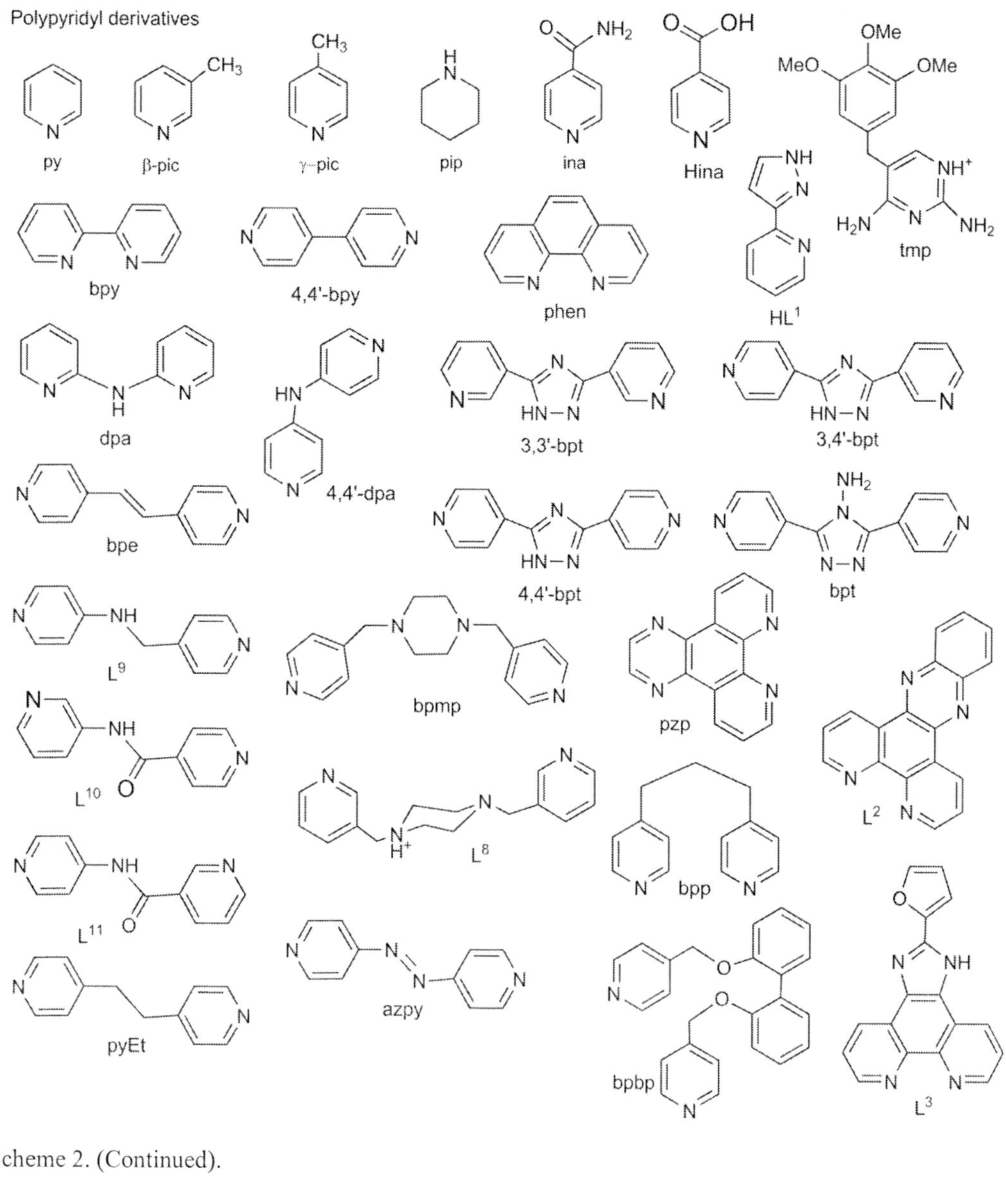

Scheme 2. (Continued).

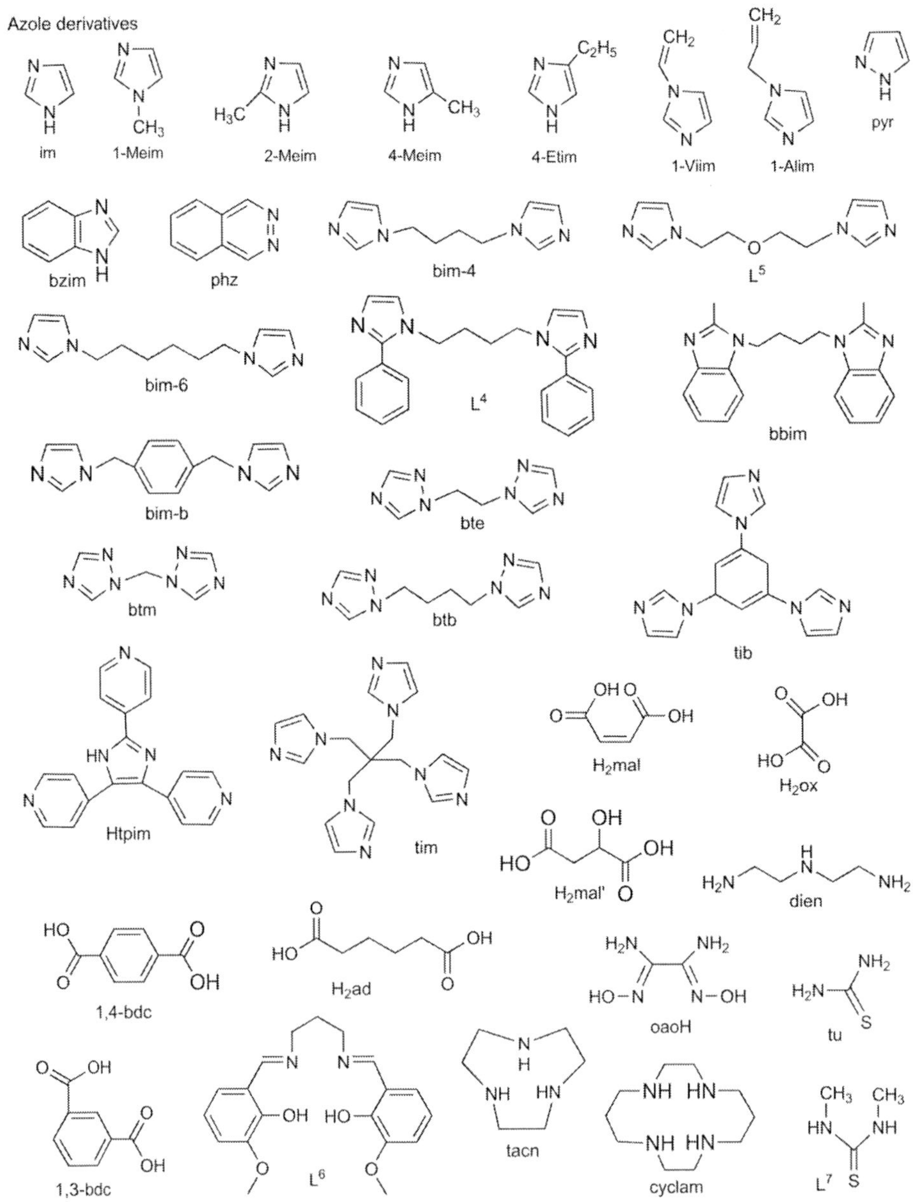

Scheme 2. Ancillary ligands discussed in this chapter.

1.1. One-Dimensional Coordination Polymers (1D)

As mentioned above, the first reported phthalate-based coordination polymers $[Hg_2(pht)]_n$ (1) [25] and $[Cu(pht)(NH_3)_2]_n$ (2) [26] display one-dimensional (1D) chain structures, where the phthalate ligands bridge metal atoms into infinite chains. Such a 1D arrangement of metal ions by the bridging phthalate ligand is the most favorable one and the resulting chains can be classified as linear, zigzag, helical, waving, ribbon and double-stranded chains, ladder-like and tube-like chains and chains composed of polynuclear clusters. Some representative examples of the one-dimensional phthalate-based chains are shown in Figures 1-5.

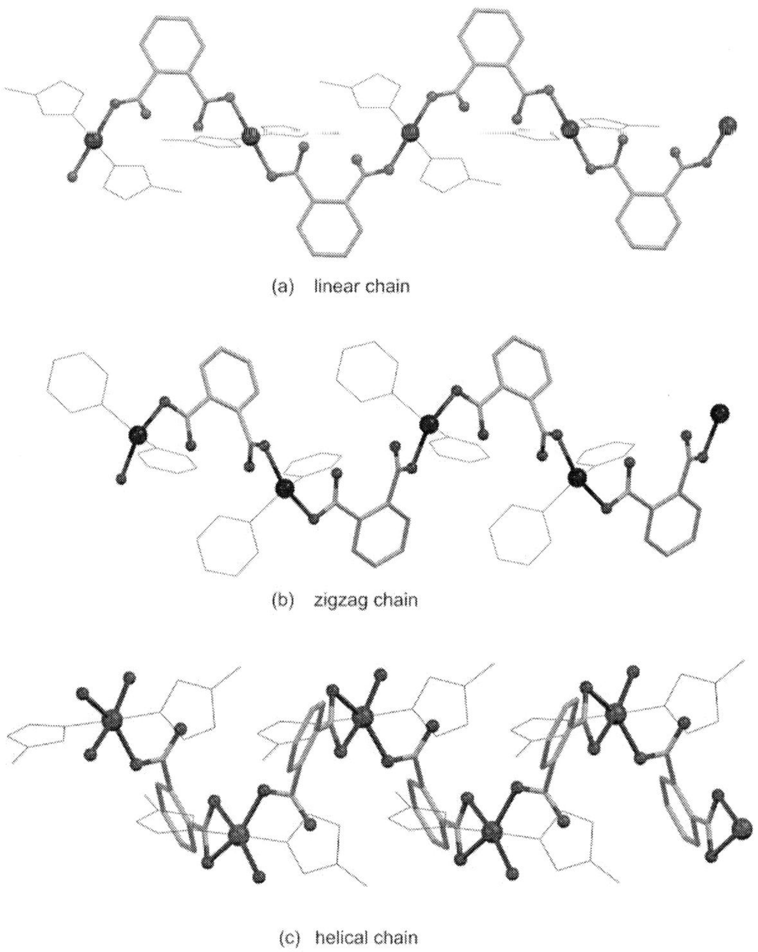

Figure 1. View of the linear (a), zigzag (b) and helical (c) chains in phthalate-based CPs (16 [40], 3 [31] and 19 [42]). Metal atoms, black spheres; O atoms, black balls; C atoms, grey sticks. Ancillary N-donor ligands are shown as thin black lines.

The main strategy in the design of 1D phthalate-based coordination polymers is to use a multi-bridging phthalate ligand and $[M(L)_n]^{2+}$ [where M = Mn^{2+}, Co^{2+}, Ni^{2+}, Cu^{2+}, Zn^{2+}, Cd^{2+}; L = additional N- and/or O-containing ligand and n = 1, 2] building blocks as the key elements. The construction of such 1D coordination polymers is influenced by different factors such as the nature of metal ions, the structural characteristics of ancillary organic ligands, the metal-acid-ligand ratio, the reaction conditions, and the presence of groups able to form non-covalent interactions such as hydrogen bonds and π-π stacking and others.

In general, cobalt(II), nickel(II), copper(II) and zinc(II) 1D coordination polymers based on *o*-phthalic acid and N-donor ligands such as pipyridine (pip), pyridine (py), pyrazole (pyr) or imidazole (im) and their derivatives [3-methylpyridine (β-pic), 4-methylpyridine (γ-pic), isonicotinamide (ina), 1-methylimidazole (1-Meim), 2-methylimidazole (2-Meim), 4-

methylimidazole (4-Meim), 4-ethylimidazole (4-Etim), 1-vinylimidazole (1-Viim) and 1-allylimidazole (1-Alim)] were prepared by the addition of *o*-phthalic acid and the corresponding amines in an aqueous or alcohol solution to the aqueous solution of metal salts. Some of them have also been prepared by dissolving the metal phthalates such as $[Co(H_2O)_6]$ • $(Hpht)_2$ [29] or $[Cu(Hpht)_2]$ • $2H_2O$ [30] in an excess of 1-methylimidazole (1-Meim) ligand. In some cases hydro(solvo)thermal conditions were used. Depending on the ratio of starting materials, different products were formed: $[M(pht)L_2]_n$ [where L = pip, py, pyr, β-pic, γ-pic, ina, im, 1-Meim, 2-Meim, 4-Meim, 4-Etim, 1-Viim and 1-Alim] and $[M(pht)L]_n$ [where L = γ-pic and 2-MeIm]. Thus, using a 1:1:2 molar ratio of the starting materials in the reaction of zinc(II) and copper(II) salts with pyridine and its derivatives leads to crystalline products with the formulae $[Zn(pht)L_2]_n$ [L = py (3); β-Pic (4); γ-pic (5) [31]], $[Cu(pht)L_2]_n$ [L = py (6) [32]; pip (7) [33]], $[Cu_2(pht)_2(\gamma\text{-pic})_4(H_2O)]_n$ (8) [34], and $\{[Cu(pht)(ina)_2]$ • $H_2O\}_n$ (9) [35] having 1D structures due to 1,6-bridging function of the dicarboxylic ligand. Both carboxylate groups adopt a μ_1-η^1:η^0-monodentate coordination mode in 3, 4, 8 and 9; the phthalate ligand also acts as a chelating one through O atoms from one carboxylate group displaying a μ_2-η^1:η^1:η^0:η^1 coordination mode in 6, whereas in 5 there are two different phthalate ligands – one pht ligand acts as bidentate 1,6-bridging, and the other behaves both as 1,6-bridging and 1,3-chelating. All coordination polymers 3–6, 8 and 9 have zigzag chains in their crystal structures as shown in Figure 1b, in which the $[ML_2]^{2+}$ building blocks are connected by phthalate bridges. Each metal atom is in a tetrahedral (3, 4) or square-planar (9) N_2O_2 donor atom environment which comprises two N atoms of two N-donor ligands and two carboxylate O atoms from two phthalates; square-pyramidal N_2O_3 in 6 arising from two N atoms and three carboxylate O atoms from different phthalates; and two independent metal atoms in 5 and 7 have different environments: for 5 - tetrahedral N_2O_2 and distorted square-pyramidal N_2O_3, and square-planar and trigonal bipyramidal geometries for 7. In a further related phthalate-based CP, $[Cu(pht)(pip)_2]_n$ (7) [33], the use of pipyridine in the synthesis led to the formation of 1D linear chains, with the $[Cu(pip)_2]$ building units linked through the bridging μ_2-η^1:η^1:η^1:η^1 phthalate ligand. Here, each Cu(II) ion is in a distorted octahedral coordination geometry being coordinated by two N atoms belonging to two pip molecules along with two O atoms of two different phthalate bridges all occupying the equatorial positions, whereas the two remaining O atoms of the same carboxylate groups occupy the axial positions. Antiferromagnetic interactions between the copper(II) ions were found in 7, and the strength of the magnetic exchange interactions depends mainly on the topology of the bridging phthalate ligand [33].

The use of heterocyclic diazole, pyrazole (pyr), imidazole (im), and their derivatives in similar reactions gave rise to a larger diversity of 1D chains depending on the nature of metal(II) ions and other factors. As a representative example, the phthalate-based CPs with imidazole of the general formula $[M(pht)(im)_2]_n$ [M = Co^{2+} (10) [17, 36]; Cu^{2+} (11) [17], (12) [37]; Zn^{2+} (13) [20, 38]] have all a one-dimensional motif consisting of $[M(im)_2]^{2+}$ units connected by phthalate ligands. However, the one-dimensional motif in CPs 10, $[Co(pht)(im)_2]_n$, and 13, $[Zn(pht)(im)_2]_n$, is a zigzag chain (Figure 1b), in 11, $\{[Cu(pht)(im)_2]$ • $1.5H_2O\}_n$, it is a helical chain as shown in Figure 1c, whereas in 12, $[Cu(pht)(im)_2]_n$, linear chains are formed (Figure 1a). Each Co(II) and Zn(II) atom in CPs 10 and 13 has a tetrahedral N_2O_2 geometry, being coordinated by a pair of imidazole ligands and a pair of monodentate carboxylate groups. Here, the phthalate ligand adopts the usual μ_2-η^1:η^0:η^0:η^1 coordination mode. In contrast, two imidazole and two carboxylate groups from two different phthalates are coordinated to each Cu(II) center in 11 and 12 to provide an approximately square planar N_2O_2 geometry. Additionally, copper(II) atoms form long contacts of 2.559–2.894 Å in 11 and 2.727–3.081 Å in 12 with uncoordinated O atoms of carboxylic groups, thus the Cu(II) atom formally is situated in an octahedral environment. Consider that the pht^{2-} ligand adopts a μ_2-η^1:η^1:η^0:η^1 and μ_2-η^1:η^1:η^1:η^1 coordination fashion in 11 and a μ_2-η^1:η^1:η^1:η^1 coordination mode in 12. Moreover, solvate water molecules present in the structure of 11 promote the formation of a 3D hydrogen-bonded network compared to the 2D hydrogen-bonded network found in 12. Higher temperature and hydrothermal condition which were applied for 12, in comparison to 11, which was prepared under reflux condition, can also influence the formation of the final product. Further, the oxidative catalytic behavior of 1D CPs 10 and 11 has been examined by Baca *et al.* [17] in the reaction of cyclohexene oxidation using the hydrogen peroxide as oxidizing agent, and the CPs demonstrate good catalytic activities. Ferroelectric measurements reveal that the coordination polymer 12 could be a promising candidate in the development of magnetically controllable ferroelectric materials [37]. Using a pyrazole ligand gave a zigzag 1D chain structure $[Zn(pht)(pyr)_2]_n$ (14) [39] with $[Zn(pyr)_2]^{2+}$ units connected by 1,6-bridging phthalates.

An extensive series of 1D phthalate-based coordination polymers with the derivatives of the imidazoles [1-methyl- [40], 2-methyl- [20, 41], 4-methyl- [42], 4-ethyl- [43], 1-vinylimidazole [44, 45] and 1-allylimidazole [46], and benzimidazole [47-49]] has been developed. The reaction of copper(II) or cobalt(II) phthalate complexes such as $[Co(H_2O)_6](Hpht)_2$ and $[Cu(Hpht)_2]$ • $2H_2O$ and *o*-phthalic acid with excess of 1-Meim in methanol solution under refluxing conditions resulted in the formation of two new coordination polymers $[Co(pht)(1\text{-Meim})_2]_n$ (15) and $[Cu(pht)(1\text{-Meim})_2]_n$ (16) [40]. Using zinc(II) acetate in a similar fashion $[Zn(pht)(1\text{-Meim})_2]_n$ (17) [40] has been prepared. Analogously to imidazole-phthalate CPs 10–13, the CPs with the derivatives of imidazoles all also exhibit 1D chain structures consisting of $[M(1\text{-Meim})_2]^{2+}$ units connected by bridging phthalate ligands. However, the one-dimensional motif in cobalt(II) and zinc(II) complexes 15 and 17 (the CPs are isostructural) is the helical chain as shown in Figure 1c with a pitch of 25.7 Å, whereas in copper(II) coordination polymer 16 linear chains are formed (Figure 1a). One helical pitch in 15 and 17 includes five metal ions, which are separated by 6.943 and 6.985 Å [two independent metal atoms are presented in the structures] and four phthalate bridges. The metal atoms are tetrahedrally N_2O_2 coordinated to a pair of pht ligands and a pair of 1-Meim molecules, whereas in complex 16, each Cu(II) center adopts a square coordination mode of N_2O_2 type by two O atoms from different phthalate ions and two N atoms of 1-Meim. There are only van der Waals interactions between the linear chains in 16, while the helical chains of complexes 15 and 17 are connected by C–H...O interactions, which lead to the formation of a supramolecular 3D structure. Slow evaporation of the methanol solution of $[Co(H_2O)_6](Hpht)_2$ and 2-methylimidazole under room temperature afforded a new 1D zigzag CP, $[Co(pht)(2\text{-Meim})_2]_n$ (18) [41]. When 4-methylimidazole reacted with metal salts and *o*-phthalic acid or potassium phthalate under heating conditions, the helical CPs with general formula $[M(pht)(4\text{-Meim})_2(H_2O)]_n$ [M = Co^{2+} (19); Cu^{2+} (20)] have been isolated by Baca *et al.* [42]. Each Co(II) atom in 19 has a distorted octahedral N_2O_4 coordination geometry being coordinated by two N atoms from two 4-Meim, three O atoms belonging to the two phthalate moieties and a water molecule. The pht ligand adopts a μ_2-

$\eta^1:\eta^1:\eta^0:\eta^1$ coordination mode using one carboxylate group to coordinate to the Co(II) atom in a monodentate fashion and the other carboxylate group to ligate the adjacent metal in a 1,3-chelating mode (Figure 1c). In 20, each Cu(II) atom has a square-pyramidal N_2O_3 coordination geometry with two N atoms of 4-Meim, two O atoms of two pht ligands and a water molecule. The pht ligand executes an usual 1,6-bridging function displaying a μ_2-$\eta^1:\eta^0:\eta^0:\eta^1$ mode. Liu and co-workers reported 1D phthalate-based CPs with 4-ethylimidazole, $[Zn(pht)(4\text{-}Etim)_2]_n$ (21) [43], 1-vinylimidazole, $[Zn(pht)(1\text{-}Viim)_2]_n$ (22) [44], and $[Cu(pht)(1\text{-}Viim)_2]_n$ (23) [45], and 1-allylimidazole, $[Zn(pht)(1\text{-}Alim)_2]_n$ (24) [46], prepared by the reaction of metal chlorides with *o*-phthalic acid and the corresponding heterocyclic diazole. In CPs 21–24, each metal atom is coordinated by two N and two O atoms in a distorted tetrahedral geometry (for Zn) or in a distorted square-planar geometry (for Cu) resulting in a 1D zigzag motif for all of them.

Cui and co-workers [47] have employed benzimidazole (bzim) as an ancillary ligand in the hydrothermal synthesis with *o*-phthalic acid and cadmium acetate to form a helical coordination polymer $[Cd(pht)(bzim)_3]_n$ (25). In CP 25, each Cd(II) atom is coordinated by three N atoms from three bzim ligands and four O atoms from two bridging pht ligands and adopts a distorted pentagonal-bipyramidal geometry. Other 1D phthalate-based CPs with this co-ligand is a zigzag coordination polymer $\{[Cd(pht)(bzim)_2(H_2O)] \bullet 0.5H_2O\}_n$ (26) [48] and a linear coordination polymer $[Co(pht)(bzim)_2]_n$ (27) [49], where only two bzim coordinate to the metal atom. The bridging pht ligand adopts different coordination modes: μ_2-$\eta^1:\eta^1:\eta^1:\eta^1$ in 25, μ_2-$\eta^1:\eta^1:\eta^0:\eta^1$ in 26, and μ_2-$\eta^1:\eta^0:\eta^0:\eta^1$ in 27.

As noticed previously, the use of a different metal: acid: ligand ratio permitted to isolate different species. For instance, the application of 4-methylpyridine (γ-pic) and 2-methylimidazole (2-MeIm) in the reaction of zinc(II) salt with *o*-phthalic acid in the molar ratio of 1:1:1 leads to 1D coordination polymers $[Zn(pht)(\gamma\text{-}pic)]_n$ (28) [31] and $[Zn(pht)(2\text{-}Meim)]_n$ (29) [20], in which the *o*-phthalic anion behaves as a tetra- and tridentate ligand, respectively. In the CP 28, one phthalate ligand adopts a μ_3-$\eta^1:\eta^1:\eta^1:\eta^1$ coordination mode (type 16 in Scheme 1), whereas other phthalate coordinates in a μ_3-$\eta^1:\eta^1:\eta^2:\eta^0$ binding mode (type 15 in Scheme 1). As a result, the 1D double-stranded chains with alternation of 6-membered and 11-membered metallocycles are formed in 28 as shown in Figure 2a. The Zn...Zn distance in the smaller 6-membered cycle is 3.470 Å and in the 11-membered ring this distance equals to 4.467 Å. The double-stranded chain in 28 includes two independent Zn(II) atoms with different coordination environments. One Zn(II) atom is coordinated by one N atom of γ-pic and three O carboxylate atoms from three different pht ligands to form a NO_3 tetrahedron, while the coordination environment around the other Zn(II) atom corresponds to a distorted tetragonal bipyramid (NO_5) with one N atom from a 4-methylpyridine and five O atoms from three pht ligands. In 29, the phthalate ligand behaves as a tridentate ligand and adopts a μ_3-$\eta^1:\eta^1:\eta^0:\eta^1$ coordination mode that promotes the formation of 8-membered and 14-membered rings: one of the zinc(II) pairs is linked by a 1,6-bridge of two pht to form a 14-membered ring with a Zn...Zn distance of 4.195 Å, while another zinc(II) pair is held together by 1,3-bridging mode from the same phthalates resulting in a smaller 8-membered ring with a shorter Zn...Zn distance of 3.594 Å (Figure 2b). Each Zn(II) atom in 29 is in a tetrahedral NO_3 environment being coordinated by one N atom from a 2-Meim and three O atoms from three phthalate ligands. The ancillary N-donor ligands, 4-methylpyridine in 28 and 2-methylimidazole in 29, lie on one side of the infinite chains in parallel fashion with strong aromatic π-π stacking interactions at a centroid-centroid distance of 3.583 Å between the γ-pic pairs and 3.596 and 3.530 Å between 2-Meim heterocycles. Alternatively, the double structural motifs with the repeating units of only 14-membered metallocycles were achieved in several copper(II) phthalate-based CPs with the general formula $\{M_2[Cu(pht)_2] \bullet nH_2O\}_n$ [M = Li^+ (30), Rb^+ (31), Cs^+ (32), K^+ (33)] synthesized by Cingi and co-workers [27, 28] without any ancillary organic ligands by dissolving copper carbonate hydroxide in an aqueous solution of the alkaline hydrogen phthalate, as well as in Mn(II) and Cu(II) hydrogen phthalate CPs, $[M(Hpht)_2(H_2O)_2]_n$ [M = Mn^{2+} (34) [50], Cu^{2+} (35) [51]], obtained from an aqueous solution of metal carbonate and an excess of *o*-phthalic acid. Note, that in 30–33 the anionic $[Cu(pht)_2]^{2-}$ chains are formed, whereas in CPs 34 and 35 the 1D chains are neutral. Each copper atom adopts a square-planar geometry being coordinated by four O atoms from four phthalate groups that act in the usual 1,6-bridging function giving rise to the formation of only 14-membered metallocycles with the Cu...Cu distance in the range 5.744 – 6.514 Å in 30–33.

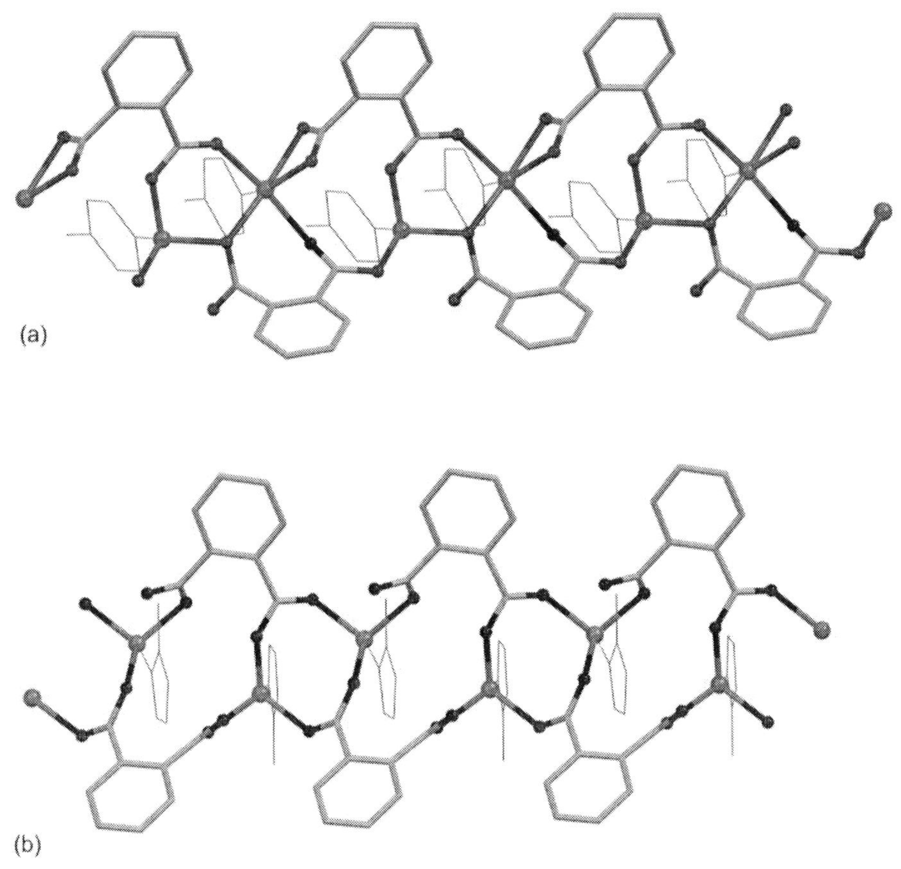

Figure 2. View of the double-stranded chains with alternation of (a) 6- and 11-membered metallocycles and (b) 8- and 14-membered metallocycles in the phthalate-based CPs (28 [31] and 29 [20]). Metal atoms, black spheres; O atoms, black balls; C atoms, grey sticks. Ancillary N-donor ligands are shown as thin black lines.

The alkaline cations differing in radius and coordination abilities have influence on the polymeric motifs formed in the CPs: the smallest Li^+ cations involve in their coordination spheres two O carboxylate atoms from two phthalates belonging to the same chain and thus promote the formation of linear double chains in 30, whereas other alkaline ions interpose between two chains and are surrounded by eight O atoms from four carboxylates and contribute to the generation of zigzag double chains.

The metal atoms, Mn(II) in 34 and Cu(II) in 35, are octahedrally coordinated to six O atoms, four from the phthalate groups and two from water molecules. A pair of pht ligands connects neighboring metal(II) centers *via* a single-bridging coordination mode with the formation of a 14-membered metallocycle with the M...M distance being 5.158 Å (Mn) and 5.111 Å (Cu), to afford the zigzag double chains in both CPs. The presence of other cations in the synthesis such as propane-1,3-diammonium (pnH_2) or trimethoprim (tmp, 2,4-diamino-5-(3′,4′,5′-trimethoxy benzyl) pyrimidine), has also produced anionic $[M(pht)_2]^{2-}$ chains composed of 14-membered cycles [M…M distance 5.453 (Zn), 5.825 Å (Cu)] in a Zn(II) phthalate-based CP $\{(pnH_2)_2[Zn(pht)_2]\}_n$ (36) reported by Padmanabhan *et al.* [52] and Cu(II) phthalate-based CP $\{(tmp)_2[Cu(pht)_2]\}_n$ (37) described by Muthiah *et al.* [53]. The cations disposed between the metal-phthalate chains interact with them through extensive hydrogen bonding. Metal atoms adopt a tetrahedral O_4 (Zn) or a distorted octahedral O_6 (Cu) coordination geometries with carboxylate O atoms from phthalates.

An extensive part of the research in the developing of 1D phthalate-based CPs has been devoted to the use of polypyridyl-like ligands such as 2,2′-bipyridine (bpy), 1,10′-phenanthroline (phen) and their derivatives as co-ligands. Numerous 1D CPs with the general formula $[M(pht)(bpy)(H_2O)_x]_n$ [M = Co^{2+} , Cu^{2+}, Zn^{2+}, Cd^{2+}; x = 1, 2] and $[M(pht)(phen)(H_2O)_x]_n$ [M = Mn^{2+}, Co^{2+}, Ni^{2+}, Cu^{2+}, Zn^{2+}, Cd^{2+}; x = 0.5, 1, 2] were synthesized. The most common one-dimensional motif seen in such CPs is a helical chain (Figure 1c). Thirumurugan and Rao [54] reported a series of cadmium(II) and zinc(II) phthalate-based 1D helical CPs of the formula $[Cd(pht)(bpy)(H_2O)]_n$ (38), $[Cd_2(pht)_2(phen)_2(H_2O)]_n$ (39) and $\{[Zn(pht)(phen)(H_2O)] \bullet H_2O\}_n$ (40) prepared by the hydrothermal reaction of cadmium acetate or zinc acetate with H_2pht acid, bpy or phen in the presence of the base, piperidine for 38 and NaOH for 39 and 40, in water at 180°C during 72 h. The 1D CPs 38 and 40 consist of one-dimensional helical chains formed by the metal building units, $[Cd(bpy)(H_2O)]$ (38) or $[Zn(phen)(H_2O)]$ (40), linked by the bridging phthalate ligands. Each metal atom is surrounded by two N atoms from a diimine ligand and four O atoms, of which three come from two phthalate groups and one from a water molecule; thus, the six-coordinated M(II) atom adopts a distorted octahedral N_2O_4 geometry. In 39, the one-dimensional motif resembles an exotic helical ribbon-like chain: four Cd(II) atoms are bridged by four phthalate groups to form a 28-membered metallocycle and the adjacent metallocycles are further linked by a carboxylate oxygen giving rise to a four-membered ring. The ribbon-like chains are stabilized by intra- and interchain π-π stacking interactions between the coordinated phen molecules at distances of 3.476–3.651 Å. The pht ligand adopts both μ_2-η^1:η^1:η^0:η^1 and μ_3-η^0:η^1:η^1:η^2 coordination modes. Other reaction conditions and lower temperature can also lead to similar helical chains which were found in several cobalt(II), zinc(II) and cadmium(II) CPs with these co-ligands. Thus, the reaction of cobalt carbonate with *o*-phthalic acid and bpy under refluxing conditions afforded a helical coordination polymer $[Co(pht)(bpy)(H_2O)]_n$ (41) [56] which co-crystallized with a mononuclear complex [Co(pht)(bpy)(H_2O)$_3$] • 3H_2O. It is interesting to note that the interaction of the similar starting materials but at 80°C, generated a hydrogen-bridged dinuclear complex [(bpy)$_2$Co(pht)H(pht)Co(bpy)$_2$](Hpht)(H_2pht) • 2H_2O [56]. The dinuclear phthalate complex consists of two [Co(pht)(bpy)$_2$] units which are connected by a very strong hydrogen bond (2.447 Å) between carboxylate groups from pht ligands into a dimer. A similar hydrogen-bonded dinuclear Zn(II) compound has also been reported [57]. The reaction of Co(NO_3)$_3$ • 6H_2O, potassium hydrogen phthalate and KOH with phen in ethanol/water solution at 60°C in an autoclave for 6 hours yielded a helical phenanthroline-phthalate CP, {[Co(pht)(phen)(H_2O)] • EtOH}$_n$ (42) [58], while heating of the methanol/water solution containing [Zn(bpy)](NO_3)$_2$ and sodium hydrogen phthalate at 55°C for 8 h gave another helical [Zn(pht)(bpy)(H_2O)]$_n$ (43) CP [59]. Hence, each metal atom in 41–43 is bonded to a chelating diimine ligand, two phthalate groups [one carboxylate group of pht coordinated to a metal ion in a chelating mode and another in a monodentate fashion] and a water molecule showing a distorted octahedral N_2O_4 geometry.

Further, it was found that 1D phthalate-based CPs with heteroaromatic N-donor ligands can also display linear, zigzag, waving or double-stranded structures. Thus, the reaction under room temperature generated a zigzag phthalate-based copper(II) CP with phen, {[Cu(pht)(phen)(H_2O)] • 0.5H_2O}$_n$ (44), firstly prepared by Krstanovic *et al.* in 1985 [60] by diffusion of an aqueous solution of sodium diaquadiphthalatocuprate(II) into an ethanol/water solution of copper nitrate and phen and later synthetized by Bi *et al.* [61] by the crystallization under room temperature from ethanol/water solution containing copper sulfate, potassium hydrogen phthalate and phen in a molar 1:1:2 ratio. Another phenanthroline-phthalate copper(II) CP containing one solvent water molecule {[Cu(pht)(phen)(H_2O)] • H_2O}$_n$ (45) [62, 63] with similar zigzag chains has also been prepared by crystallization from an aqueous solution of copper chloride, H_2pht and phen at pH = 4 adjusted with aqueous ammonia under room temperature. The Cu(II) atoms in 44 and 45 adopt a square pyramidal N_2O_3 geometry in which two N atoms of chelating phen and two O atoms from different pht ligands construct the equatorial plane, and a coordinating water molecule occupies the apical position. The same coordination environment was found in a zigzag phenanthroline-phthalate nickel(II) CP, {[Ni(pht)(phen)(H_2O)] • H_2O}$_n$ (46), prepared by Tian *et al.* [64]. It can be noted, that a week antiferromagnetic interaction between Ni(II) ions has been reveled in 46.

Utilization of the 2,2′-bipyridyl ligand has resulted in linear, zigzag, waving and double-stranded 1D CPs. The bipyridine-phthalate {[Cu(pht)(bpy)(H_2O)] • 3.5H_2O}$_n$ (47) CP described by Li *et al.* [65] was prepared under room temperature from ethanol-water solution of copper(II) perchlorate hexahydrate, H_2pht acid and bpy in equimolar 1:1:1 ratio at pH = 5. Here, the [Cu(bpy)(H_2O)] units are linked by the 1,6-bridging phthalate ligands into a waving chain. The copper(II) atoms display a square pyramidal coordination geometry. The base plane of the square pyramid is provided by two N atoms from bpy and two O atoms from different phthalates. The apical position of the square pyramid is occupied by a water molecule. Prelesnik *et al.* [66] have reported a double-stranded 1D phthalate-based CP, {[Cu_2(pht)$_2$(bpy)$_2$(H_2O)] • 3H_2O}$_n$ (48), with the same ligand. Crystallization from an aqueous solution of copper nitrate, bpy and sodium phthalate at a pH of approximately 2.5 under room temperature yielded CP 48. One phthalate in 48 behaves as a bidentate ligand bridging two Cu(II) atoms through O atoms from both carboxylate groups and adopts a μ_2-η^1:η^0:η^0:η^1 mode. The other phthalate acts as a tridentate ligand bridging three Cu atoms, one of which belongs to an adjacent chain in a μ_3-η^1:η^1:η^0:η^1 feature with the formation of double-stranded chains. As a result, the double-stranded chains consist of the alternating 14-membered and 22-membered metallocycles. The copper(II) atoms adopt a square pyramidal geometry. Another related zigzag bipyridine-phthalate CP with zinc(II), {[Zn_2(pht)$_2$(bpy)$_2$(H_2O)] • H_2O}$_n$ (49), reported by Song *et al.* [67] was synthesized by hydrothermal reaction of zinc acetate, H_2pht and bpy. In 49, there are two crystallographically unique Zn(II) atoms: one Zn atom is five-coordinated and displays a distorted trigonal bipyramidal environment completed by two N atoms of bpy, two carboxylate O atoms and one aqua ligand, whereas another Zn atom is four-coordinated by two N atoms from bpy and two O atoms from two bridging pht groups adopting a distorted tetrahedron geometry. The pht ligand shows an usual 1,6-bridging mode in 49. Liu *et al.* [68] and later Zhou *et al.* [69] have reported a 1D phthalate-based manganese(II) CP, [Mn(pht)(bpy)(H_2O)$_2$]$_n$ (50), in which [Mn(bpy)(H_2O)$_2$] units are linked by the 1,6-bridging phthalate into linear chains. Both, refluxing of a EtOH/H_2O mixture of manganese acetate, H_2pht acid, potassium hydroxide and bpy taken in a 1:1:2:1 ratio for 4 hours [68] and the reaction of dichloro bisbipyridine manganese and *o*-phthalic acid in EtOH/H_2O solution at 60°C for 12 hours [69] afforded the same product. Each Mn(II) atom is in a distorted octahedral N_2O_4 geometry being coordinated by two N atoms of a chelating bpy and four O atoms, two of them from phthalate groups and other two from two aqua ligands. All coordinated bpy ligands are situated on one side of the linear chain. Weak antiferromagnetic interactions between Mn(II) ions mediated through phthalate ligands have been found in 50. A similar linear phthalate-based CP with cadmium(II) ions, [Cd(pht)(bpy)(H_2O)$_2$]$_n$ (51), has also been reported [70].

Diverse derivatives of 2,2′-bipyridine and 1,10′-phenanthroline chelating ligands have been proven to be also good candidates as terminal ligands for obtaining low-dimensional phthalate-based CPs. Bu *et al.* [71] used the chelating bipyridine-like 3-(2-pyridyl)pyrazole ligand (HL^1) in the hydrothermal synthesis with metal salts, H_2pht acid and NaOH at 140°C for two days to produce two new one-dimensional CPs, {[Zn_2(pht)(L^1)$_2$(H_2O)]$_n$ (52) and {[Cd_2(pht)$_2$(HL^1)$_2$] • H_2O}$_n$ (53). Slow evaporation of the mother liquor of CP 53 under room temperature gave {[Cd(pht)(HL^1)(H_2O)]$_n$ (54). In all these CPs, the phthalate ligand shows a great variety in its coordination modes to metal centers that results in the different structural 1D motifs formed. In CP 54, the phthalate ligand links [Cd(HL^1)(H_2O)] units in an usual 1,6-bridging mode adopting a bidentate-chelating μ_2-η^1:η^1:η^0:η^1 coordination mode to generate linear 1D chains. In CP 52, four Zn(II) atoms are connected through four L^1 to form a tetrazinc cluster, the phthalate ligand displaying a μ_4-η^2:η^1:η^1:η^1 coordination mode (type 26 in Scheme 1) coordinates *via* one of its carboxylate group to three Zn atoms from the tetrazinc cluster and further bounds to another Zn atom from the adjacent tetranuclear zinc cluster by another carboxylate group to propagate a 1D cluster-based CP as shown in Figure 3a. Within the tetranuclear zinc cluster there are two crystallographically unique Zn(II) atoms. Both metal atoms are six-coordinated; one of them is surrounded by four N atoms from two L^1 ligands and two O atoms from two different phthalates adopting a distorted octahedral N_4O_2 geometry, while the second Zn atom is coordinated to two N atoms of two L^1 ligands, three O atoms of two phthalates and a water molecule with the formation of a distorted octahedral N_2O_4 geometry. Similar to 52, the formation of tetranuclear cadmium clusters within one-dimensional chains is a remarkable feature of 53, where the phthalate exhibits two different coordination modes. Four Cd(II) atoms in the cluster bridged by two phthalates adopting a μ_3-

$\eta^1:\eta^2:\eta^2:\eta^1$ coordination mode (type 21, Scheme 1) and four HL^1 ligands are chelated to the Cd(II) atoms. Two other phthalate ligands connect the tetranuclear cadmium clusters coordinated to Cd(II) atoms in a 1,3-bridging and chelating mode, μ_3-$\eta^1:\eta^1:\eta^1:\eta^1$ (type 16, Scheme 1), to develop 1D chains consisting of the tetracadmium clusters.

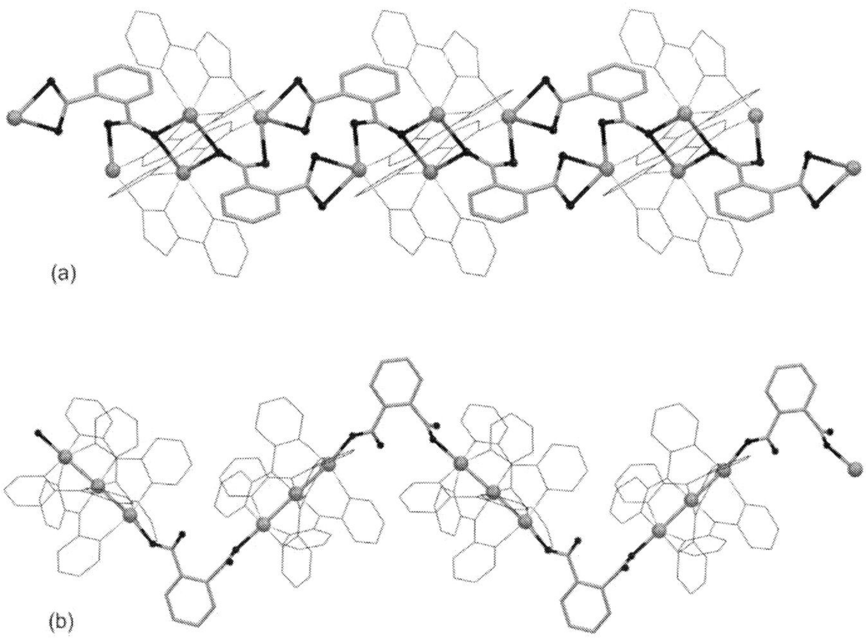

Figure 3. View of the 1D cluster-based chains in CPs 52 [71] (a) and 55 [72] (b). Metal atoms, black spheres; O atoms, black balls; C atoms, grey sticks. Ancillary N-donor ligands are shown as thin black lines.

Two crystallographically independent Cd(II) atoms adopt different coordination geometries: one Cd atom is seven-coordinated by two N atoms from chelating HL^1 and five O atoms from three phthalates, while the second Cd atom is six-coordinated and surrounded by two N atoms from chelating HL^1 and four O atoms of three phthalates. The investigation of luminescent properties of 52–54 showed the existence of strong emission in the solid-state at room temperature.

Recently, Zhu and Zhang [72] have shown that the trinuclear nickel cluster $[Ni_3(dpa)_4(ClO_4)_2]$ (dpa = 2,2′-dipyridylamine) can be also successfully employed as building blocks along with the bridging phthalate ligand in the construction of 1D CP. Self-assembly of the $[Ni_3(dpa)_4(ClO_4)_2]$ clusters with potassium hydrogen phthalate gives rise to the formation of a zigzag 1D cluster-based CP, $\{[Ni_3(pht)(dpa)_4] \cdot 1.5H_2O\}_n$ (55) (Figure 3b). The CP 55 can also be prepared from a discrete $[Ni_3(Hpht)_2(dpa)_4] \cdot CH_2Cl_2 \cdot H_2O$ cluster under treatment with KOH in MeOH/H_2O solution. In the trinickel cluster two terminal Ni(II) atoms are penta-coordinated and the equatorial positions are occupied with four N atoms from four dpa and the apical ones with an O atom from a phthalate ligand, whereas a central Ni(II) atom is tetra-coordinated by four N atoms from four dpa ligands. Note, that Poleti and Karanovic [73] have synthesized a linear 1D phthalate-based cobalt(II) CP, $[Co(pht)(dpa)]_n$ (56), with the same 2,2′-dipyridylamine ligand.

A very interesting 1D zinc(II) CP with the formula $\{[Zn_2(pht)_2(bpe)] \cdot MeOH\}_n$ (57) has been prepared by Tong *et al.* [74] by exploring another bipyridyl-like ligand, 1,2-(bis(4-pyridyl)ethene (bpe), in the reaction with zinc acetate and *o*-phthalic acid in MeOH/water solution at room temperature. Here, both bridging ligands, *o*-phthalate and 1,2-(bis(4-pyridyl)ethane, act as bifunctional spacers and bridge metal ions to generate a ladder-like chain. The phthalate ligands coordinate to Zn atoms in different ways while adopting μ_3-η^1:η^1:η^1:η^1 and μ_3-η^0:η^2:η^2:η^1 modes (types 16 and 20 respectively, Scheme 1), and bridge the metal atoms to give a linear 1D chain with the Zn...Zn separations across the carboxylate bridges lying in the range 3.48 to 4.45 Å. Two such parallel chains are further connected by bpe (the "rungs" of the ladder; Zn...Zn distance of 13.44 Å) with the formation of the ladder-like chain. Two crystallographically unique Zn atoms in 57 are six- and four-coordinated while adopting distorted octahedral and tetrahedral geometries, respectively. The Zn atom with the octahedral NO_5 donor set is bounded to one N atom from bpe ligand and five O atoms from four carboxylate groups. The Zn atom in the tetrahedral NO_3 geometry is surrounded by one N atom of bpe and three O atoms from three carboxylate groups. It is interesting to note, that the 1D phthalate-based CP 57 exhibits a solvochromic effect upon reversible removal and adsorption of linear alcohols.

Using other spacer for "rungs" such as 1*H*-3-(3-pyridyl)-5-(4-pyridyl)-1,2,4-triazole (3,4′-bpt) a similar ladder-like 1D CP, $[Zn_2(pht)_2(3,4'\text{-bpt})]_n$ (58), has been synthesized by Yan *et al.* [75] under hydrothermal conditions from zinc nitrate, *o*-phthalic acid, 3,4′-bpt and NaOH in ethanol/water solution at 140°C for 3 days. In contrast to CP 57, the phthalate ligand in 58 adopts a μ_3-η^1:η^1:η^0:η^1 coordination mode (type 12, Scheme 1) and links the adjacent Zn(II) centers into infinite {Zn-pht} chains with the Zn...Zn distance of 3.66 and 4.25 Å. The neighboring chains are further bridged by bpt ligands to furnish ladder-like chains with a Zn...Zn distance of 13.68 Å as shown in Figure 4a. The Zn(II) atoms adopt a distorted tetrahedral geometry and are coordinated by one N atom of bpt and three O atoms from two phthalate ligands. Jiang and co-workers [76] reported an intriguing ladder CP with an extremely long bipyridine-like ligand 2,2′-bis(4-pyridylmethyleneoxy)-1,1′-biphenylene (bpbp), $\{[Zn(pht)(bpbp)] \cdot H_2O\}_n$ (59). The CP 59 consists of two-leg ladders; in contrast to 57 and 58, the bpbp ligands connecting the adjacent Zn centers serve as the rails of the ladder, and phthalates further bridge these chains and act as the cross rungs (Figure 4b). Bu *et al.* [77] and later Ma *et al.* [78] used pyrazino[2,3-f][1,10]phenanthroline (pzp) as an additional terminal co-ligand in the preparation of double-stranded 1D zinc(II) and lead(II) CPs with the formula $\{[Zn_2(pht)_2(pzp)_2] \cdot H_2O\}_n$ (60) [77] and $\{[Pb(pht)(pzp)] \cdot H_2O\}_n$ (61) [78], respectively, by the hydrothermal synthesis of metal nitrates [$Zn(NO_3)_2 \cdot 6H_2O$ for 60 and $Pb(NO_3)_2$ for 61], *o*-phthalic acid and pzp with the base [NaOH, 60 or triethylamine, 61] in water at 160°C during six (60) or three (61) days. Here, the phthalate ligands bridge two Zn atoms with the formation of a 14-membered dinuclear macrocycle, that is similar to zinc(II) phthalate-based CP 29 with 2-methylimidazole. The 14-membered metallocycles are connected with each other by 1,3-bridging carboxylate groups to furnish a 1D double chain; thus, the pht adopts a μ_3-η^1:η^1:η^0:η^1 coordination mode. The double chain is decorated with pzp ligands. It is interesting to mention that using of an excess of *o*-phthalic acid in the hydrothermal reactions with zinc or nickel nitrates and pzp at 160°C during 3 days gave discrete dinuclear $[M_2(pht)_2(pzp)_2(H_2O)_4] \cdot 2H_2O$ (M = Zn, Ni) clusters [79] where two phthalate ligands linked two $[M(pzp)(H_2O)_2]$ units to form the same 14-membered metallocycle. Zhang and Sun [80] have prepared the similar double-stranded 1D phthalate-based zinc(II) CP, $[Zn(pht)(L^2)]_n$ (62), composed of 8- and 14-membered cycles by the reaction of zinc chloride and phthalic acid with dipyrido[3,2-*a*:2′,3′-*c*]phenazine ligand (L^2) by refluxing conditions in $EtOH/H_2O$. The Zn…Zn distances in smaller cycles are 4.435 (61), 4.323 Å (62), and in the larger ones 5.213 (61) and 5.323 Å (62).

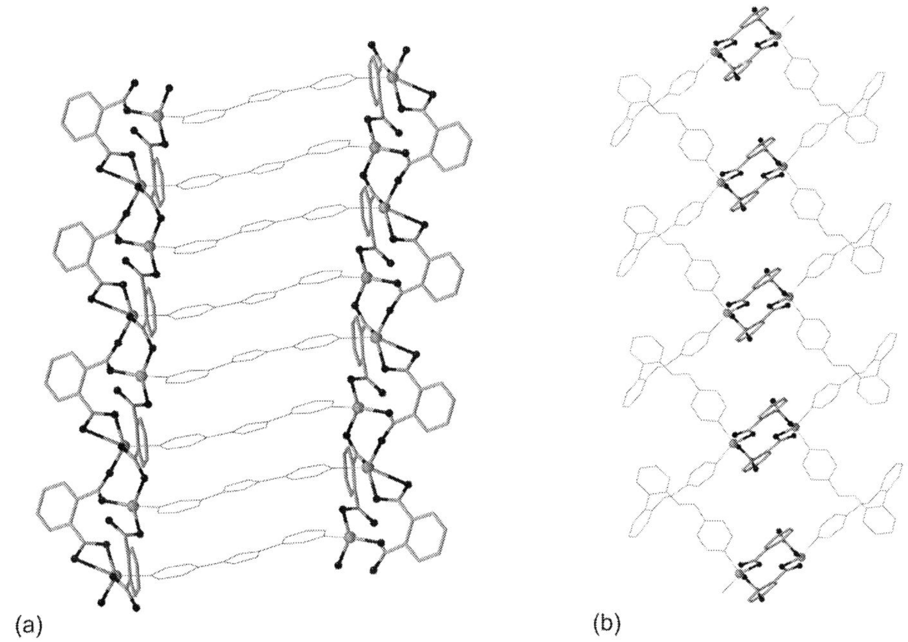

Figure 4. View of the ladder-like chains in the phthalate-based CPs 58 [75] (a) and 59 [76] (b). Metal atoms, black spheres; O atoms, black balls; C atoms, grey sticks. Ancillary N-donor ligands are shown as thin black lines.

Another phenanthroline-like ligand, 2-(furan-2-yl)-1*H*-imidazo[4,5-*f*] [1,10] phenanthroline (L^3), used in the hydrothermal reaction with $ZnCl_2$, H_2pht in water solution adjusted with NaOH to pH = 7.5, afforded a linear 1D phthalate-based CP, $[Zn(pht)(L^3)]_n$ (63) [81], where Zn(II) atoms are connected by bridging phthalate. In 60, 62 and 63, the Zn atoms are surrounded by two N atoms from the phenanthroline-like ligand and three O atoms from two phthalates adopting a distorted N_2O_3 trigonal bipyramid.

The *exo*-polypyridyl ligands have been proven to be good candidates for obtaining 1D phthalate-based CPs. For instance, Yao *et al.* [82] used 1,3-bis(4-pyridyl)propane (bpp) in the hydrothermal reaction with phthalic anhydride and zinc sulfate in $EtOH/H_2O$ solution at 160°C for 60 hours to produce a double stranded $[Zn_2(pht)_2(bpp)]_n$ (64) CP. The double phthalate-based Zn(II) chain, $\{[Zn(pht)(L^4)] \cdot 2H_2O\}_n$ (65), has been also prepared with the substituted bis(imidazole) ligand, 1,1′-(1,4-butanediyl)bis(imidazole-2-phenyl) (L^4), by Ma *et al.* [83]. In these double-stranded chains both the pht and *N,N′*-donor ligands playing the same bridging function connect the neighboring metal atoms with the formation of different metallocycles. In CP 64, the Zn(II) atoms are doubly bridged by phthalate ligands forming 6-membered cycles with a Zn…Zn distance of 3.458 Å and 11-membered cycles where the Zn…Zn distance equals 4.400 Å. The pht ligand acts as μ_3-bridge to link three zinc atoms displaying μ_3-η^1:η^1:η^1:η^1 and μ_3-η^1:η^1:η^2:η^0 coordination patterns. The bpp co-ligands additionally link the metal centers. In contrast to 64, a pair of pht dianions connects neighboring Zn(II) centers *via* a μ_2-η^1:η^0:η^0:η^1 coordination mode to furnish the formation of a 14-membered metallocycle with the Zn...Zn distance being 6.196 Å in 65. Each metal ion is further connected through two bridging L^4 ligands to form a 22-membered ring with Zn atoms being at 9.571 Å from each other. As a result, the 1D double-stranded chains with alternation of 14- and 22-membered metallocycles are formed in 65 as shown in Figure 5a. The Zn(II) atom is coordinated by two N atoms of different L^4 ligands and two O carboxylate atoms from two different phthalates to form a distorted N_2O_2 tetrahedron. The other *N,N'*-donor linker, 1,1'-(2,2'-oxybis(ethane-2,1-diyl))bis(1*H*-imidazole) (L^5) used in the hydrothermal synthesis with $CoCl_2 \cdot 2H_2O$ and phthalic acid gave a similar 1D double-stranded chain $\{[Co_2(pht)_2(L^5)_2(H_2O)_2] \cdot H_2O\}$ (66) that consists of alternating 14-membered and 22-membered metallocycles [84]. Similar to 65, the 14-membered cycle is formed by two phthalates coordinated in an usual 1,6-bridging fashion to two Co(II) centers [Co…Co, 5.254 Å], while the larger 22-membered rings formed by two flexible bis(imidazole) ligands with the Co…Co distance being 9.807 Å. The Co atoms exhibit a five-coordinated environment with two O atoms from different phthalates and two N atoms from two L^5 ligands and one water molecule.

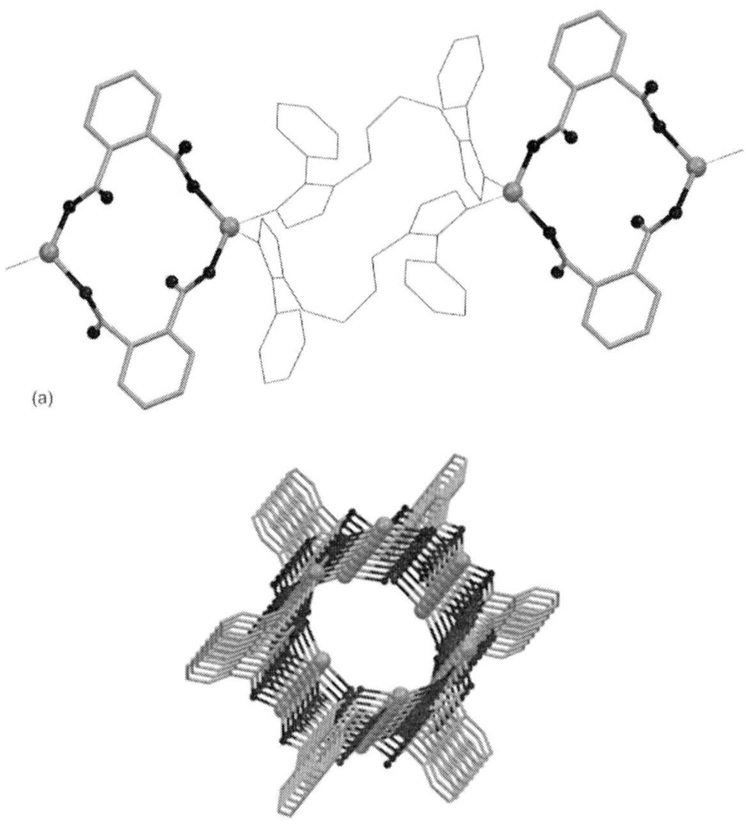

Figure 5. View of (a) double-stranded chain in CP 65 [83] and (b) tube-like chain in CP 97 [108]. Metal atoms, black spheres; O atoms, black balls; C atoms, grey sticks. Ancillary N-donor ligands are shown as thin black lines.

Other polydentate organic co-ligands such as *N*-, *N,O*-containing ligands with potential di-, tri- and tetradentate coordination sites have also been employed in the construction of phthalate-based CPs. For instance, Bermejo *et al.* [85] have described a linear 1D phthalate-based Mn(III) CP with a tetradentate N_2O_2 Schiff base ligand [L^6, 1,3-bis(3-methoxysalicylideneimino)propane], $[Mn(Hpht)_2(L^6)]_n$ (67), where 1,6-bridging hydrogen phthalates connect $[Mn(L^6)]$ units into linear chains. Each Mn(III) atom is octahedrally coordinated by two N and two O atoms of L^6 ligand forming an equatorial plane, and two O atoms from two different phthalate ligands occupying the axial positions. The interaction of a tridentate 1,4,7-triazacyclononane macrocyclic ligand (tacn) with phthalic acid and copper salt under refluxing conditions in MeOH/H_2O solution afforded a zigzag 1D phthalate-based CP with the formula $\{[Cu_4(pht)_2(tacn)_4](ClO_4)_4 \cdot 2H_2O\}_n$ (68), as reported by Cheng *et al.* [86]. The main feature of the CP 68 is the formation of anionic chains composed of [Cu(tacn)] units bridged by phthalates. The phthalate ligand adopts a μ_3-η^1:η^1:η^1:η^1 coordination pattern (type 14, Scheme 1) and chelates to one [Cu(tacn)] and simultaneously bridges other two [Cu(tacn)] units to propagate a one-dimensional chain. In 68, the square-pyramidal Cu atoms are five-coordinated by three N atoms from tacn and two carboxylate O atoms from different phthalates. A very interesting 1D phthalate-based CP has been prepared by Sadler *et al.* [15] with a tetradentate macrocyclic cyclam ligand, a potential anti-HIV agent. Self-assembly of the cyclam with zinc oxide and phthalic acid in water under room temperature followed by recrystallization from methanol gave $\{[Zn(pht)(cyclam)] \cdot 2MeOH\}_n$ (69) in which the [Zn(cyclam)] blocks are connected by 1,6-bridging phthalate ligands into linear chains. The coordination geometry around the Zn atoms is distorted N_4O_2 octahedral: the equatorial plane is formed by four N atoms of macrocyclic cyclam; two carboxylate atoms from the phthalate dianions occupying the axial positions complete the coordination sphere around Zn atoms. The investigation of the CP 69 in both solid state by X-ray crystallography and in solution by NMR spectroscopy showed that in solid state the cyclam adopts the most stable configuration *trans*-III (*S,S,R,R*), in which six-membered rings are in the chair conformation and five-membered rings in the gauche conformation, whereas in solution the phthalate groups stabilized the *cis*-V (*R,R,R,R*) configuration; thus, carboxylates can exert a strong influence over configurational selectivity and play an important role in the anti-HIV activity.

Burrows *et al.* [87, 88], who extensively studied the influence of intermolecular hydrogen bonding and substituents of the terminal co-ligands on the topology of CPs, have prepared several phthalate-based CPs with *S*-donor co-ligands, thiourea (tu) and its derivatives. They showed that the thiourea substitution can affect the formation of discrete compounds or CPs. The 1D thiourea-phthalate zinc(II) CPs $[Zn(pht)(tu)_2]_n$ (70) [87] and $[Zn(pht)(L^7)_2]_n$ (71) [88] crystallized from an aqueous solution containing zinc thiourea complex ($[Zn(tu)_4]Cl_2$ or $[Zn(L^7)_4]Cl_2$, L^7 = *N,N*-dimethylthiourea) and sodium phthalate at room temperature. The one-dimensional chains in both CPs comprise $[Zn(tu)_2]$ (70) or $[Zn(L^7)_2]$ (71) building blocks connected by 1,6-bridging phthalates, however, the simple thiourea ligand afforded the formation of helical chains in 70, whereas in $[Zn(pht)(L^7)_2]_n$ (71) with methyl substitutes on the thiourea ligand, linear chains are generated. It is worth to note that the presence of intensive intra-molecular and inter-chain hydrogen bonding in 70 where 3D hydrogen-bonded networks formed by the interaction between thiourea NH groups and carboxylate atoms, stabilized the helical chains formed in 70 compared to 71 (only 2D hydrogen-bonded layer). In both CPs, the zinc adopts a distorted tetrahedral S_2O_2 coordination geometry coordinated by two S atoms from thiourea ligands and two carboxylate atoms. The pitch of the helix in 70 includes three Zn atoms and two bridging phthalates and equals 10.919 Å. Using a oxamide oxime ligand (oaoH) Endres [89] has synthesized the similar helix phthalate-based nickel(II) CP, $\{[Ni(pht)(oaoH)_2] \cdot 4H_2O\}_n$ (72), where intra- and intermolecular O...O and N...O interactions stabilize the helical chains $[Ni(pht)(oaoH)_2]$ with a helical pitch of 11.324 Å.

In all above mentioned phthalate-based CPs, the phthalate ligand executes a bridging function and coordinates to metal ions through both carboxylate groups. However, there are rare examples where the connection between metal atoms is *via* only one carboxylate group of phthalate while its other group remains pendant. Using 1,10′-phenanthroline as a co-ligand, such an unusual 1D linear CP, $\{[Mn(pht)(phen)(H_2O)_2] \bullet H_2O\}_n$ (73), consisting of the $[Mn(phen)(H_2O)_2]$ units bridged *via* one carboxylate group of pht ligand and the other participated in a hydrogen bonding, has been reported by Zhang *et al.* [90]. The CP 73 was crystallized from an aqueous-alcohol solution containing equimolar amounts of phen, H_2pht acid and $MnSO_4 \bullet H_2O$ at room temperature. The structure of 73 possesses isotactic chains where all aromatic rings are located on the same side of the polymeric backbone. Mn(II) centers are coordinated by two N atoms of phen, two carboxylate O atoms from different pht groups and two water molecules adopting a distorted octahedral N_2O_4 geometry. Such coordination of the phthalate ligand resembles those found in another phthalate-based copper(II) CP with diethylenetriamine co-ligand (dien), $\{[Cu(pht)(dien)] \bullet 0.5MeOH\}_n$ (74), synthesized by Bakalbassis *et al.* [91] by slow diffusion of methanol into an aqueous solution of copper acetate and piperidinium phthalate. However, in the latter case, the carboxylate group joined [Cu(dien)] units into a zigzag one-dimensional chain. The copper(II) atoms in 74 adopt a distorted square pyramidal N_3O_2 environment: one O carboxylate atom and three N atoms from one dien molecule form an equatorial plane and one O atom from the carboxylate group of another pht moiety occupies the apical position. Magnetic properties of this CP reveal both a nearest-neighbor antiferromagnetic coupling between copper atoms through one carboxylate group within the chain and very weak interchain antiferromagnetic interactions through pendant carboxylate groups. It is interesting to note, that using in the same reaction cobalt perchlorate instead of cobalt acetate afforded a CP, $\{[Cu_2(pht)(ClO_4\ (dien)_2)](ClO_4)\}_n$ (75) [92], with the similar chains where the neighboring [Cu(dien)] building units are connected by one carboxylate group, however, the pendant carboxylate group of phthalate ligands additionally attach $[Co(ClO_4)(dien)]$ units to the skeleton forming cationic one-dimensional chains.

The position of nitrogen donor atoms in bipyridine-like ligands can influence the formation of the resulting 1D CPs. LaDuca *et al.* have used *exo*-bidentate 4,4′-dipyridylamine (4,4′-dpa) [93] or *N,N′*-bis(3-pyridyl-methyl)piperazine (L^8) [94] in the hydrothermal reactions to produce linear 1D CPs with the general formula $\{[M_2(pht)_2(4,4'\text{-dpa})_2(H_2O)_4] \cdot H_2O\}_n$ [M = Co^{2+} (76); Ni^{2+} (77)] and a waving 1D CP $\{[Cu(pht)(Hpht)(L^8)(H_2O)] \cdot 3H_2O\}$ (78). In CPs 76 and 77, the polymeric motif is formed with $[M(H_2O)_4]$ and $[M(pht)_2]$ building blocks which are alternately linked by the bridging 4,4′-dpa to generate linear chains with M...M distances of 11.308 Å (Co) and 11.124 Å (Ni), whereas in 78, the $[Cu(pht)_2(H_2O)_2]$ and $[Cu(Hpht)_2]$ units are bridged through the *exo*-bidentate L^8 ligand with a Cu...Cu distance of 10.087 Å. Here, the phthalate ligand does not execute a bridging function and coordinates to metal atoms in a chelating μ_1-η^0:η^1:η^1:η^0 coordination mode (type 3, Scheme 1) in CPs 76 and 77, and adopts a monodentate coordination in 78. Liao *et al.* [95] also reported a 1D cobalt(II) CP, $[Co(pht)(azpy)(H_2O)_3]_n$ (79), that consists of the alternating $[Co(H_2O)_4]$ and $[Co(pht)_2(H_2O)_2]$ building blocks bridged by *exo*-bidentate 4,4′-azopyridine (azpy). The CP 79 was precipitated from the aqueous solution containing equimolar amounts of $Co(ClO_4)_2 \cdot 6H_2O$ and *o*-phthalate piperidine salt. In the chain, the Co(II) atoms belonging to different building units have a different distorted octahedral coordination environment: one Co atom is coordinated by two N atoms from different azpy and four water molecules, while the other one is bonded to two N atoms from bridging azpy ligands, two O atoms from monodentate coordinated phthalate ligands and two water molecules. Magnetic studies reveal very weak antiferromagnetic interactions between Co(II) ions through azpy bridges. Das *et al.* [96] have described 1D phthalate-based cobalt(II) CPs with other N-donor *exo*-bidentate ligands such as *N*-(4-pyridyl)isonicotinamide (L^9), *N*-(3-pyridyl)isonicotinamide (L^{10}), and *N*-(4-pyridyl)isonicotinamide (L^{11}), $\{[Co(pht)(L^9)(H_2O)_2] \cdot 2.36H_2O\}_n$ (80), $\{[Co_2(pht)_2(L^{10})(H_2O)_6] \cdot H_2O\}_n$ (81), and $\{[Co_2(pht)_2(L^{11})(H_2O)_6] \cdot 1.12H_2O\}_n$ (82), prepared by the reaction of N-donor ligand (L^9, L^{10} or L^{11}) with cobalt nitrate and disodium phthalate in $EtOH/H_2O$ solution. In all CPs, one-dimensional chains are formed. The chains consist of $[Co(pht)(H_2O)]$ building units bridged by L^9 ligand in 80; and $[Co(H_2O)_4]$ and $[Co(pht)_2(H_2O)_2]$ units linked by L^{10} in 81 or L^{11} in 82. Depending on the ligating topology of bridging ligands [either linear in L^9, or angular in L^{10} and L^{11}] the linear one-dimensional motif is formed in 80, while in 81 and 82 the 1D motif is a zigzag chain. The Co atoms in all CPs are hexa-coordinated displaying a distorted N_2O_4 octahedral geometry, where the equatorial plane is completed by the O atoms from phthalate and water molecules, and axial positions are occupied by the N atoms of the bidentate ligands L^9, L^{10} and L^{11}. Similar to above-mentioned CPs 76–79, the phthalate anion behaves only as a chelating ligand in 80, or a monodentate ligand in 81 and 82.

The coordination ability of the phthalate ligand towards other metal ions in the construction of 1D phthalate-based CPs, such as yttrium, silver, tin and lanthanides was also explored. Thirumurugan and Natarajan [97] have reported a 1D phthalate-based CP $\{[Y_2(pht)_3(phen)_2] \cdot H_2O\}_n$ (83) with yttrium(III) and a terminal 1,10′-phenanthroline co-ligand employing a hydrothermal method. In the formed CP 83, the dimeric $[Y_2(phen)_2]$ building units are bridged by phthalates into double stranded chains. There are two different building units: in one of them two neighboring [Y(phen)] subunits are connected by two carboxylate anions forming an eight-membered cycle, while in the other, two neighboring [Y(phen)] subunits are bridged by four phthalates forming a paddle-wheel like arrangement. The dimeric units alternate along the chains and are bridged by the carboxylate groups of phthalate ligands. Here, the phthalate ligands link three Y(III) atoms showing different connectivities: μ_3-η^1:η^1:η^1:η^1 (type 14 and 16, Scheme 1) and μ_3-η^1:η^1:η^2:η^1 (type 18, Scheme 1) coordination modes. The Y(III) atoms are in a distorted dodecahedral N_2O_6 geometry by two nitrogen atoms from a chelating phen and six carboxylate oxygen atoms. The 1D phthalate-based CP with silver(I) $[Ag_2(pht)(phz)_2(H_2O)]_n$ (84) prepared from the reaction of disilver phthalate $Ag_2(pht)$ with phthalazine (phz) in water has been reported by Whitcomb and Rogers [98]. In 84, the phthalazine behaves as an *exo*-dentate bridging ligand and is coordinated to two Ag(I) ions forming dimeric units, which are further connected by bridging phthalate ligands to generate a linear 1D CP. The two crystallographically unequivalent Ag(I) atoms are three- and four-coordinated by two nitrogen atoms from two phz ligands and one oxygen atom from carboxylate group of pht, and two N atoms of different phz and two O atoms from phthalate, respectively. Using hydrothermal conditions and *N,O*-bidentate isonicotinic acid (Hina) Li *et al.* [99] have isolated another 1D silver(I) CP $[Ag(Hpht)(ina)]_n$ (85), in which the isonicotinic acid executes a bridging function to connect the neighboring Ag(I) atoms into one-dimensional chains. The adjacent chains are linked by Ag...Ag interactions of 3.212 Å in a double motif. Each Ag(I) atoms adopts a T-shaped geometry by one N atom from one isonicotinate and O atom from other ina ligand, the phthalate dianion completes the coordination sphere around Ag(I) atom acting as a monodentate ligand in 85.

Usually, the preparation of CPs with lanthanides that have high coordination numbers furnishes the formation of high-dimensional 2D or 3D coordination polymers. For the design and synthesis of CPs with one-dimensional structures, the additional termini ligands which block certain metal coordination sites, have been extensively used. Several research groups [100-105] employed 1,10′-phenanthroline and 2,2′-bipyridine as terminal ligands to assemble 1D phthalate-based CPs with lanthanides (La^{3+}, Pr^{3+}, Nd^{3+}, Sm^{3+}, Gd^{3+}, Eu^{3+}, Ho^{3+}, and Yb^{3+}) by hydrothermal synthesis at 90-180°C during 3-6 days. Thirumurugan and Natarajan [100] and Yan and Song [101] have used lanthanide nitrates in the hydrothermal synthesis to prepare 1D CPs with the general formula $\{[Ln_3(pht)_4(NO_3)(phen)_2] \cdot xH_2O\}_n$ [x = 1, La^{3+} (86), Pr^{3+} (87) [100]; x = 0, Sm^{3+} (88) [101]], whereas employing of lanthanide chlorides gave 1D CPs $\{[Eu_3(pht)Cl(phen)_3(H_2O)_6]Cl_2 \cdot 4H_2O\}_n$ (89) reported by Yue *et al.* [102] and $\{[Ln_2(pht)_3(phen)(H_2O)] \cdot xH_2O\}_n$ [x = 1, Nd^{3+} (90); x = 3, Ho^{3+} (91) [103]; x = 3.5, Yb^{3+} (92) [104] described by Jin and co-workers. The hydrothermal reaction of 2,2′-bipyridine with *o*-phthalic acid and GdO_2 afforded $\{[Gd_2(pht)_3(bpy)(H_2O)_2] \cdot 3H_2O\}_n$ (93) [105]. All CPs 86 – 93 have one-dimensional structures with phthalate ligands connecting adjacent metal ions to generate double stranded chains decorated with phen/bpy and phthalate ligands. The metal atoms are coordinated by phen/bpy or nitrate/chloride/water groups along with the bridging phthalates resulting in the different coordination geometry. In 86 and 87, the metal atoms are eight- and ten-coordinated adopting a distorted dodecahedral N_2O_6 environment or a distorted bicapped square antiprismatic N_2O_8 and O_{10} coordination geometry; nine-coordinated Sm(III) centers - a distorted square monoantiprism O_9 in 88; Eu(III) atoms adopt N_2O_6 and N_2O_6Cl coordination environments in 89; nine-coordinated Nd(III) atoms in 90 – O_9 and N_2O_7 coordination geometries; eight and nine-coordinated Ho(III) atoms in 91 – O_8 and N_2O_7 coordination environments; in 92 Yb(III) atoms are eight-coordinated and have O_8 and N_2O_6 donor sets; in 93 Gd(III) atoms are eight and nine-coordinated with distorted dodecahedral O_8 and capped square antiprismatic N_2O_7 geometries. The phthalate ligand coordinated to two or three lanthanide atoms shows the great diversity of coordination modes such as types 10, 18 and 21 in 86 and 87, types 14, 18 and 21 in 88, types 12 and 18 in 89, types 16, 18 and 25 in 90, types 9, 14 and 16 in 91 and 92, types 9, 16 and 18 in 93 (Scheme 1). Additionally, Wang *et al.* [106] used isonicotinic acid (Hina) in the hydrothermal synthesis with lanthanide oxides and *o*-phthalic acid at 170°C for 6 days to isolate 1D wave-like chain CPs with a general formula $\{[Ln_2(pht)(OH)_2(ina)_2(H_2O)] \bullet H_2O\}_n$ [Ln = Er^{3+} (94); Tm^{3+} (95)].

The employment of the phthalate ligand in the construction of 1D CPs with actinides is still rare and only a limited number of examples are presented in the literature with U^{6+}, $Np^{5+/6+}$ and Pu^{6+} ions. Several research groups of Charushnikova [107], Grigoriev [108] and Loiseau [109] have reported one-dimensional uranyl and neptunium phthalate polymers with a general formula $\{[AnO_2(pht)(H_2O)] \bullet xH_2O\}_n$ [x = 1, An = U^{6+} (96) [107]; x = 0.33, An =

Np^{6+} (97) [108]; $x = 0.32$, An = U^{6+} (98) [109]] built up from double-stranded chains in 96 or very interesting nanotubes-like chains in 97 and 98. The CP 97 has been prepared by the heating of the mixture which contained neptunium nitrate and lithium phthalate at 70°C for 3 hours with a pH = ~4, whereas the hydrothermal conditions have been employed for the isolation of CPs 96 and 98 from $UO_2(NO_3)_2 \bullet 6H_2O$: for 96 the hydrothermal reaction of uranyl nitrate and lithium phthalate occurred at 125°C for 2 hours with the pH = ~3, and for 98 - at 200°C for 24 hours in the presence of KOH with the pH = 0.9 – 1.5 and *o*-phthalic acid instead of lithium phthalate. In 96, each phthalate ligand coordinates to three uranium atoms adopting a μ_3-η^1:η^1:η^1:η^1 (14 in Scheme 1) coordination fashion: two phthalates bridge two uranium atoms through three O atoms from each ligand with the formation of 7-membered and 8-membered metallocycles, the adjacent dinuclear $[UO_2(pht)(H_2O)]_2$ units are connected *via* the forth O atom from the same phthalates into double-stranded chains. The formation of similar double chains has also been found in the related uranyl phthalate CP with urea (ur), $\{[UO_2(pht)(ur)] \bullet 0.5H_2O\}_n$ (99), reported by Shchelokov *et al.* [110]. Different from 96 and 99, the phthalate ligand in 97 and 98 is coordinated to four actinide atoms (U or Np) displaying a μ_4-η^1:η^1:η^1:η^1 mode (22 in Scheme 1) to arrange them into nanotubes-like chains as shown in Figure 5b with an approximately ring cross section. Each section consist of 24-membered macrocyclic $[An_6O_{12}(pht)_6(H_2O)_6]$ rings with the An...An distance of ca. 9.5 Å in diameter; these are staked vertically *via* carboxylate bridges to give an infinite tubular array, that is a remarkable feature of these CPs. The internal cavities of the nanotubes encapsulate the water molecules. In 96–99, the actinide atoms are seven-coordinated and adopt pentagonal bipyramidal environments whose equatorial planes are occupied by four oxygen atoms from three (96 and 99) or four (97 and 98) different carboxylates and one water (96–98) or urea (99) molecule. Charushnikova *et al.* [111] have reported heterometallic cobalt-neptunium phthalate 1D CP, $\{[Co(NH_3)_6][NpO_2(Pht)_2] \bullet 2H_2O\}_n$ (100), which starts immediately to precipitate after addition of aqueous $[Co(NH_3)_6]Cl_3$ solution to the aqueous solution containing NpO_2NO_3 and $(NH_4)_2(pht)$. Here, the phthalate ligands adopt different coordination modes: one of them is coordinated to Np(V) atom in a bidentate fashion (3 in Scheme 1) with the formation of a 7-membered chelate ring, whereas other executes a bridging function (type 6 in Scheme 1) to connect the neighboring metal atoms into anionic infinite $[NpO_2(pht)_2]^{3-}$ chains; the chains are separated by $[Co(NH_3)_6]^{3+}$ cations.

1.2. Two-Dimensional Coordination Polymers (2D)

Owing to a huge variety of coordination modes of the *o*-phthalic acid displayed in metal complexes and possibility to connect several metal atoms simultaneously, this dicarboxylic acid has been extensively used in the design of two-dimensional coordination polymers. Two strategies have mainly been explored. One of them is based on the interaction of *o*-phthalic acid with a range of metal salts (Li^{I}, Na^{I}, K^{I}, Cs^{+}, Mg^{2+}, Sr^{2+}, Co^{2+}, Cu^{2+}, Ag^{+}, Cd^{2+}, In^{3+}, Tl^{+}, Pb^{2+}, Ln^{3+} and $An^{5+/6+}$) in water or alcohol solution at room temperature or hydro(solvo)thermal conditions, and the other one relies on the exploration of bridging functions displayed by the ancillary polydentate ligands used in the same reactions. As shown by the pioneering works of Cingi *et al.* and Prout *et al.* on the construction of copper(II) phthalate-based CPs, the use of the *o*-phthalic ligand and metal salts in aqueous solution leads to the formation of 2D polymeric structures in both the copper(II) phthalate CPs, $[Cu(pht)(H_2O)_x]_n$ [x = 1 (101) [112, 113]; x = 2 (102) [114]], and in copper(II) phthalate CPs with additional alkali and alkaline earth metals: $[Na_2Cu(pht)_2(H_2O)_2]_n$ (103) [28, 115-117] and $[MgCu(pht)_2(H_2O)_4]_n$ (104) [118], as well as with Ag^+ ions, in $\{[AgCu_2(Pht)_2(OH)(H_2O)] \cdot 3H_2O\}_n$ CP (105) [119]. In copper(II) phthalate monohydrate CP 101, the phthalate ligand is coordinated to four Cu(II) atoms adopting a μ_4-η^1:η^0:η^1:η^2 coordination fashion (24 in Scheme 1): one of its carboxylate group is coordinated to three Cu(II) atoms in the formed $\{Cu(\mu\text{-}H_2O)\}$ chains, whereas the other carboxylate group is coordinated to the Cu(II) atoms from the neighboring chain to furnish the formation of 2D layer structure with 4^4-sql topology. In copper(II) phthalate dihydrate CP 102, the phthalate ligand adopts a μ_3-η^1:η^0:η^1:η^1 coordination mode (12 in Scheme 1) to link $\{Cu(H_2O)_2\}$ units into a 2D layer structure with the same 4^4-**sql** net topology and with the dimension of the formed grids of 4.226 x 6.795 Å compare to that of 6.443 x 6.670 Å in 101. The insertion of metal cations such as Na^+, Mg^{2+}, and Ag^+ into the above mentioned copper(II) phthalate CP, extensively studied by group of Cingi and others, leads to the formation of diverse 2D layer structures, in which the phthalate ligands adopt different coordination modes to link three (104, 12 in Scheme 1), four (105, 22 and 24 in Scheme 1) or five (103, 42 and 43 in Scheme 1) metal atoms.

The formation of 2D layer structures has also been found in the phthalate-based CPs only with s-block metals: with alkali metals in $[Li_4(pht)_2(H_2O)_3]_n$ (106) [120] and $[Li_4(pht)_2(EtOH)(H_2O)]_n$ (107) [23] prepared by the crystallization of lithium salt and *o*-phthalic acid from aqueous or ethanol solutions at room temperature, respectively, in $[K(Hpht)]_n$ (108) [121-124] and $[K(Hpht)(H_2O)]_n$ (109) [125] grown from aqueous solution; and with alkaline earth metals in $[Sr_5(pht)_4(NO_3)_2(H_2O)_2]_n$ (110) [126] CP. Similar to CPs 103-105, the phthalate ligand connecting three, four, five and six metal atoms displays various coordination modes: μ_5-η^2:η^1:η^2:η^1 and μ_5-η^2:η^2:η^1:η^0 in 106 (types 37 and 38 in Scheme 1); μ_3-η^1:η^2:η^0:η^0 and μ_3-η^1:η^1:η^2:η^0 in 107 (types 13 and 15 in Scheme 1); μ_4-η^2:η^3:η^1:η^0 and μ_5-η^0:η^3:η^2:η^3 in 108 (types 33 and 40 in Scheme 1); μ_4-η^1:η^1:η^2:η^2 in 109 (type 32) and μ_6-η^3:η^2:η^3:η^2 in 110 (type 47). Note, in all CPs 106-110 the phthalate ligands and metal ions are interconnected giving complex layered architectures.

Under hydrothermal conditions at 200°C for 60 h, the reaction between cobalt nitrate and potassium hydrogen phthalate in aqueous solution give rise to an interesting cobalt(II) 2D CP, $[Co_3(pht)_2(OH)_2]_n$ (111) [127], with a honeycomb layered structure. The use of KOH or Cs_2CO_3 in the solvothermal reaction of cobalt acetate with *o*-phthalic acid in methanol/water solution at 160°C for 3 days afforded the heterometallic 2D CPs $[MCo(pht)(OH)]_n$ (M = K^+ (112); Cs^+ (113) [24]) where the 1D zigzag chains of {Co(pht)} are linked by the alkali metal ions to form two-dimensional layers. In 111, the phthalate ligand adopts a μ_6-η^2:η^1:η^1:η^2 coordination fashion (48 in Scheme 1) bridging six Co atoms into 2D 6^3-**hcb** net, whereas, in 112 and 113, it exhibits μ_7-η^2:η^3:η^3:η^1 (for M = K) and μ_7-η^2:η^2:η^3:η^2 (for M = Cs) (51 and 53, Scheme 1) coordination modes linking four Co(II) atoms and three K^+ or Cs^+ ions into 2D layers with the shortest intrachain Co...Co distance of 3.387 Å. The Co...Co distance between the adjacent Co(II) chains equals 6.504 Å. It is interesting to mention, that both phthalate-based CPs 112 and 113 with the Co(II) chains separated by the alkali ions show spin canting antiferromagnetic behaviors. Moreover, superparamagnetic dynamic relaxations are observed for these compounds.

Robl [128] and Faure *et al.* [129] have reported cadmium(II) phthalate-based 2D CP $[Cd(pht)(H_2O)]_n$ (114) prepared by the crystallization of cadmium nitrate and sodium phthalate or *o*-phthalic acid from aqueous solution at room temperature. In addition, the hydrothermal heating of an aqueous mixture containing potassium hydrogen phthalate, cadmium nitrate and oxalic acid at 160°C for 5 days gave the same cadmium(II) phthalate hydrate CP 114 [130, 131]. In 114, the phthalate adopts the uncommon bridging mode 28 in Scheme 1 to connect four Cd(II) atoms and further generates 2D layers showing a honeycomb 6^3-**hcb** net with cavity dimensions of 9.492 x 5.788 Å. The hydrothermal reaction of phthalyl-γ-glutaminyl-histidine with $Cd(ClO_4)_2 \bullet 6H_2O$ in aqueous ethanol solution at 120°C for 3 days afforded an anhydrous cadmium(II) phthalate $[Cd(pht)]_n$ (115) [132] CP with layers consisting of square grids with dimension of 4.216 x 4.552 Å. Similar to 114, the phthalate in 115 also coordinates to four metal atoms, but instead show a μ_4-η^2:η^2:η^2:η^1 (34 in Scheme 1) coordination modes leading to 2D 4^4-**sql** net topology. Both CPs 114 and 115 demonstrate a strong fluorescent emission in the solid state at room temperature and can be candidates for potential photoactive materials.

Wang *et al.* [133] have reported the first example of 2D CP with In^{3+} ions, $[In(pht)(OH)(H_2O)]_n$ (116), which was prepared by the hydrothermal reaction of an aqueous solution containing *o*-phthalic acid and 2-picoline with $InCl_3$ at 160°C for 4 days. The In(III) atoms are linked by hydroxyl groups into zigzag {In(OH)} chains which are further connected by phthalate ligands (it adopts an usual 1,3- and 1,6-bridging mode) to furnish a 2D layer structure with a (4,4)-connected 4^4-**sql** net in 116. It was shown that thallium(I) nitrate reacts with *o*-phthalic acid or sodium phthalate in ethanol or methanol solution at room temperature resulting in the formation of two 2D phthalate-based CPs with the Tl^+ ions, $[Tl_2(pht)]_n$ (117) [22] and $[Tl(Hpht)]_n$ (118) [134, 135], where the phthalate ligands coordinate to seven (117) or four (118) Tl(I) atoms (Tl–O distances are in the range 2.509 – 2.969 Å) showing μ_7-η^4:η^2:η^2:η^3 (54 in Scheme 1) or μ_4-η^2:η^3:η^1:η^0 (33 in Scheme 1) coordination modes. The formation of 2D networks with a 4^4-**sql** topology has also been found in the phthalate-based CP with lead(II) ions, $[Pb(pht)(H_2O)]_n$ (119) [136].

Recently, Huang and co-workers [137] have described the synthesis of two 2D phthalate-based CPs with Ag^+ metal ions, $\{[Ag_2(pht)]_4 \bullet MeCN\}_n$ (120) and $[Ag_2(pht)(H_2O)]_n$ (121), which have been prepared under ultrasonic conditions (160W, 4 KHz) by the reaction of Ag_2O and *o*-phthalic acid in acetonitrile at 50°C for 20 min. Both CPs 120 and 121 display the same 2D layer structures with 4^4-**sql** nets, but in 120 the cavities are occupied by acetonitrile molecules, whereas in 121, the 2D layer does not include any solvents. These coordination polymers display luminescence in the solid state at room temperature, the origin of which has been assigned to the ligand-to-metal charge transfer (LMCT) mainly arising from Ag(I)-pht transitions. Other silver phthalate 2D CP, $[Ag_4(pht)_2(Hac)]_n$ (122), with acetic acid (Hac) acting as a termini ligand has been reported by Whitcomb and Rajeswaran [138].

Similar to 1D phthalate-based CPs, the other strategy in the preparation of two-dimensional phthalate-based CPs relies on the use of a combination of bridging ligands in the same material. Depending on the nature of the metal ions, the geometry of the additional bridging co-ligands (linear, T-shaped) used and synthetic conditions the rectangular grids, rhombic grids, honeycomb grids, brick wall, herringbone-like and others two-dimensional layers were identified in 2D phthalate-based CPs.

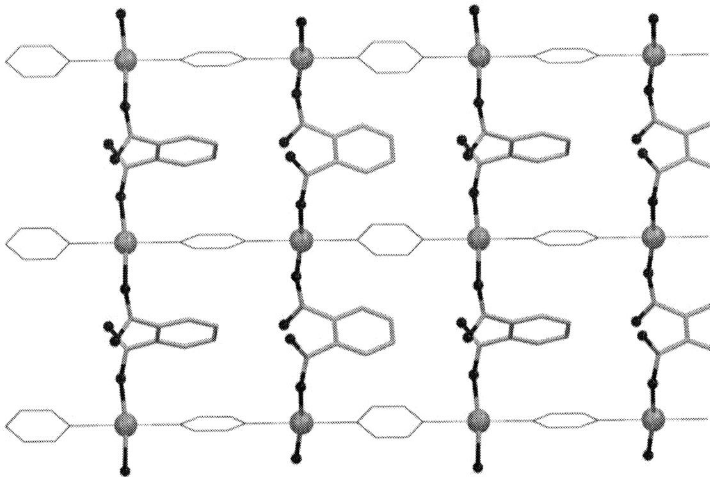

(a)

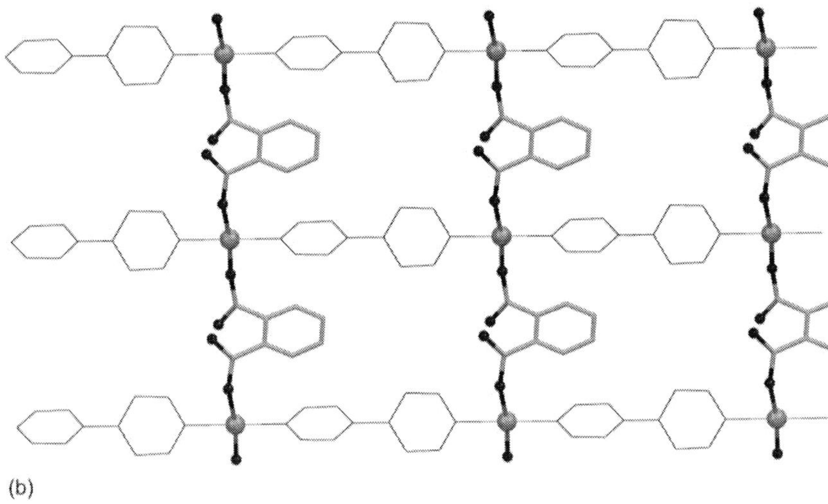

(b)

Figure 6. View of (a) square grids in CP 123 [140] and (b) rectangular grids in CP 127 [144]. Metal atoms, black spheres; O atoms, black balls; C atoms, grey sticks. Ancillary N-donor ligands are shown as thin black lines.

Linear, exo-bidentate and rigid ligands such as pyrazine (pyz) or 4,4′-bipyridine (4,4′-bpy) have been used by numerous research groups to develop an extensive series of rectangular grid-like phthalate-based 2D CPs with 3d metals (Mn^{2+}, Fe^{2+}, Co^{2+}, Ni^{2+}, Cu^{2+}, Zn^{2+}, and Cd^{2+}). In general, metal atoms in such CPs are bridged by phthalate groups into one-dimensional {M(pht)} chains, linear or zigzag, which are further connected by the ancillary bridging ligand, pyz or 4,4′-bpy, to form a two-dimensional 4-connected 4^4-**sql** nets with cavities, dimensions of which depend on both the length of diimine ligands used and the coordination modes displayed by phthalate anions. For instance, in $[Mn(pht)(pyz)(H_2O)_2]_n$ (123) prepared by the hydrothermal reaction of manganese acetate and H_2pht with pyrazine in $MeOH/H_2O$ solution at 140°C for 4 days [139] or under refluxing [140], the 2D layer consists of rectangular grids with size of ca. 7.496 x 7.628 Å through pyz and pht bridges, respectively as shown in Figure 6a. By using a longer spacer, 4,4′-bipyridine, the 2D layers with elongated grids were found in {[M(pht)(4,4′-bpy)(H_2O)$_2$] • 2H_2O}$_n$ (M = Mn^{2+} (124) [141, 142]; Co^{2+} (125) [142, 143]; Cd^{2+} (126) [143]; Zn^{2+} (127) [144, 145]; Cu^{2+} (128) [146]; Fe^{2+} (129) [147]; Ni^{2+} (130) [148]) prepared under both ambient and hydrothermal conditions.

The size of the grids was tuned by the 4,4′-bpy linker (Figure 6b). The different metal ions impact a very little change in the grid size: 11.586 x 7.704 Å (Mn); 11.390 x 7.650 Å (Fe); 11.300 x 7.596 Å (Co, [143]) and 11.372 x 7.616 Å (Co, [142]); 11.372 x 7.616 Å (Ni); 11.103 x 7.696 Å (Cu); 11.316 x 7.635 Å (Zn); 11.647 x 7.790 Å (Cd). Notably, in 124–130, every phthalate ligand acts in an usual 1,6-bridging mode and links two pairs of adjacent metal atoms with M...M separations of 7.596 – 7.760 Å, however, when the ligand connects the adjacent metals in a chelating-bridging mode (8 in Scheme 1) as it was found in grid polymer {[Ni(pht)(4,4′-bpy)(H_2O)] • 3H_2O}$_n$ (131) [149], the M...M distance through the phthalate linker is shorter and the size of the rectangular grids in 131 corresponds ca. 11.265 x 5.134 Å. Additionally, Jin *et al.* have reported a rectangular grid polymer {[Co(pht)(pyEt)(H_2O)] • H_2O}$_n$ (132) [150] prepared with a longer polypyridyl ligand, 1,2-bis(4-pyridyl)ethane (pyEt), where the phthalate ligand displays the similar coordination mode to metal atoms bridging them into a zigzag 1D polymer. The connection of these zigzag chains by pyEt leads to longer rectangular grids with the dimensions of 13.627 x 5.290 Å. In all CPs 124–132 the cavities accommodate water molecules which form intensive hydrogen bonding with carboxylate groups to connect the 2D layers into multi-layer 3D networks. Recently, research groups of Gao [151] and Yan [75, 152] have introduced the 1*H*-1,2,4-triazole moiety between the two pyridyl groups to design positional isomeric *N,N′*-donor ligands, 3,5-bis(4-pyridyl)-1,2,4-triazole (4,4′-bpt) and 3,5-bis(3-pyridyl)-1,2,4-triazole (3,3′-bpt), which have been used in the hydrothermal reactions with metal nitrates, H_2pht and NaOH in ethanol (140°C, 3 days) to build grid-like coordination polymers with the formula [M(pht)(4,4′-bpt)(H_2O)$_x$]$_n$ [x = 0, Zn^{2+} (133) [75]; x = 1, Co^{2+} (134) [151, 152]] and [Co(pht)(3,3′-bpt)(H_2O)]$_n$ (135) [152]. Similar to 131 and 132, the phthalate ligand displays the chelating-bridging μ_2-η^1:η^1:η^1:η^0 mode connecting Co(II) atoms to furnish the polymeric chain {Co(pht)} with a Co...Co separation of 5.266 Å in 134 and 5.281 Å in 135; while the adjacent Zn(II) centers in the formed {Zn(pht)} chains in 133 are connected by 1,3- and 1,6-bridging phthalate with a Zn...Zn separation of 7.622 Å. These chains are further interlinked through 3,3′-bpt (135) or 4,4′-bpt (133 and 134) spacers to generate 2D wave-like layers. The M...M separation by the 4,4′-bpt linker is 13.893 Å in 133, and 14.325 Å in 134, and it is 12.092 Å in 135, where 1D chains are bridged by 3,3′-bpt linker. Two related cobalt(II) and cadmium(II) phthalate-based CPs have been reported by Du *et al.* [153] with 4-amino-3,5-bis(4-pyridyl)-1,2,4-triazole (bpt), {[M(pht)(bpt)(H_2O)] • H_2O}$_n$ [M = Co^{2+} (136); Cd^{2+} (137)]. The CPs 136 and 137 have also layered architectures formed by the linking of one-dimensional {M(pht)} chains by the polypyridyl bpt organic linker. The 2D layer, which can be described as a 4-connected 4^4-**sql** net has the M...M separation by bpt of 14.842 Å (Co) and 15.149 Å (Cd), and through the phthalate bridge it is 5.263 Å (Co) and 5.391 Å (Cd). Notably, owing to the flexibility of the angular dipyridyl derivatives the layers in 133–137 are more corrugated compared to CPs 124–131 with rodlike bpy ligands.

LaDuca and co-workers [154] have reported a series of copper(II) and cobalt(II) phthalate-based 2D CPs with bis(4-pyridylmethyl)piperazine (bpmp) prepared by solvent diffusion or hydrothermal method, {[Cu_2(pht)$_2$(Hbpmp)$_2$(H_2O)$_2$](NO_3)$_2$ • H_2O}$_n$ (138),

{[Cu_2(pht)$_2$(Hbpmp)$_2$(H_2O)$_2$](SO_4) • $2H_2O$}$_n$ (139) and {[Co(pht)(bpmp)(H_2O)] • $5.5H_2O$}$_n$ (140). Slow diffusion of an aqueous mixture of copper nitrate or copper sulfate and H_2pht with a methanolic solution of bpmp afforded phthalate-based CPs 138 and 139, whereas hydrothermal conditions, e.g. the heating of the mixture with Co(ClO_4)$_2$ • $6H_2O$, phthalic acid and bpmp in an oven at 120°C for 2 days, are required for the isolation of 140. In 138 and 139, the neighboring Cu(II) atoms are linked by 1,6-bridging phthalate ligands into {Cu(pht)(H_2O)} chains with Cu...Cu distances of 7.532 and 7.483 Å. These chains are further joined by Hbpmp$^+$ ligands to generate cationic {Cu(pht)(Hbpmp)(H_2O)}$^+$ layers, the Cu...Cu separations through bpmp ligands are 13.924 and 13.978 Å. Thus, the layer motifs adopt a 4^4-**sql** topology. Charge-balancing NO_3^- (138) or SO_4^{2-} (139) counterions are situated between the cationic layers. Different from 138 and 139, a pair of phthalate ligands in 140 is coordinated to two Co(II) atoms in a μ_2-η^1:η^1:η^1:η^0 coordination mode (8 in Scheme 1) with the formation of a dinuclear [Co_2(pht)$_2$(H_2O)$_2$] cluster, the adjacent dinuclear clusters are further connected by bpmp ligands into 2D layers with rhombic grids, Co...Co distances *via* bpmp are 16.514 and 16.576 Å.

Alternatively, grid-like copper(II) coordination polymer {[Cu_4(pht)$_4$(btm)$_6$] • 4dmf • $3H_2O$}$_n$ (141) has been made using a bis(triazole) ligand bearing methyl spacer, bis(1,2,4-triazole-1-yl)methane (btm), in the reaction with copper perchlorate and potassium hydrogen phthalate in dimethylformamide-water solution [155]. In 141, Cu(II) atoms are linked through single and double btm bridges with the Cu...Cu distances of 8.974 and 8.797 Å, respectively, to form alternating one-dimensional chains, which are further linked by the bis(monodentate) phthalates to generate 2D layers with cavities [Cu...Cu distance through pht is 7.684 Å]. Varying the length of the alkyl spacer in bis(triazole) derivatives, e.g. using of bis(1,2,4-triazole-1-yl)ethane (bte) or 1,4-bis(1,2,4-triazol-1-yl)butane (btb), alters the structure considerably. A two-dimensional network in {[Zn_2(pht)$_2$(bte)$_2$] • $4H_2O$}$_n$ (142) prepared by Li *et al.* [156] is formed through helical chains linking {Zn_2(pht)$_2$} dinuclear units by the bte ligands with the formation of 50-membered macrometallocyclic rings. Thus, two 1,6-bridging phthalates connect Zn centers forming a 14-membered ring with a Zn...Zn distance of 5.326 Å. A flexible bte ligand bridges the dinuclear units to generate a helical chain, and finally the adjacent helical chains are connected by the *exo*-bidentate bte linkers to furnish the formation of 2D structure containing extremely large 50-membered rings. Each metallocycle consists of four bpe bridged four {Zn_2(pht)$_2$} dinuclear units, two of them belong to the neighboring helixes. The authors extended their research and used a more flexible long chain ligand such as 1,4-bis(1,2,4-triazol-1-yl)butane in the reaction with cadmium nitrate and H_2pht in methanol/water solution adjusted to pH = 6 with dilute Et_3N to generate {[Cd_4(pht)$_4$(btb)$_2$(H_2O)$_6$] • $9H_2O$}$_n$ (143) [157]. In 143, the phthalate anions adopt two different coordination modes (9 and 16 in Scheme 1) to connect Cd atoms resulting in a 2D structure, in which the Cd...Cd distances are 5.538 and 6.528 Å. The flexible btb ligands additionally bridge the Cd centers with a Cd...Cd separation of 11.369 Å and decorate the 2D networks. The lattice water molecules occupy the cavities of the 2D layers and form a hydrogen bonding network to furnish a 3D hydrogen-bonded architecture in 143.

For the construction of 2D phthalate-based CPs the flexible bis(imidazole) derivatives holding different alkyl spacers between imidazole rings have also been extensively explored. Ma and co-workers [158, 159] have reported four new zinc(II), cadmium(II) and cobalt(II) 2D CPs, namely [[Zn(pht)(bim-6)]$_n$ (144), [Zn(pht)(bim-5)]$_n$ (145) and {[Cd(pht)(bim-

5)$(H_2O)_{1.5}$] • $0.5H_2O\}_n$ (146) and [Co(pht)(bim-5)]$_n$ (147), prepared by hydrothermal reactions of 1,1′-(1,5-pentanedidyl)bis(imidazole) (bim-5) and 1,1′-(1,6-hexanedidyl)bis(imidazole) (bim-6) with metal salts and phthalic anhydride in water at 170-180°C for 4 days. Upon hydrothermal reaction of bim-6 with cobalt chloride and phthalic acid in aqua solution at pH = 5 (150°C for 3 days), a 2D CP [Co(pht)(bim-6)]$_n$ (148) formed [160]. Owing to the flexible nature of the ancillary bridging co-ligands, the different 2D motifs are observed in 144–148. For instance, two phthalates adopting bis(monodentate) modes in zinc(II) CP 144 and bis(chelating) coordination modes in cadmium(II) CP 146 connect the adjacent metal atoms with the formation of 14-membered metallocyclic dinuclear units, which are further extended by bis(imidazole) ligands into grid-like 2D architectures with a 4^4-**sql** topology. The longer bim-6 ligand gives the rectangular grids with dimensions of 16.485 x 14.189 Å compared to rhombic grids with 15.116 x 15.116 Å, as found in 146. The similar rhombic grid 2D network was observed in silver(I) phthalate-based CP, [Ag_2(pht)(bbim)(H_2O)]$_n$ (149), prepared recently by Ma *et al.* [161] upon the crystallization from ethanol solution containing Ag_2CO_3, H_2pht and 1,1′-(1,4-butanediyl)bis(2-methylbenzimidazole) (bbim). In 149, two phthalate ligands connect four Ag(I) ions in a $\mu_4:\eta^1:\eta^1:\eta^1:\eta^1$ coordination fashion (22 in Scheme 1) to form tetranuclear {$Ag_4(pht)_2$} units, which are further joined by the bbim linkers into a 2D layer with 4^4-**sql** net. The formed rhombic grids have a dimension of 13.961 x 13.961 Å through bbim bridges. Alternatively, in CPs 145, 147 and 148, the phthalate ligand linked metal atoms in an usual 1,6-bridging mode to generate a 1D chain of {M(pht)} and these chains are further connected together by bim-5 or bim-6 to construct a hexagonal honeycomb 6^3-**hcb** layer or an undulated grid (4^4-**sql**) 2D network. In another study by Ma and co-workers [162], a copper(II) phthalate-based CP [Cu(pht)(bie)]$_n$ (150) with the related flexible N-donor bis(imidazole) ligand, 2,2′-bis(1*H*-imidazolyl)ether (bie), was reported. The CP 150 also shows a 4-connected two-dimensional 4^4-**sql** net formed by the 1D {Cu(pht)} chains which are linked through bie bridges.

Several potential tripodal ligands, e. g. 2,4,5-tris(4-pyridyl)imidazole (Htpim) [163] and 1,3,5-tris(1-imidazolyl)benzene (tib) [164], have also been used for the construction of 2D polymeric phthalate-based networks. Solvothermal reaction of cadmium nitrate with Htpim and H_2pht in DMF/H_2O solution at 140°C for 3 days afforded a 2D {[$Cd_3(pht)_2(H_{0.5}tpim)_4(H_2O)_2$] • $5H_2O\}_n$ (151) CP. In 151, two phthalate ligands in a bidentate-chelating fashion connect two Cd(II) atoms to form a [$Cd_2(pht)_2$] dinuclear unit with a 14-membered metallocycle. $H_{0.5}$tpim ligand *via* N atom from its one pyridyl group coordinates to one of the Cd(II) atoms from this dimer, whereas its other pyridyl group ligates the third Cd(II) atom; thus three Cd atoms are linked by $H_{0.5}$tpim and phthalate ligands to produce the {[$Cd_3(pht)_2(H_{0.5}tpim)_4(H_2O)_2$} trinuclear unit which extends in four directions to form a 2D layer with $(5^3)(5^4)$ net topology. The Cd...Cd distances separated by the pht and $H_{0.5}$tpim ligands are 5.141 and 10.115 Å, respectively. Two crystallographically independent octahedral Cd atoms have different coordination environments: one of them is surrounded by four pyridyl N atoms from $H_{0.5}$tpim ligands in the equatorial plane, and two water molecules occupy the axial positions; whereas in the other one the equatorial plane is formed by three carboxylate O atoms and one pyridyl N atom of $H_{0.5}$tpim ligand, and the axial positions are filled with O atom from phthalate and N atom from $H_{0.5}$tpim ligands. Sun *et al.* [164] have reported a zinc(II) phthalate-based 2D CP, [$Zn_4(pht)_4(tib)_2$] • $2H_2O\}_n$ (152), with another auxiliary tripodal imidazole-containing ligand, 1,3,5-tris(1-imidazolyl)benzene (tib). An interesting feature of the CP 152 is the formation of a tetranuclear Zn_4 unit in which the Zn atoms are linked by four phthalate ligands. The phthalate ligand adopts three different μ_2- and μ_3-bridging coordination modes: μ_2-η^1:η^0:η^0:η^1, μ_3-η^1:η^1:η^2:η^0 and μ_3-η^0:η^2:η^2:η^1 modes (5, 15 and 20 in Scheme 1). Further connection of the $\{Zn_4(pht)_4\}$ units *via* μ_3-bridging tib ligands leads to the formation of a 2D framework.

Escuer and co-workers [165] have reported an interesting phthalate-based CP, $\{[Cu(Hpht)(N_3)] \bullet H_2O\}_n$ (153), having 2D layers following the strategy of mixing different potential bridging ligands with the copper salt. In CP 153, azide ligands in μ_2-end-on fashion bridge the neighboring copper(II) atoms to build a one-dimensional linear chain $\{Cu(N_3)\}$, the chains are further extended by the phthalate connectors to give rise to a 2D 4^4-**sql** motif possessing rectangular grids with dimensions of 3.284 x 7.383 Å. Furthermore, when the same starting materials were heated under autogenous pressure at 120°C for 2 days, an unprecedented 2D CP $\{[Cu(H_2O)_6][Cu_{12}(pht)_6(OH)_6(N_3)_8]\}_n$ (154) with the considerably different 2D framework is generated [166]. Compared to the 2D layer in 153 built up from 1D $(Cu$-μ-$N_3)\}$ chains which were interconnected by the phthalate groups, the coordination polymer 154 has a layered architecture formed by linking large metallocycles through μ_3-azide and hydroxyl groups. The metallocycles consist of twelve Cu atoms bridged by six μ_2-azide and six μ_4-phthalate ligands with ca. 12 x 12 Å pores, wherein charge balancing $[Cu(H_2O)_6]^{2+}$ ions reside. Magnetic measurements of 153 indicate strong ferromagnetic coupling through the carboxylate-azido-bridged chains and very weak antiferromagnetic interactions *via* phthalate ligands which link the chains at longer distance, whereas in 154, the antiferromagnetic interactions dominate over the ferromagnetic ones.

Several other *N*-donor ligands, such as ethylenediamine (eda) and imidazole (im), have also been used in the preparation of copper(II) and cadmium(II) 2D phthalate–based CPs with the formula $[Cu(pht)(eda)]_n$ (155) [167], $\{[Cu_2(pht)_2(im)_4(H_2O)] \bullet H_2O\}_n$ (156) [168] and $\{[Cd_2(pht)_2(im)_4] \bullet H_2O\}_n$ (157) [169]. The CP 155 was prepared from the reaction of $[Cu(eda)(H_2O)]^{2+}$ with sodium phthalate. Refluxing of aqueous solution containing a basic copper(II) carbonate, *o*-phthalic acid and imidazole for 30 min afforded the 2D copper(II) phthalate-based CP 156, whereas the hydrothermal reaction of cadmium acetate with imidazole and *o*-phthalic acid in aqueous solution at 160°C for 4 days gave the 2D cadmium(II) phthalate-based CP 157. Notably, different from above mentioned CPs with a range of polydentate N-donor ligands, the bridging function in 155–157 is only executed by the phthalate ligand, and N-donor amine or azole ligands act as terminal ones. Although these ligands block the coordination sites around metal ions, the μ_3-bridging phthalate ligand manages the $\{Cu(eda)\}$ or $\{M(im)_2\}$ (M = Cu, Cd) units in such way that 2D grids with 4-connected 4^4-**sql** nets in 155 and 157, and a 6^3-**hcb** net topology in 156 are formed.

Recent efforts have also been devoted to using additional carboxylate ligands with appropriate donor sites in the preparation of two-dimensional phthalate-based CPs. For example, Thirumurugan and Natarajan [170] have reported the hydrothermal reaction between yttrium nitrate, *o*-phthalic acid and ancillary dicarboxylic acids – 1,3-benzenedicarboxylic (1,3-bdc, isophthalic) or 1,4-benzenedicarboxylic (1,4-bdc, terephthalic) acids, in the presence of 1,10′-phenanthroline, which gave rise two 2D CPs, $\{[Y_2(pht)_2(1,3$-$bdc)(phen)_2]\}_n$ (158) and $\{[Y_2(pht)_2(1,4$-$bdc)(phen)_2]\}_n$ (159). Both CPs consist of one-dimensional double-stranded chains formed from $\{Y(phen)\}$ units interconnected by the μ_3-bridging phthalate ligand which adopts a μ_3-η^1:η^1:η^1:η^1 coordination mode (16 in Scheme 1) with the formation of 8-membered and 14-membered metallocycles (Y...Y distances through pht bridges are 7.517 and 5.316 Å for 158, and 7.172 and 5.395 Å for 159). The chains are additionally connected by isophthalate (158) or terephthalate (159) connectors to furnish 2D layer architectures with 4^4-**sql** nets. The longer terephthalate linker in 159 provides the longer Y...Y distance of 11.359 Å between the double chains, compared to 10.963 Å found in 158 through the isophthalates. In CP $[Pb_5(pht)_4(ac)_2]_n$ (160), reported by Maginn and co-workers [171], the adjacent lead atoms are linked *via* either acetate or phthalate ligands to generate an complicated 2D layer. Here, the phthalate ligand adopts diverse μ_3-, μ_4- and μ_5-bridging coordination modes: μ_3-η^1:η^1:η^1:η^1, μ_4-η^1:η^1:η^1:η^1, and μ_5-η^2:η^2:η^1:η^0 (14, 22 and 38 in Scheme 1).

A series of two-dimensional copper(II) phthalate-based CPs have been prepared by exploring the several mixed *N,O*-donor ligands such as *N,N'*-substituted oxamide: *N,N'*-bis(3-aminopropyl)oxamide (L^9); *N,N'*-bis(2-aminoethyl)oxamide (L^{10}) and *N,N'*-bis(2-aminopropyl)oxamide (L^{11}), with the general formula $\{[Cu_2(pht)(L)] \cdot xH_2O\}_n$ [$x = 2$, L = L^9 (161) [172], L^{10} (162) [173]; $x = 4$, L = L^{11} (163) [174]]. The reaction of CuL (L = L^9, L^{10}, and L^{11}) with equimolar amounts of copper bromide or perchlorate salts and sodium phthalate in aqueous or aqueous-methanol solution at room temperature afforded CPs 161 – 163, where the *N,N'*-substituted oxamide bridges two Cu(II) atoms to form dimeric $\{Cu_2L\}$ units with the Cu...Cu distance through the oxamide ligand of 5.234 – 5.283 Å. These dimeric building units are then connected by the phthalate spacers to generate the resulting 2D grid-like polymers with a 4-connected 4^4-*sql* net topology. Each grid consists of two dinuclear Cu-L units and two phthalates adopting a unique coordination mode μ_3-η^0:η^1:η^1:η^2 (17 in Scheme 1), with one carboxylate group ligating in a monodentate coordination fashion and another acts as a μ_2-bridge, in 161 and 162, or an usual 1,6-bridging mode in 163. As a result, the dimensions of the grids are different: 7.609 x 7.511 (161), 7.888 x 7.486 (162) and 7.978 x 8.593 Å (163) through oxamide and phthalate connectors. Magnetic analysis reveals strong antiferromagnetic coupling interactions between copper(II) ions in 161 and 162.

The coordination ability of the phthalate ligand towards other metal ions such as lanthanides in the generation of 2D polymers was also extensively explored. Examples of 2D phthalate-based coordination polymers with lanthanides are very abundant. Jin *et al.* [175] have reported a series of lanthanides 2D phthalate-based CPs with the formula $[Ln_2(pht)_3(H_2O)]_n$ [Ln = La^{3+} (164); Eu^{3+} (165); Tb^{3+} (166); Yb^{3+} (167)] prepared by the reaction of $LnCl_2 \bullet xH_2O$ with *o*-phthalic acid at pH = 3,5 adjusted with NaOH aqueous solution under hydrothermal conditions at 170°C for 80 h. Using dysprosium nitrate instead of lanthanide chlorides in the hydrothermal reactions at 160°C for 3 days Yan *et al.* [176] and later Song *et al.* [177] have prepared the related phthalate-based CP with dysprosium(III), $[Dy_2(pht)_3(H_2O)]_n$ (168). Lu *et al.* [178] have described the hydrothermal synthesis of gadolinium oxide with *o*-phthalic acid in the presence of L-glutamic acid at 170°C for 3 days to give $[Gd_2(pht)_3(H_2O)]_n$ (169). Li *et al.* [179] hydrothermally prepared a 2D CP with samarium ions, $[Sm_2(pht)_3(H_2O)]_n$ (170), from the reaction of samarium oxide with *o*-phthalic acid at 160°C for 3 days, and Wang *et al.* [180] with thulium ions, $[Tm_2(pht)_3(H_2O)]_n$ (171). In all CPs, the phthalate ligand coordinates to three or four metal ions and adopts diverse coordination modes resulting in the different 2D motifs. In 164, the phthalate displays two different coordination fashions, types 22 and 28 in Scheme 1, linking four metal ions into 2D grids with 4^4-*sql* nets in which the La...La distances range from 4.434 to 6.450 Å. Different from 164, the phthalate ligand in 165, 166, and 168–170 acts as μ_3- and μ_4-bridges linking three or four metal ions through its four carboxylate O atoms (18, 25 and 28 in Scheme 1), and each metal atoms connects four phthalate ligands to form a distorted honeycomb layer (6^3-*hcb* net) with Ln...Ln distances 3.815, 4.455 and 5.290 Å (Sm); 3.786, 4.440 and 5.277 Å (Eu); 3.785, 4.469 and 5.314 Å (Tb); 3.725, 4.412 and 5.262 Å (Dy) and 3.767, 4.437 and 5.276 Å (Gd). In 167 and 171, the phthalate ligand acts also as μ_3- and μ_4-bridges and adopts three kinds of coordination modes to connect Yb^{3+} ions (16, 28 and 29 in Scheme 1) or Tm^{3+} ions (16, 22 and 29 in Scheme 1), however, here the formed 2D motif is a 4-connected 4^4-*sql* coordination network. Utilization of 1,10′-phenanthroline as a terminal ligand to block certain metal coordination sites in the hydrothermal reaction of $EuCl_3 \bullet 6H_2O$ with the same starting materials but heated at 150°C for 5 days provides a 2D phthalate-based CP, $[Eu(pht)(Hpht)(phen)(H_2O)]_n$ (172) [181], with a 2D herringbone architecture having the larger grid dimensions of 12.554 x 8.509 Å as shown in Figure 7a.

Interestingly, Natarajan and Thirumurgam [182] used piperazine (piz) in the combination with $Gd(NO_3)_3$, *o*-phthalic acid, 1,10′-phenanthroline and NaOH in water under hydrothermal conditions at 180°C for 3 days to produce 2D CP $\{(piz)[Gd_2(pht)_4(H_2O)_2]\}_n$ (173) consisting of a 2D anionic coordination network $[Gd_2(pht)_4(H_2O)_2]$ with large pores containing the protonated piperazine molecules. A (6,3)-honeycomb topology results in this case (Figure 7b).

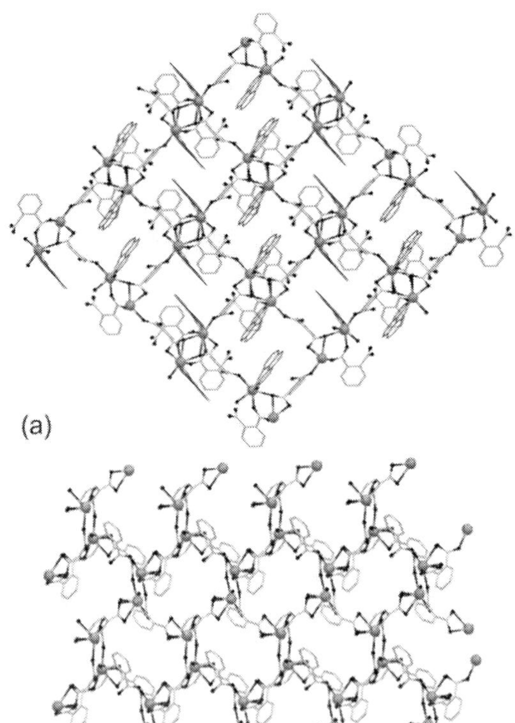

Figure 7. View of (a) herringbone-like layer in CP 172 [181] and (b) honeycomb-like layer in CP 173 [182]. Metal atoms, black spheres; O atoms, black balls; C atoms, grey sticks. Ancillary N-donor ligands are shown as thin black lines.

The use of different flexible dicarboxylates as co-ligands has allowed the isolation of many 2D phthalate-based CPs with lanthanides. Several research groups [179, 183-186] have recently investigated the influence of ancillary aliphatic dicarboxylate ligands such as oxalic (H_2ox), maleic (H_2mal) and adipic (H_2ad) acid on the topology of coordination networks formed in an extensive series of lanthanide CPs with the general formula $\{[Ln_2(pht)_2(L)(H_2O)_4] \bullet xH_2O\}_n$ [L = ox, x = 1, Ln = La^{3+} (174), Ce^{3+} (175), Pr^{3+} (176), Nd^{3+} (177), Sm^{3+} (178), Eu^{3+} (179), Gd^{3+} (180), Tb^{3+} (181); L = mal, x = 2, Ln = Sm^{3+} (182), Eu^{3+} (183); L = ad, x = 0, Eu^{3+} (184), Sm^{3+} (185)]. Remarkably, all lanthanide CPs have been synthetized under hydrothermal conditions. In the reaction with the oxalic acid for the preparation of phthalate-based CPs, Xu *et al.* [184] used lanthanide nitrate salts (Ce, Pr, Nd, Sm, Eu, and Gd), and the pH values were controlled in the range 3-4 by addition of KOH solution; the reaction mixture were heated at 160°C for 3 days, whereas lanthanide chlorides (Sm and Tb) were used by Xing *et al.* [185] in the hydrothermal reaction at 180°C for 4 days, and pH was adjusted by ethylenediamine. Li *et al.* [186] used lanthanide oxides in the similar hydrothermal reaction (180°C, 3 days, without any base) to produce La- and Nd-containing phthalate-based CPs, 174 and 177.

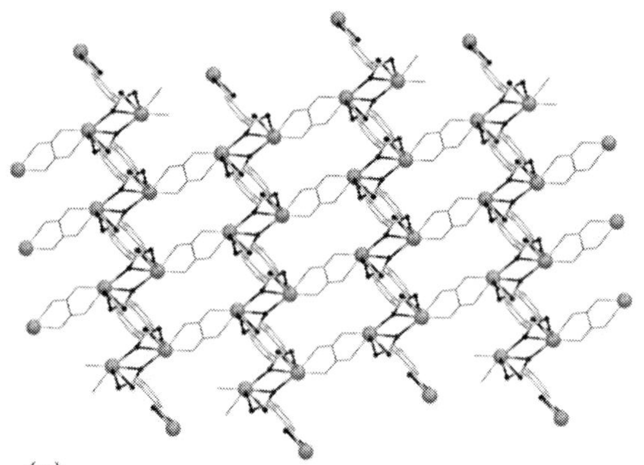

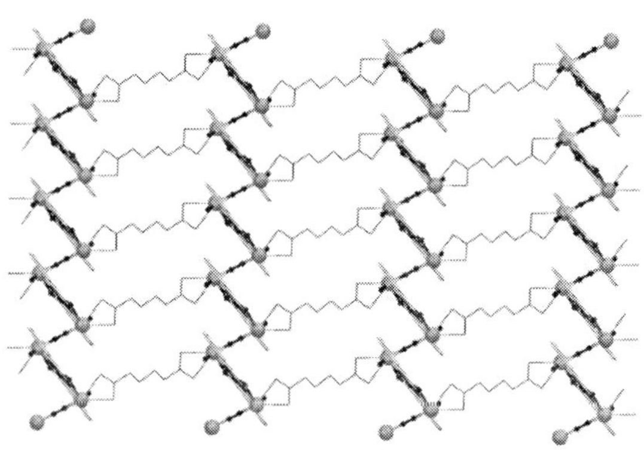

Figure 8. View of brick wall grids in CP 175 [184] (a) and CP 184 [185] (b). Metal atoms, black spheres; O atoms, black balls; C atoms, grey sticks. Ancillary N-donor ligands are shown as thin black lines.

Although, the hydrothermal conditions were different as well as starting materials (lanthanide salts and a base), the final products 174–181 have the similar composition and structures. Here, two phthalate ligands coordinate to two Ln ions forming a $[Ln_2(pht)_2(H_2O)_4]$ dimer, the adjacent dimeric units are further connected by the O atom from the carboxylate group of the same phthalate ligand [pht adopts a μ_3-η^1:η^2:η^1:η^1 coordination mode (19 in Scheme 1)] to generate a staircase chain. The neighboring staircase-like chains are linked by oxalate ligands into a 2D layer architecture with the (4,3) net containing rectangular cavities in 174–181 (Figure 8). The dimensions of these cavities range from 6.639 x 10.550 Å for lanthanum(III) phthalate-based CP (174) to 6.505 x 10.200 Å for terbium(III)-containing CP [181].

When maleic acid was used by Xing *et al.* [185] in the hydrothermal reactions with samarium and europium chlorides, the 2D $\{[Sm_2(pht)_2(mal)(H_2O)_4] \bullet 2H_2O\}_n$ (182) and $\{[Eu_2(pht)_2(mal)(H_2O)_4] \bullet 2H_2O\}_n$ (183) CPs have been successfully synthesized. Similar to CPs 174–181 with the oxalic acid, the phthalate ligand in 182 and 183 adopts a μ_3-bridging mode connecting Sm(III) or Eu(III) atoms into {Ln-pht} stairs-like chains, which are further bridged by the auxiliary dicarboxylate ligands to furnish 2D layers. As expected, the formed rectangular grids through the alkyl dicarboxylate ligands are longer with the dimensions 6.385 x 13.139 Å (Sm) and 6.378 x 13.111 Å (Eu). Interesting to note is that *cis*-maleate anion under this condition is transformed to *trans*-fumarate in 182 and 183, the conversion which is used for the industrial synthesis of fumaric acid. Moreover, utilization of praseodymium chloride in the similar reaction leads to the chemical rearrangement from maleic acid to malic acid (H_3mal') and the formation of 2D $\{[Pr_3(pht)_3(mal')(H_2O)_3] \bullet H_2O\}_n$ (186) CP, where the phthalate ligand displays different coordination modes 16, 18 and 34 in Scheme 1. The use of a longer adipic acid in the reactions with europium or samarium chlorides and *o*-phthalic acid in the presence of triethylamine at 160°C for 3 days gave directly 2D phthalate-based CPs 184 and 185, $[Ln_2(pht)_2(ad)(H_2O)_4]_n$ (Ln = Eu, Sm), with similar 2D structural motifs, which consist of {Ln-pht} stairs-like chains bridged by adipate ligands (Figure 8b). In 184 and 185, the dimensions of the formed cavities are ca. 6.519 x 14.632 Å for Eu and 6.526 x 14.631 Å for Sm.

Relative to 2D phthalate-based CPs with actinides, only several examples have been reported in the literature. Charushnikova *et al.* [187] have described a 2D phthalate-based CP with Np^{5+} ions, $\{[NpO_2)_2(pht)] \cdot 4H_2O\}_n$ (187), prepared by the hydrothermal reaction of neptunium nitrate with ammonium phthalate and *o*-phthalic acid at 160°C for 5 hours. In 187, the $\{NpO_2\}^+$ dioxocations are bonded to each other through O atoms to form cationic two-dimensional layers with rhombic grids (Np…Np, 4.165 x 4.202 Å). The phthalate ligand additionally connects the dioxocations showing a μ_3-bridging mode (12 in Scheme 1) and decorates the formed neutral 2D layer. The coordination polyhedron around two unique Np centers is a pentagonal bipyramid: one of them is coordinated by four O atoms from adjacent dioxocations, two water molecules and one O atom from the phthalate ligand, whereas another is bonded to O atoms from four neighboring $\{NpO_2\}^+$ dioxocations, two O atoms from different phthalates and a single water molecule. Under mild synthetic conditions, e.g. refluxing of plutonium nitrate with lithium phthalate at 70°C for 3 hours with a pH = ~4, Grigoriev *et al.* [108] have prepared CP, $\{[PuO_2(pht)(H_2O)] \bullet H_2O\}_n$ (188), which exhibits a 2D honeycomb framework. Other 2D phthalate-based CPs with uranyl has been reported by Loiseau *et al.* [109]. The presence of potassium or ammonium cations in the hydrothermal reactions of uranyl nitrate with *o*-phthalic acid at pH adjusted by KOH or NH_3 bases up to 3 at 200°C for 24 hours led to the isolation of 2D $[K(UO_2)_3O(pht)_2(OH)(H_2O)]_n$ (189) and $[NH_3(UO_2)_3O(pht)_2(OH)(H_2O)]_n$ (190) CPs, respectively. In these polymers, the formed 1D anionic $[(UO_2)_3O(pht)_2(OH)(H_2O)]^-$ motif consists of the interconnected by phthalate bridges di- and tetranuclear uranyl building blocks. The chains are further interlinked by potassium or ammonium cations into 2D architectures.

1.3. Three-Dimensional Coordination Polymers (3D)

Three-dimensional CPs, also known as metal-organic frameworks (MOFs) – the term "MOF" was introduced by Yaghi *et al.* [188] in 1995, represent the class of materials with porous structures that is currently of great interest due to their potential applications for gas storage and separation, ion exchange, luminescence, magnetism and catalysis, in medicine as drug delivery devices and contrast agents for magnetic resonance imaging, as active components in sensing devices [2]. For the design and construction of phthalate-based MOFs the main efforts were focused on the exploration of both the mixed ligand systems, e.g. the use of ancillary bridging *N,N'*-, *O,O'*- or *N,O*-donor ligands, and transition metals (Co^{2+}, Ni^{2+}, Cd^{2+}, and Ag^+), and lanthanide metals (Y^{3+}, La^{3+}, Pr^{3+}, Nd^{3+}, Sm^{3+}, Gd^{3+}, and Dy^{3+}) or d-f (Cu–Ln where Ln = La^{3+}, Ce^{3+}, Nd^{3+}, Sm^{3+}, Eu^{3+}, Tb^{3+}, and Er^{3+} ions, and Ag–Nd) heterometallic systems that provide this type of framework. It is noticeable that almost all phthalate-based MOFs have been synthesized under hydrothermal conditions.

Lightfoot and Snedden [189] reported the first cobalt(II) phthalate-based MOF with 4,4′-bipyridine (4,4′-bpy), $[Co(Hpht)_2(4,4\text{-bpy})]_n$ (191), in 1999, which was prepared hydrothermally from the mixture containing a 1:2:1 molar ratio of cobalt nitrate, *o*-phthalic acid and 4,4′-bpy at 100°C for 3 days. The CP 191 exhibits a three-dimensional framework structure built up from linear {Co(4,4′-bpy)} chains with Co...Co distance of 11.534 Å, which are further linked to each other by 1,6-bridging phthalate ligands (Co...Co distance through the pht is 9.187 Å) to generate 3D frameworks as shown in Figure 9. The coordination environment around Co atoms is octahedral, being composed of two N atoms from 4,4′-bpy and four O atoms from four phthalate ligands. Wang *et al.* [190] have designed a similar 3D phthalate-based CP with cadmium ions, $[Cd(Hpht)_2(4,4'\text{-bpy})]_n$ (192).

Zheng and co-workers [160] have explored the derivatives of bis(imidazole) ligands such as 1,1′-(1,4-butanediyl)bis(imidazole) (bim-4) and 1,4-bis(imidazol-1-ylmethyl)benzene (bim-b), to construct nickel(II) and cobalt(II) coordination polymers, $[Ni(pht)(bim\text{-}4)]_n$ (193) and $[Co(pht)(bim\text{-}b)]_n$ (194). The CPs 193 and 194 have been prepared from the hydrothermal reaction of metal chlorides with *o*-phthalic acid and bis(imidazole) ligands at 160°C (193) or 150°C (194) for 3 days and the pH was adjusted by NaOH to 5.5 for 193 or 6.0 for 194. The resulting polymers have 3D structures consisting of infinite {M(pht)} chains, linear in 193 and zigzag in 194, interlinked by the auxiliary bis(imidazole) ligands (Ni...Ni, 13.891 Å; Co...Co, 13.029 Å) to give a 3D framework.

The Ni(II) atoms in 193 adopt a octahedral N_2O_6 geometry, which is provided by two N atoms from different bim-4 ligands and four O atoms from two phthalates. The phthalate dianions link Ni(II) centers in a chelating-bridging mode (9 in Scheme 1) with the Ni...Ni distance of 6.364 Å.

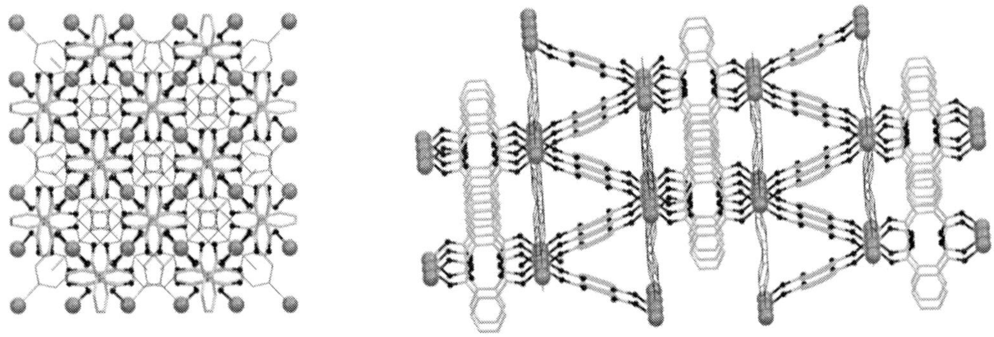

Figure 9. View of a 3D framework in CP 191 [189]. Metal atoms, black spheres; O atoms, black balls; C atoms, grey sticks. Ancillary N-donor ligands are shown as thin black lines.

In contrast, each Co(II) center in 194 is coordinated to two N atoms of two independent bim-b and two O atoms from two phthalate ligands in a tetrahedral N_2O_2 coordination geometry, and the phthalate adopts an usual 1,6-bridging mode (5 in Scheme 1, Co...Co, 6.726 Å). As a result, the CP 193 holds a 4-connected 3D $6^5{\cdot}8$ framework, while in 194 the 4-connected 3D $6^5{\cdot}10$ framework is formed. Additionally, Sun *et al.* [191] have reported a similar zinc(II) phthalate-based MOF with bim-b ligand, $\{[Zn_2(pht)_2(bim\text{-}b)] \cdot 2H_2O\}_n$ (195), prepared from the hydrothermal reaction of zinc sulfate with *o*-phthalic acid and bim-b ligand in the presence of NaOH with a molar ratio of 2:1:2:4 at 160°C for 3 days. Here, the phthalate ligand connects three Zn(II) atoms using its two carboxylate groups with a μ_3-η^1:η^1:η^0:η^1 (12 in Scheme 1) coordination mode to form a 4-connected 2D grid layer with Zn...Zn edge distances 5.047 and 6.299 Å. Furthermore, the 2D layers with a 4^4-*sql* net are linked together by the bis(imidazole) ligands (Zn...Zn, 13.768 Å) to generate a 3D architecture with a bimodal (3,4)-connected net. The photoluminescent properties of 195, together with the bim-b and pht ligands, were studied in the solid state at room temperature. The 195 exhibits emission at 485 nm (λ_{ex} = 390 nm), that can be assigned to intraligand transition of coordinated bim-b ligands. Ma *et al.* [161] have reported a zinc(II) phthalate MOF with 1,1′-(1,4-butanediyl)bis(2-methylbenzimidazole) (bbim), $[Zn(pht)(bbim)]_n$ (196), prepared by the solvothermal reaction of $Zn(OH)_2$ with H_2pht and bbim in EtOH/H_2O at 140°C for 3 days. In 196, the Zn(II) atoms are linked by the bridging pht and bbim ligands with the Zn...Zn distances of 6.405 and 14.171 Å to give a 3D diamond network. In contrast, slow mixing a MeOH/H_2O solution of H_2pht and 1,4-bis(1,2,4-triazol-1-yl)butane (btb) with pH = 6 adjusted by NEt_3 with an aqueous solution of copper nitrate into "H-shape" tube afforded a copper(II) phthalate-based CP, $[Cu(pht)(btb)]_n$ (197) [192], in which a {Cu(pht)} chain connects four other {Cu(pht)} chains through btb connectors to build a 3D frameworks, the Cu...Cu separations *via* btb bridges are 13.550 Å.

An extensive series of phthalate-based MOFs has been developed by introducing into a phthalate-metal system the multidentate tris- and tetrakis(imidazole) ligands by groups of Sun [164] and Ma [193]. Sun and co-workers [164] have investigated the hydrothermal reaction of different zinc salts ($ZnCl_2$, $Zn(OAc)_2 \bullet 2H_2O$, $Zn(NO_3)_2 \bullet 6H_2O$, ZnI_2) with *o*-phthalic acid and 1,3,5-tris(1-imidazolyl)benzene (tib). Under similar synthetic conditions, e.g. heating the mixture which contains zinc(II) salt, tib, H_2pht and NaOH with a ratio of 1:1:1:2 at 180°C for 3 days, the zinc(II) phthalate-based MOFs with different framework structures were isolated, $[Zn_2(pht)Cl_2(tib)]_n$ (198), $\{[Zn_2(pht)(tib)_2](NO_3)_2\}_n$ (199), and $\{[Zn_2(pht)(tib)_2]I_2\}_n$ (200); that emphasized the influence of different anions (Cl^-, NO_3^- and I^-) used in the reactions on the structure of the CPs. All CPs 198 – 200 have three-dimensional structures, however in 198 the 3D architecture contains infinite 2D networks pillared by the phthalates resulting in a trinodal (3,4) connected net, while 199 and 200 feature unusual 3D frameworks with a binodal (3,4)-connected net where 2D layers with a $4,8^2$ net topology are linked by the phthalate ligands. Moreover, in 198, the Cl^- anions are terminal and coordinate to Zn atoms, whereas NO_3^- and I^- anions occupy the 1D channels formed in 199 and 200. In addition, each tib ligand in 198 using its three imidazole arms links three Zn(II) atoms to form 1D chains which are joined by carboxylates into 2D layers with 38-membered metallocycles formed by three tib and two pht ligands together with five Zn atoms. Such 2D layers are further connected by phthalates to generate a final 3D network. In 199 and 200, each tib ligand also links three Zn(II) atoms, but organizes them into a 2D network with 20-membered and 40-membered metallocycles; thus the 2D layers have a $4,8^2$ net topology. The 20-membered cycle is formed by two tib and two metal atoms, whereas four tib ligands connect four Zn atoms to form the 40-membered metallocycle. The final 3D architecture is completed by the 1,6-bridging phthalates that join together these 2D layers. The Zn atoms in 198 – 200 all have the same tetrahedral coordination geometry, but a different coordination environment. In 198, two unique Zn atoms have $NOCl_2$ and N_2O_2 donor sets – with two Cl^- ions, and one N atom of tib and one phthalate carboxylate O atom for the first one, and two imidazole N donor atoms of two tib and two carboxylate O atoms from different phthalates for the other one. The tetrahedral N_3O geometry of Zn atoms in 199 and 200 is completed by three imidazole N atoms from three tib and one O atom from phthalate. Ma and co-workers [193] used the flexible tetrakis(imidazole) ligand such as tetrakis(imidazol-1-ylmethyl)methane (tim), to produce zinc(II) and cobalt(II) CPs, $\{[Zn_2(pht)_2(tim)] \bullet 3.5H_2O\}_n$ (201) and $\{[Co_2(pht)_2(tim)] \bullet 2.5H_2O\}_n$ (202), with a trinodal (3,4)-connected 3D framework. Both MOFs have been prepared by the heating of the mixture containing metal carbonate, H_2pht and tetrakis(imidazole) ligand in the ratio of 1:1:0.5 under hydrothermal conditions at 170°C for 3 days. Remarkably, each tim ligand using all its four imidazole arms links metal atoms to generate 2D layers, which are further bridged by two kinds of phthalate anions to form a 3D network. In 201, the Zn atoms are four- and five-coordinated, exhibiting a tetrahedral N_2O_2 or trigonal bipyramidal N_2O_3 coordination environment through two N atoms from two tim ligands and two or three carboxylate O atoms from two phthalates. The Co atoms in 202 are four- and six-coordinated and have trigonal N_2O_2 and octahedral N_2O_4 donor sets, respectively, being surrounded by two imidazole N atoms from different tim, and two or four carboxylate O atoms from two phthalate ligands.

Replacement of flexible *N,N'*-donor derivatives of azole ligands with the rigid dicarboxylate such as terephthalic acid leads to an interesting cobalt(II) 3D CP, $[Co_3(pht)_2(1,4\text{-bdc})(H_2O)_4)]_n$ (203), reported by Zheng's group [194]. Heating of the mixture containing cobalt chloride, *o*-phthalic and terephthalic acids in a ratio of 1.5:1:1 with the pH = 7 adjusted by NaOH solution at 170°C for 3 days gave rise to CP 203. The remarkable feature of 203 is the formation of isosceles trinuclear Co_3 units, in which Co atoms are held together by μ-H_2O molecules and carboxylate groups from phthalates to construct anionic 2D $\{Co_3(pht)_2(H_2O)_4)\}^{2-}$ nets, that are further supported by the terephthalates, thus creating the (3,8)-connected 3D pillar-layered frameworks with the Co…Co distance of 11.266 Å between adjacent sheets. Two crystallographically independent Co atoms are both six-coordinated and show a O_6 octahedral geometry: for one of them the coordination environment is completed by three O atoms from three phthalates, one O atom from terephthalate and two water molecules, and the other one through four O atoms from phthalates and two water molecules; and here the phthalate ligand adopts a μ_5-η^1:η^1:η^2:η^1 (39 in Scheme 1) coordination mode. The study of magnetic properties of 203 reveals the presence of antiferromagnetic interactions ($\theta = -13.2$ K) between Co^{2+} ions with a Curie constant of 6.57 cm^3 K mol^{-1}.

The terephthalic acid has also been used by Thirumurugan and Natarajan [195, 196] in the preparation of an extensive series of phthalate-based MOFs with lanthanides of the general formula $[Ln_2(pht)_2(1,4\text{-bdc})(H_2O)_4]_n$ [Ln = Y^{3+} (204); La^{3+} (205); Pr^{3+} (206); Nd^{3+} (207); Sm^{3+} (208); Gd^{3+} (209); Dy^{3+} (210)]. All CPs 204–210 have been hydrothermally synthesized from the reaction of lanthanide chloride with 1,2-dicyanobenzene and terephthalic acid with pH adjusted to 5 by NaOH at 180°C for 3 days. Notably, under these conditions, the 1,2-dicyanobenzene is oxidized to *o*-phthalic acid. The prepared polymers all have a three-dimensional structure built up from 2D $[Ln_2(pht)_2(H_2O)_4]$ corrugated layers, in which Ln atoms are connected through the phthalate dianions, and these layers are further cross-linked by bridging terephthalate ligands to furnish a 3D framework. However, the coordination modes of the phthalate ligand as well as the coordination environment around Ln atoms are different. In 205 – 207, the lanthanide (La, Pr, and Nd) centers are nine-coordinated while binding to seven carboxylate O atoms and two water molecules with an overall geometry being a triply capped trigonal prism; the phthalate coordinates to four Ln atoms adopting a μ_4-η^1:η^1:η^2:η^1 (25 in Scheme 1) coordination mode. Different from 205 – 207, the coordination geometry of Ln (Y, Sm, Gd, and Dy) atoms in other MOFs (204, and 208 – 210) is a distorted dodecahedron and the Ln atoms are eight-coordinated, surrounded by six carboxylate O atoms and two water molecules; here the phthalate acts as a μ_4-bridge, linking four lanthanide atoms through its four carboxylate O atoms and it adopts a μ_4-η^1:η^1:η^1:η^1 (22 in Scheme 1) coordination fashion.

A series of mixed lanthanide-transition metal phthalate-based MOFs has been synthesized by research groups of Yang [197-199] and Wang [200, 201], who used in the heterometallic (Ln–Cu) systems an ancillary *N,O*-donor ligand such as isonicotinic acid (Hina). Hydrothermal reactions of lanthanide oxides and copper halides with isonicotinic and phthalic acids at 170°C for 7 [197-199] or 9 [200] days lead to 3D CPs with the general formulae $[Ln_3Cu_7(pht)(ina)_8Br_6(H_2O)_4]_n$ [Ln = La^{3+} (211) [200]; Ce^{3+} (212) [197]] and $[Er_4Cu_8(pht)_2(ina)_8I_7(OH)(H_2O)_5)]_n$ (213) [197]. Using in the similar hydrothermal reactions $HClO_4$ [198] or $NaClO_4$ [201] with a ratio of starting materials Ln_2O_3: $CuCl_2$/CuCl: Hina: H_2pht: $HClO_4$/$NaClO_4$ as 0.5:0.2:2:1:0.7/0.5, afforded 3D CPs with the general formula $\{[Ln_2Cu(pht)_2(ina)_2(H_2O)_2]X\}_n$ [X = ClO_4^-, Ln = Nd^{3+} (214), Sm^{3+} (215), Eu^{3+} (216) [198, 201], Tb^{3+} (217) [201]; X = Cl^-, Ln = Nd^{3+} (218) [198]]. The remarkable feature of 211 – 218, is the formation of two-dimensional lanthanide-carboxylate layers and Cu-containing chains. In 211 and 212, the Ln (La in 211 and Ce in 212) atoms are connected by isonicotinate and phthalate ligands to generate two-dimensional cationic Ln-carboxylate layers, $\{Ln_3(pht)(ina)_6(H_2O)_4\}^+$. Such Ln-containing 2D layers are further extended by the bridging copper(I)-isonicotinate units, $\{Cu(ina)_2\}^-$, forming a 3D motif that possesses 1D channels. The inorganic $\{Cu_6Br_6\}$ chains are located inside these channels. Note, although the Cu^{2+} ions were used as starting materials, under these conditions the reduction to the oxidation state of +1 takes place.

Table 1.

N	CPs	Coordination modes displayed by the phthalate ligand (type in Scheme 1)	Structural features	Ref. ccde in CSD	Ref.
1	$[Hg_2(pht)]_n$	μ_2-η^1:η^0:η^0:η^1 (5)	1D linear chain	HGOPHT	[25]
2	$[Cu(pht)(NH_3)_2]_n$	μ_2-η^1:η^0:η^0:η^1 (5)	1D linear chain	PALCUA10	[26]
3	$[Zn(pht)(py)_2]_n$	μ_2-η^1:η^0:η^0:η^1 (5)	1D zigzag chain	RAJREZ	[31]
4	$[Zn(pht)(\beta\text{-pic})_2]_n$	μ_2-η^1:η^0:η^0:η^1 (5)	1D zigzag chain	RAJRID	[31]
5	$[Zn(pht)(\gamma\text{-pic})_2]_n$	μ_2-η^1:η^0:η^0:η^1 (5) μ_2-η^1:η^1:η^0:η^1 (6)	1D zigzag chain	RAJROJ	[31]
6	$[Cu(pht)(py)_2]_n$	μ_2-η^1:η^1:η^0:η^1 (6) μ_2-η^1:η^1:η^1:η^1 (9)	1D zigzag chain	BARPOZ	[32]
7	$[Cu(pht)(pip)_2]_n$	μ_2-η^1:η^1:η^1:η^1 (9)	1D linear chain	HEHPEP	[33]
8	$[Cu_2(pht)_2(\gamma\text{-pic})_4(H_2O)]_n$	μ_2-η^1:η^0:η^0:η^1 (5)	1D zigzag chain	LITMOQ	[33]
9	$\{[Cu(pht)(ina)_2] \cdot H_2O\}_n$	μ_2-η^1:η^0:η^0:η^1 (5)	1D zigzag chain	YIQRIA	[35]
10	$[Co(pht)(im)_2]_n$	μ_2-η^1:η^0:η^0:η^1 (5)	1D zigzag chain	UHUXAW UHUXAW01	[36] [17]
11	$\{[Cu(pht)(im)_2] \cdot 1.5\ H_2O\}_n$	μ_2-η^1:η^1:η^0:η^1 (6) μ_2-η^1:η^1:η^1:η^1 (9)	1D helical chain	CEHFUR	[17]
12	$\{[Cu(pht)(im)_2]_n$	μ_2-η^1:η^1:η^1:η^1 (9)	1D linear chain	COWEOG	[37]
13	$[Zn(pht)(im)_2]_n$	μ_2-η^1:η^0:η^0:η^1 (5)	1D zigzag chain	LUKHEE LUKHEE01	[38] [20]
14	$[Zn(pht)(pyr)_2]_n$	μ_2-η^1:η^0:η^0:η^1 (5)	1D zigzag chain	BEZXOU	[39]
15	$[Co(pht)(1\text{-Meim})_2]_n$	μ_2-η^1:η^0:η^0:η^1 (5)	1D helical chain	XAHKIB	[40]
16	$[Cu(pht)(1\text{-Meim})_2]_n$	μ_2-η^1:η^0:η^0:η^1 (5)	1D linear chain	XAHKOH	[40]
17	$[Zn(pht)(1\text{-Meim})_2]_n$	μ_2-η^1:η^0:η^0:η^1 (5)	1D helical chain	XAHLAU	[40]
18	$[Co(pht)(2\text{-Meim})_2]_n$	μ_2-η^1:η^0:η^0:η^1 (5)	1D zigzag chain	ECIDID	[41]
19	$[Co(pht)(4\text{-Meim})_2(H_2O)]_n$	μ_2-η^1:η^1:η^0:η^1 (6)	1D helical chain	OBICAE	[42]
20	$[Cu(pht)(4\text{-Meim})_2(H_2O)]_n$	μ_2-η^1:η^0:η^0:η^1 (5)	1D helical chain	OBICEI	[42]
21	$[Zn(pht)(4\text{-Etim})_2]_n$	μ_2-η^1:η^0:η^0:η^1 (5)	1D zigzag chain	NIQGOK	[43]
22	$[Zn(pht)(1\text{-Viim})_2]_n$	μ_2-η^1:η^0:η^0:η^1 (5)	1D zigzag chain	NIQGIE	[44]
23	$[Cu(pht)(1\text{-Viim})_2]_n$	μ_2-η^1:η^0:η^0:η^1 (5)	1D zigzag chain	XILWOF	[45]
24	$[Zn(pht)(1\text{-Alim})_2]_n$	μ_2-η^1:η^0:η^0:η^1 (5)	1D zigzag chain	RUSMUO	[46]

Table 1. (Continued)

N	CPs	Coordination modes displayed by the phthalate ligand (type in Scheme 1)	Structural features	Ref. code in CSD	Ref.
25	$[Cd(pht)(bzim)_3]_n$	μ_2-η^1:η^1:η^1:η^1 (9)	1D helical chain	YERRUJ	[47]
26	$\{[Cd(pht)(bzim)_2(H_2O)] \cdot 0.5H_2O\}_n$	μ_2-η^1:η^1:η^0:η^1 (6)	1D zigzag chain	IZULIR	[48]
27	$[Co(pht)(bzim)_2]_n$	μ_2-η^1:η^0:η^0:η^1 (5)	1D linear chain	HAVFAM	[49]
28	$[Zn(pht)(\gamma\text{-pic})]_n$	μ_3-η^1:η^1:η^2:η^0 (15) μ_3-η^1:η^1:η^1:η^1 (16)	1D double chain	RAJRUP	[31]
29	$[Zn(pht)(2\text{-Meim})]_n$	μ_3-η^1:η^1:η^0:η^1 (12)	1D double chain	LUWVUU	[20]
30	$\{Li_2[Cu(pht)_2] \cdot 4H_2O\}_n$	μ_3-η^0:η^2:η^0:η^1 (11) μ_3-η^1:η^1:η^0:η^1 (12)	1D double chain	LICUPH	[27]
31	$\{Rb_2[Cu(pht)_2] \cdot 2H_2O\}_n$	μ_2-η^1:η^0:η^0:η^1 (5)	1D double chain	RBCUPH	[27]
32	$\{Cs_2[Cu(pht)_2] \cdot 2H_2O\}_n$	μ_2-η^1:η^0:η^0:η^1 (5)	1D double chain	ZZZBQS	[27]
33	$\{K_2[Cu(pht)_2] \cdot 2H_2O\}_n$	μ_2-η^1:η^0:η^0:η^1 (5)	1D double chain	KPHTCU	[28]
34	$[Mn(Hpht)_2(H_2O)_2]_n$	μ_2-η^1:η^0:η^0:η^1 (5)	1D double chain	MNHPHD	[50]
35	$[Cu(Hpht)_2(H_2O)_2]_n$	μ_2-η^1:η^0:η^0:η^1 (5)	1D double chain	DICWER	[51]
36	$\{(pnH_2)_2[Zn(pht)_2]\}_n$	μ_2-η^1:η^0:η^0:η^1 (5)	1D double chain	SIJNOP	[52]
37	$\{(tmp)_2[Cu(pht)_2]\}_n$	μ_2-η^1:η^1:η^0:η^1 (6)	1D double chain	HAGRUD	[53]
38	$[Cd(pht)(bpy)(H_2O)]_n$	μ_2-η^1:η^1:η^0:η^1 (6)	1D helical chain	JAVQAZ	[54]
39	$[Cd(pht)(phen)(H_2O)]_n$	μ_2-η^1:η^1:η^0:η^1 (6) μ_3-η^0:η^1:η^1:η^2 (17)	1D helical chain	JAVQED	[54]
40	$\{[Zn(pht)(phen)(H_2O)] \cdot H_2O\}_n$	μ_2-η^1:η^1:η^0:η^1 (6)	1D helical chain	WAFDAJ WAFDAJ01	[55] [54]
41	$[Co(pht)(bpy)(H_2O)]_n$	μ_2-η^1:η^1:η^0:η^1 (6)	1D helical chain	LATQED	[56]
42	$\{[Co(pht)(phen)(H_2O)] \cdot EtOH\}_n$	μ_2-η^1:η^1:η^0:η^1 (6)	1D helical chain	MEPFET	[58]
43	$[Zn(pht)(bpy)(H_2O)]_n$	μ_2-η^1:η^1:η^0:η^1 (6)	1D helical chain	EGUNAV	[59]
44	$\{[Cu(pht)(phen)(H_2O)] \cdot 0.5H_2O\}_n$	μ_2-η^1:η^0:η^0:η^1 (5)	1D zigzag chain	COWFUP COWFUP01	[60] [61]
45	$\{[Cu(pht)(phen)(H_2O)] \cdot H_2O\}_n$	μ_2-η^1:η^0:η^0:η^1 (5)	1D zigzag chain	YACTUS YACTUS01	[62] [63]

N	CPs	Coordination modes displayed by the phthalate ligand (type in Scheme 1)	Structural features	Ref. code in CSD	Ref.
46	$\{[Ni(pht)(phen)(H_2O)] \cdot H_2O\}_n$	μ_2-η^1:η^0:η^0:η^1 (5)	1D zigzag chain	DEZLOK	[64]
47	$\{[Cu(pht)(bpy)(H_2O)] \cdot 3.5H_2O\}_n$	μ_2-η^1:η^0:η^0:η^1 (5)	1D waving chain	VEZGIQ	[65]
48	$\{[Cu_2(pht)_2(bpy)_2(H_2O)] \cdot 3H_2O\}_n$	μ_2-η^1:η^0:η^0:η^1 (5) μ_3-η^1:η^1:η^0:η^1 (6)	1D double chain	LALGEK	[66]
49	$\{[Zn_2(pht)_2(bpy)_2(H_2O)] \cdot H_2O\}_n$	μ_2-η^1:η^0:η^0:η^1 (5)	1D zigzag chain	REBPIY	[67]
50	$[Mn(pht)(bpy)(H_2O)_2]_n$	μ_2-η^1:η^0:η^0:η^1 (5)	1D linear chain	ABIFAT ABIFAT01	[68] [69]
51	$[Cd(pht)(bpy)(H_2O)_2]_n$	μ_2-η^1:η^0:η^0:η^1 (5)	1D linear chain	WIHQAG	[70]
52	$\{[Zn_2(pht)(L^1)_2(H_2O)]_n$	μ_4-η^2:η^1:η^1:η^1 (26)	1D cluster-based chain	TIYYEG	[71]
53	$\{[Cd_2(pht)_2(HL^1)_2] \cdot H_2O\}_n$	μ_3-η^1:η^1:η^1:η^1 (16) μ_3-η^1:η^2:η^2:η^1 (21)	1D cluster-based chain	TIYYIK	[71]
54	$\{[Cd(pht)(HL^1)(H_2O)]_n$	μ_2-η^1:η^1:η^0:η^1 (6)	1D linear chain	TOBQIL	[71]
55	$\{[Ni_3(pht)(dpa)_4] \cdot 1.5H_2O\}_n$	μ_2-η^1:η^0:η^0:η^1 (5)	1D cluster-based chain	UJOYOI	[72]
56	$[Co(pht)(dpa)]_n$	μ_2-η^1:η^1:η^1:η^1 (9)	1D linear chain	GELWUP	[73]
57	$\{[Zn_2(pht)_2(bpe)] \cdot MeOH\}_n$	μ_3-η^1:η^1:η^1:η^1 (16) μ_3-η^0:η^2:η^2:η^1 (20)	1D ladder chain		[74]
58	$[Zn_2(pht)_2(3,4'\text{-}bpt)]_n$	μ_3-η^1:η^1:η^0:η^1 (12)	1D ladder chain	NUWWUY	[75]
59	$\{[Zn(pht)(bpbp)] \cdot H_2O\}_n$	μ_2-η^1:η^0:η^0:η^1 (5)	1D ladder chain	SIPHIJ	[76]
60	$\{[Zn_2(pht)_2(pzp)_2] \cdot H_2O\}_n$	μ_3-η^1:η^1:η^0:η^1 (6)	1D double chain	MIXSES	[77]
61	$\{[Pb(pht)(pzp)] \cdot H_2O\}_n$	μ_3-η^1:η^2:η^2:η^1 (12)	1D double chain	MOXYII	[78]
62	$[Zn(pht)(L^2)]_n$	μ_2-η^1:η^1:η^0:η^1 (6)	1D double chain	PESGEA	[80]
63	$[Zn(pht)(L^3)]_n$	μ_2-η^1:η^1:η^0:η^1 (6)	1D linear chain	MEQBIU	[81]
64	$[Zn_2(pht)_2(bpp)]_n$	μ_3-η^1:η^1:η^1:η^1 (16) μ_3-η^1:η^1:η^2:η^0 (15)	1D double chain	ICEMAF	[82]
65	$\{[Zn(pht)(L^4)] \cdot 2H_2O\}_n$	μ_2-η^1:η^0:η^0:η^1 (5)	1D double chain	COCTAQ	[83]
66	$\{[Co_2(pht)_2(L^5)_2(H_2O)_2] \cdot H_2O\}$	μ_2-η^1:η^0:η^0:η^1 (5)	1D double chain	POCSUW	[84]
67	$[Mn(Hpht)(L^6)]_n$	μ_2-η^1:η^0:η^0:η^1 (5)	1D linear chain	COXQOV	[85]
68	$\{[Cu_4(pht)_2(tacn)_4](ClO_4)_4 \cdot 2H_2O\}_n$	μ_3-η^1:η^1:η^1:η^1 (14)	1D zigzag chain	EBABAL	[86]
69	$\{[Zn(pht)(cyclam)] \cdot 2MeOH\}_n$	μ_2-η^1:η^0:η^0:η^1 (5)	1D linear chain	ALOJOA	[15]
70	$[Zn(pht)(tu)_2]_n$	μ_2-η^1:η^0:η^0:η^1 (5)	1D helical chain	QIKXIR	[87]

Table 1. (Continued)

N	CPs	Coordination modes displayed by the phthalate ligand (type in Scheme 1)	Structural features	Ref. code in CSD	Ref.
71	$[Zn(pht)(L^7)_2]_n$	μ_2-η^1:η^0:η^0:η^1 (5)	1D linear chain	ULIHIG	[88]
72	$\{[Ni(pht)(oaoH)_2] \cdot 4H_2O\}_n$	μ_2-η^1:η^0:η^0:η^1 (5)	1D helical chain	CIWTIL	[89]
73	$\{[Mn(pht)(phen)(H_2O)_2] \cdot H_2O\}_n$	μ_2-η^1:η^1:η^0:η^0 (4)	1D linear chain	FEYQAB	[90]
74	$\{[Cu(pht)(dien)] \cdot 0.5MeOH\}_n$	μ_2-η^1:η^1:η^0:η^0 (4)	1D linear chain	HIRXAH	[91]
75	$\{[Cu_2(pht)(ClO_4\ (dien)_2)](ClO_4)\}_n$	μ_3-η^1:η^1:η^0:η^1 (12)	1D linear chain	JIWVAM	[92]
76	$\{[Co(pht)(4,4'\text{-}dpa)_4(H_2O)_2] \cdot H_2O\}_n$	η^0:η^1:η^1:η^0 (3)	1D linear chain	XEZLEU	[93]
77	$\{[Ni(pht)(4,4'\text{-}dpa)_4(H_2O)_2] \cdot H_2O\}_n$	η^0:η^1:η^1:η^0 (3)	1D linear chain	XEZLIY	[93]
78	$\{[Cu(pht)(Hpht)(L^8)(H_2O)] \cdot 3H_2O\}$	η^1:η^0:η^0:η^0 (1)	1D waving chain	JOFKUL	[94]
79	$[Co(pht)(azpy)(H_2O)_3]_n$	η^1:η^0:η^0:η^0 (1)	1D linear chain	IKUKOO	[95]
80	$\{[Co(pht)(L^9)(H_2O)_2] \cdot 2.36H_2O\}_n$	η^0:η^1:η^1:η^0 (3)	1D linear chain	DEPSOH	[96]
81	$\{[Co_2(pht)_2(L^{10})(H_2O)_6] \cdot H_2O\}_n$	η^1:η^0:η^0:η^0 (1)	1D zigzag chain	DEPSUN	[97]
82	$\{[Co_2(pht)_2(L^{11})(H_2O)_6] \cdot 1.12H_2O\}_n$	η^1:η^0:η^0:η^0 (1)	1D zigzag chain	DEPTAU	[96]
83	$\{[Y_2(pht)_3(phen)_2] \cdot H_2O\}_n$	μ_3-η^1:η^1:η^1:η^1 (14) μ_3-η^1:η^1:η^1:η^1 (16) μ_3-η^1:η^1:η^2:η^1 (18)	1D double chain	RAFLEQ	[97]
84	$[Ag_2(pht)(phz)_2(H_2O)]_n$	μ_2-η^1:η^1:η^1:η^0 (8)	1D linear chain	NARRUT	[98]
85	$[Ag(Hpht)(ina)]_n$	η^1:η^0:η^0:η^0 (1)	1D double chain	TOBHOI	[99]
86	$\{[La_3(pht)_4(NO_3)(phen)_2] \cdot H_2O\}_n$	μ_2-η^0:η^1:η^2:η^1 (10) μ_3-η^1:η^2:η^2:η^1 (18) μ_3-η^1:η^1:η^2:η^1 (21)	1D double chain	LAXWIR	[100]
87	$\{[Pr_3(pht)_4(NO_3)(phen)_2] \cdot H_2O\}_n$	μ_2-η^0:η^1:η^2:η^1 (10) μ_3-η^1:η^2:η^2:η^1 (18) μ_3-η^1:η^1:η^2:η^1 (21)	1D double chain	LAXWOX	[100]
88	$[Sm_3(pht)_4(NO_3)(phen)_2]_n$	μ_3-η^1:η^1:η^1:η^1 (14) μ_3-η^1:η^1:η^2:η^1 (18) μ_3-η^1:η^2:η^2:η^1 (21)	1D double chain	FIDXAS	[101]

N	CPs	Coordination modes displayed by the phthalate ligand (type in Scheme 1)	Structural features	Ref. code in CSD	Ref.
89	$\{[Eu_3(pht)Cl(phen)_3(H_2O)_6]Cl_2 \cdot 4H_2O\}_n$	μ_3-η^1:η^1:η^0:η^1 (12) μ_3-η^1:η^1:η^2:η^1 (18)	1D double chain	CIHNOX	[102]
90	$\{[Nd_2(pht)_3(phen)(H_2O)] \cdot H_2O\}_n$	μ_3-η^1:η^1:η^1:η^1 (16) μ_3-η^1:η^1:η^2:η^1 (18) μ_4-η^1:η^1:η^2:η^1 (25)	1D double chain	HUVWUQ	[103]
91	$\{[Ho_2(pht)_3(phen)(H_2O)] \cdot 3H_2O\}_n$	μ_2-η^1:η^1:η^1:η^1 (9) μ_2-η^1:η^1:η^1:η^1 (14) μ_3-η^1:η^1:η^1:η^1 (16)	1D double chain	HUVXAX	[103]
92	$\{[Yb_2(pht)_3(phen)(H_2O)_2] \cdot 3.5H_2O\}_n$	μ_2-η^1:η^1:η^1:η^1 (9) μ_2-η^1:η^1:η^1:η^1 (14) μ_3-η^1:η^1:η^1:η^1 (16)	1D double chain	IKAYIC	[104]
93	$\{[Gd_2(pht)_3(bpy)(H_2O)_2] \cdot 3H_2O\}_n$	μ_2-η^1:η^1:η^1:η^1 (9) μ_3-η^1:η^1:η^1:η^1 (16) μ_3-η^1:η^1:η^2:η^1 (18)	1D double chain	ICIMUD	[105]
94	$\{[Er_2(pht)(OH)_2(ina)_2(H_2O)] \cdot H_2O\}_n$	μ_3-η^0:η^2:η^2:η^1 (20)	1D wave-like chain	GIMMOF	[106]
95	$\{[Tm_2(pht)(OH)_2(ina)_2(H_2O)] \cdot H_2O\}_n$	μ_3-η^0:η^2:η^2:η^1 (20)	1D wave-like chain	GIMMUL	[106]
96	$\{[UO_2(pht)(H_2O)] \cdot H_2O\}_n$	μ_3-η^1:η^1:η^1:η^1 (14)	1D double chain	XATRIU	[107]
97	$\{[NpO_2(pht)(H_2O)] \cdot 0.33H_2O\}n$	μ_4-η^1:η^1:η^1:η^1 (22)	1D tube-like chain	NAJHEM	[108]
98	$\{[UO_2(pht)(H_2O)] \cdot 0.32H_2O\}_n$	μ_4-η^1:η^1:η^1:η^1 (22)	1D tube-like chain		[109]
99	$\{[UO_2(pht)(ur)] \cdot 0.5H_2O\}_n$	μ_3-η^1:η^1:η^1:η^1 (14)	1D double chain	DIFZAT	[110]
100	$\{[Co(NH_3)_6][NpO_2(Pht)_2] \cdot 2H_2O\}_n$	η^0:η^1:η^1:η^0 (3) μ_2-η^1:η^1:η^0:η^1 (6)	1D linear chain	OKOCAS	[111]
101	$[Cu(pht)(H_2O)]_n$	μ_4-η^1:η^0:η^1:η^2 (24)	2D rectangular grids	CUPHAL CUPHAL02	[112] [113]
102	$[Cu(pht)(H_2O)_2]_n$	μ_3-η^1:η^0:η^1:η^1 (12)	2D rectangular grids	PHALCV	[114]
103	$[Na_2Cu(pht)_2(H_2O)_2]_n$	μ_5-η^2:η^1:η^2:η^2 (42) μ_5-η^1:η^2:η^2:η^2 (43)	2D herringbone layer	NAPHCU NAPHCU01 NAPHCU01 NAPHCU01	[28] [115] [116] [117]

Table 1. (Continued)

N	CPs	Coordination modes displayed by the phthalate ligand (type in Scheme 1)	Structural features	Ref. code in CSD	Ref.
104	$[MgCu(pht)_2(H_2O)_4]_n$	μ_3-η^1:η^0:η^1:η^1 (12)	2D rectangular grids	MGPHCU	[118]
105	$\{Ag[Cu_2(Pht)_2(OH)(H_2O)] \bullet 3H_2O\}_n$	μ_4-η^1:η^1:η^1:η^1 (22) μ_5-η^1:η^2:η^2:η^2 (24)	2D layers	AGPHCU	[119]
106	$[Li_4(pht)_2(H_2O)_3]_n$	μ_5-η^2:η^2:η^1:η^0 (37) μ_5-η^2:η^1:η^2:η^1 (38)	2D layers	VUYNOS	[120]
107	$[Li_4(pht)_2(EtOH)(H_2O)]_n$	μ_3-η^1:η^2:η^0:η^0 (13) μ_3-η^1:η^1:η^2:η^0 (15)	2D layers	GOLKOI	[23]
108	$[K(Hpht)]_n$	μ_4-η^2:η^3:η^1:η^0 (33) μ_5-η^0:η^3:η^2:η^3 (40)	2D layers	KHPHAL KHPHAL01 KHPHAL02 KHPHAL03 KHPHAL04 KHPHAL05 KHPHAL06	[121] [122] [123] [123] [123] [124]
109	$[K(Hpht)(H_2O)]_n$	μ_4-η^1:η^1:η^2:η^2 (32)	2D layers	EROBES	[125]
110	$[Sr_5(pht)_4(NO_3)_2(H_2O)_2]_n$	μ_6-η^3:η^2:η^3:η^2 (47)	2D layers	FENRUM	[126]
111	$[Co_3(pht)_2(OH)_2]_n$	μ_6-η^2:η^1:η^1:η^2 (48)	2D honeycomb layer	BEBCUH	[127]
112	$[KCo(pht)(OH)]_n$	μ_7-η^2:η^3:η^3:η^1 (51)	2D layers	HOGXIL	[24]
113	$[CsCo(pht)(OH)]_n$	μ_7-η^2:η^2:η^3:η^2 (53)	2D layers	HOGXOR	[24]
114	$[Cd(pht)(H_2O)]_n$	μ_4-η^1:η^2:η^2:η^1 (28)	2D honeycomb layer	KACBOF KACBOF01 KACBOF02 KACBOF03	[128] [129] [130] [131]
115	$[Cd(pht)]_n$	μ_4-η^2:η^2:η^2:η^1 (34)	2D rectangular grids	BIFCUP	[132]
116	$[In(pht)(OH)(H_2O)]_n$	μ_3-η^1:η^1:η^0:η^1 (12)	2D rectangular grids	PELTOQ	[133]
117	$[Tl_2(pht)]_n$	μ_7-η^4:η^2:η^2:η^3 (54)	2D layers	XASVET	[22]

N	CPs	Coordination modes displayed by the phthalate ligand (type in Scheme 1)	Structural features	Ref. code in CSD	Ref.
118	$[Tl(Hpht)]_n$	μ_4-η^2:η^3:η^1:η^0 (33)	2D layers	DOKRIE02 DOKRIE03	[134] [135]
119	$[Pb(pht)(H_2O)]_n$	μ_5-η^2:η^1:η^2:η^2 (45)	2D rectangular grids	YIGZAQ	[136]
120	$\{[Ag_2(pht)]_4 \cdot MeCN\}_n$	μ_7-η^3:η^2:η^2:η^1	2D rectangular grids		[137]
121	$[Ag_2(pht)(H_2O)]_n$	μ_7-η^3:η^2:η^2:η^1	2D rectangular grids		[137]
122	$[Ag_4(pht)_2(Hac)]_n$	μ_5-η^1:η^1:η^2:η^1 (39) μ_6-η^2:η^2:η^2:η^1	2D layers	FIYNOR	[138]
123	$[Mn(pht)(pyz)(H_2O)_2]_n$	μ_2-η^1:η^0:η^0:η^1 (5)	2D rectangular grids	ABIFEX	[139] [140]
124	$\{[Mn(pht)(4,4'\text{-}bpy)(H_2O)_2] \cdot 2H_2O\}_n$	μ_2-η^1:η^0:η^0:η^1 (5)	2D rectangular grids	BASMIS BASMIS01	[141] [142]
125	$\{[Co(pht)(4,4'\text{-}bpy)(H_2O)_2] \cdot 2H_2O\}_n$	μ_2-η^1:η^0:η^0:η^1 (5)	2D rectangular grids	NEDXOJ NEDXOJ01	[143] [142]
126	$\{[Cd(pht)(4,4'\text{-}bpy)(H_2O)_2] \cdot 2H_2O\}_n$	μ_2-η^1:η^0:η^0:η^1 (5)	2D rectangular grids	NEDXEZ	[143]
127	$\{[Zn(pht)(4,4'\text{-}bpy)(H_2O)_2] \cdot 2H_2O\}_n$	μ_2-η^1:η^0:η^0:η^1 (5)	2D rectangular grids	JEKTUP	[144] [145]
128	$\{[Cu(pht)(4,4'\text{-}bpy)(H_2O)_2] \cdot 2H_2O\}_n$	μ_2-η^1:η^0:η^0:η^1 (5)	2D rectangular grids	TEJKUP	[146]
129	$\{[Fe(pht)(4,4'\text{-}bpy)(H_2O)_2] \cdot 2H_2O\}_n$	μ_2-η^1:η^0:η^0:η^1 (5)	2D rectangular grids	QEHCUC	[147]
130	$\{[Ni(pht)(4,4'\text{-}bpy)(H_2O)_2] \cdot 2H_2O\}_n$	μ_2-η^1:η^0:η^0:η^1 (5)	2D rectangular grids	WIPFOR	[148]
131	$\{[Ni(pht)(4,4'\text{-}bpy)(H_2O)] \cdot 3H_2O\}_n$	μ_2-η^1:η^1:η^1:η^0 (8)	2D rectangular grids	KAGQEP	[149]
132	$\{[Co(pht)(pyEt)(H_2O)] \cdot H_2O\}_n$	μ_2-η^1:η^1:η^1:η^0 (8)	2D rectangular grids	XEPDAY	[150]
133	$[Zn(pht)(4,4'\text{-}bpt)]_n$	μ_2-η^1:η^1:η^0:η^1 (6)	2D rectangular grids	NUWWAE	[75]
134	$[Co(pht)(4,4'\text{-}bpt)(H_2O)]_n$	μ_2-η^1:η^1:η^1:η^0 (8)	2D rectangular grids	SOVQOK SOVQOK01	[151] [152]
135	$[Co(pht)(3,3'\text{-}bpt)(H_2O)]_n$	μ_2-η^1:η^1:η^1:η^0 (8)	2D rectangular grids	HUZNIA	[152]
136	$\{[Co(pht)(bpt)(H_2O)] \cdot H_2O\}_n$	μ_2-η^1:η^1:η^1:η^0 (8)	2D rectangular grids	UDIQCO	[153]
137	$\{[Cd(pht)(bpt)(H_2O)] \cdot H_2O\}_n$	μ_2-η^1:η^1:η^1:η^0 (8)	2D rectangular grids	UDIQLU	[153]
138	$\{[Cu_2(pht)_2(Hbpmp)_2(H_2O)_2](NO_3)_2 \cdot H_2O\}_n$	μ_2-η^1:η^0:η^0:η^1 (5)	2D rectangular grids	TAHRAX	[154]

Table 1. (Continued)

N	CPs	Coordination modes displayed by the phthalate ligand (type in Scheme 1)	Structural features	Ref. code in CSD	Ref.
139	$\{[Cu_2(pht)_2(Hbpmp)_2(H_2O)_2](SO_4)\cdot 2H_2O\}_n$	μ_2-$\eta^1:\eta^0:\eta^0:\eta^1$ (5)	2D rectangular grids	TAHREB	[154]
140	$\{[Co(pht)(bpmp)(H_2O)]\cdot 5.5H_2O\}_n$	μ_2-$\eta^1:\eta^1:\eta^1:\eta^0$ (8)	2D rhombic grids	TAHQUQ	[154]
141	$\{[Cu_4(pht)_4(btm)_6]\cdot 4dmf\cdot 3H_2O\}_n$	μ_2-$\eta^1:\eta^0:\eta^0:\eta^1$ (5)	2D rectangular grids	NOQHAD	[155]
142	$\{[Zn_2(pht)_2(bte)_2]\cdot 4H_2O\}_n$	μ_2-$\eta^1:\eta^0:\eta^0:\eta^1$ (5)	2D layers	FAWKIY	[156]
143	$\{[Cd_4(pht)_4(btb)_2(H_2O)_6]\cdot 9H_2O\}_n$	μ_2-$\eta^1:\eta^1:\eta^1:\eta^1$ (9) μ_3-$\eta^1:\eta^1:\eta^1:\eta^1$ (16)	2D layers	XOJVOI	[157]
144	$[[Zn(pht)(bim\text{-}6)]_n$	μ_2-$\eta^1:\eta^0:\eta^0:\eta^1$ (5)	2D rectangular grids	QOHKII	[158]
145	$[Zn(pht)(bim\text{-}5)]_n$	μ_2-$\eta^1:\eta^0:\eta^0:\eta^1$ (5)	2D honeycomb layer	QOHKOO	[158]
146	$\{[Cd(pht)(bim\text{-}5)(H_2O)_{1.5}]\cdot 0.5H_2O\}_n$	μ_2-$\eta^1:\eta^1:\eta^1:\eta^1$ (9)	2D rhombic grids	QOHKUU	[158]
147	$[Co(pht)(bim\text{-}5)]_n$	μ_2-$\eta^1:\eta^0:\eta^0:\eta^1$ (5)	2D honeycomb layer	QUFVUJ	[159]
148	$[Co(pht)(bim\text{-}6)]_n$	μ_2-$\eta^1:\eta^0:\eta^0:\eta^1$ (5)	2D undulated grids	RIWVID	[160]
149	$[Ag_2(pht)(bbim)(H_2O)]_n$	μ_4:$\eta^1:\eta^1:\eta^1:\eta^1$ (22)	2D rhombic grids		[161]
150	$[Cu(pht)(bie)]_n$	μ_2-$\eta^1:\eta^1:\eta^0:\eta^1$ (6)	2D undulated grids	MOHMAY	[162]
151	$\{[Cd_3(pht)_2(H_{0.5}tpim)_4(H_2O)_2]\cdot 5H_2O\}_n$	μ_2-$\eta^1:\eta^1:\eta^1:\eta^1$ (9)	2D layers	COWWUH	[163]
152	$[Zn_4(pht)_4(tib)_2]\bullet 2H_2O\}n$	μ_2-$\eta^1:\eta^0:\eta^0:\eta^1$ (5) μ_3-$\eta^1:\eta^1:\eta^2:\eta^0$ (15) μ_3-$\eta^0:\eta^2:\eta^2:\eta^1$ (20)	2D layers	GUCZAG	[164]
153	$\{[Cu(Hpht)(N_3)]\cdot H_2O\}_n$	μ_4-$\eta^1:\eta^1:\eta^0:\eta^2$ (23)	2D rectangular grids	RIHZEN	[165]
154	$\{[Cu(H_2O)_6][Cu_{12}(pht)_6(OH)_6(N_3)_8]\}_n$	μ_4:$\eta^1:\eta^1:\eta^1:\eta^1$ (22)	2D layers	POJLIK	[166]
155	$[Cu(pht)(eda)]_n$	μ_3-$\eta^1:\eta^1:\eta^0:\eta^1$ (12)	2D rectangular grids	BUGJAO	[167]
156	$\{[Cu_2(pht)_2(im)_4(H_2O)]\cdot H_2O\}_n$	μ_3-$\eta^1:\eta^2:\eta^1:\eta^1$ (19) μ_2-$\eta^1:\eta^1:\eta^0:\eta^1$ (6)	2D honeycomb grids	KELVED	[168]
157	$\{[Cd_2(pht)_2(im)_4]\cdot H_2O\}_n$	μ_3-$\eta^1:\eta^2:\eta^1:\eta^1$ (19) μ_3-$\eta^1:\eta^1:\eta^1:\eta^1$ (16)	2D rectangular grids	QOGWOZ	[169]
158	$\{[Y_2(pht)_2(1,3\text{-}bdc)(phen)_2]\}_n$	μ_3-$\eta^1:\eta^1:\eta^1:\eta^1$ (16)	2D rectangular grids	BETTEA	[170]
159	$\{[Y_2(pht)_2(1,4\text{-}bdc)(phen)_2]\}_n$	μ_3-$\eta^1:\eta^1:\eta^1:\eta^1$ (16)	2D rectangular grids	BETTIE	[170]
N	CPs	Coordination modes displayed by the phthalate ligand (type in Scheme 1)	Structural features	Ref. code in CSD	Ref.
160	$[Pb_5(pht)_4(ac)_2]_n$	μ_3-$\eta^1:\eta^1:\eta^1:\eta^1$ (14) μ_4-$\eta^1:\eta^1:\eta^1:\eta^1$ (22)	2D layers	WIFVUC	[171]

		μ_5-η^2:η^2:η^1:η^0 (38)			
161	$\{[Cu_2(pht)(L^9)] \cdot 2H_2O\}_n$	μ_3-η^0:η^1:η^1:η^2 (17)	2D rectangular grids	JAPZOP	[172]
162	$\{[Cu_2(pht)(L^{10})] \cdot 2H_2O\}_n$	μ_3-η^0:η^1:η^1:η^2 (17)	2D rectangular grids	SUYCUK	[173]
163	$\{[Cu_2(pht)(L^{11})] \cdot 4H_2O\}_n$	μ_2-η^1:η^0:η^0:η^1 (5)	2D rectangular grids	HAVGAN	[174]
164	$[La_2(pht)_3(H_2O)]_n$	μ_4-η^1:η^1:η^1:η^1 (22) μ_4-η^1:η^1:η^2:η^1 (28)	2D rectangular grids	WUJSID	[175]
165	$[Eu_2(pht)_3(H_2O)]_n$	μ_3-η^1:η^1:η^2:η^1 (18) μ_4-η^1:η^1:η^2:η^1 (25) μ_4-η^1:η^1:η^2:η^1 (28)	2D honeycomb grids	WUJSOJ	[175]
166	$[Tb_2(pht)_3(H_2O)]_n$	μ_3-η^1:η^1:η^2:η^1 (18) μ_4-η^1:η^1:η^2:η^1 (25) μ_4-η^1:η^1:η^2:η^1 (28)	2D honeycomb grids	WUJSUP	[175]
167	$[Yb_2(pht)_3(H_2O)]_n$	μ_4-η^1:η^1:η^1:η^1 (16) μ_4-η^1:η^1:η^2:η^1 (28) μ_4-η^1:η^1:η^2:η^1 (29)	2D rectangular grids	WUJTAW	[175]
168	$[Dy_2(pht)_3(H_2O)]_n$	μ_3-η^1:η^1:η^2:η^1 (18) μ_4-η^1:η^1:η^2:η^1 (25) μ_4-η^1:η^1:η^2:η^1 (28)	2D honeycomb grids	FIBWOD FIBWOD01	[176] [177]
169	$[Gd_2(pht)_3(H_2O)]_n$	μ_3-η^1:η^1:η^2:η^1 (18) μ_4-η^1:η^1:η^2:η^1 (25) μ_4-η^1:η^1:η^2:η^1 (28)	2D honeycomb grids	LEJPIA	[178]
170	$[Sm_2(pht)_3(H_2O)]_n$	μ_3-η^1:η^1:η^2:η^1 (18) μ_4-η^1:η^1:η^2:η^1 (25) μ_4-η^1:η^1:η^2:η^1 (28)	2D honeycomb grids	KUGPUY	[179]
171	$[Tm_2(pht)_3(H_2O)]_n$	μ_4-η^1:η^1:η^1:η^1 (16) μ_4-η^1:η^1:η^1:η^1 (22) μ_4-η^1:η^1:η^2:η^1 (29)	2D rectangular grids	KIWPOW	[180]
172	$[Eu(pht)(Hpht)(phen)(H_2O)]_n$	μ_2-η^1:η^1:η^0:η^0 (4) μ_3-η^1:η^1:η^0:η^1 (12)	2D herringbone grids	IKAXUN	[181]
173	$\{(piz)[Gd_2(pht)_4(H_2O)_2]\}_n$	μ_2-η^1:η^1:η^0:η^1 (6) μ_3-η^1:η^1:η^1:η^1 (16)	2D honeycomb grids	CAXCIO	[182]

Table 1. (Continued)

N	CPs	Coordination modes displayed by the phthalate ligand (type in Scheme 1)	Structural features	Ref. code in CSD	Ref.
174	$\{[La_2(pht)_2(ox)(H_2O)_4] \cdot H_2O\}_n$	μ_3-η^1:η^2:η^1:η^1 (19)	2D brick wall grids	YIHMOS01 YIHMOS	[179] [183]
175	$\{[Ce_2(pht)_2(ox)(H_2O)_4] \cdot H_2O\}_n$	μ_3-η^1:η^2:η^1:η^1 (19)	2D brick wall grids	NULWUN	[184]
176	$\{[Pr_2(pht)_2(ox)(H_2O)_4] \cdot H_2O\}_n$	μ_3-η^1:η^2:η^1:η^1 (19)	2D brick wall grids	NULXAU	[184]
177	$\{[Nd_2(pht)_2(ox)(H_2O)_4] \cdot H_2O\}_n$	μ_3-η^1:η^2:η^1:η^1 (19)	2D brick wall grids	NULXEY NULXEY01	[184] [186]
178	$\{[Sm_2(pht)_2(ox)(H_2O)_4] \cdot H_2O\}_n$	μ_3-η^1:η^2:η^1:η^1 (19)	2D brick wall grids	KUGQEJ KUGQEJ01 KUGQEJ02	[179] [184] [185]
179	$\{[Eu_2(pht)_2(ox)(H_2O)_4] \cdot H_2O\}_n$	μ_3-η^1:η^2:η^1:η^1 (19)	2D brick wall grids	NULXOI	[184]
180	$\{[Gd_2(pht)_2(ox)(H_2O)_4] \cdot H_2O\}_n$	μ_3-η^1:η^2:η^1:η^1 (19)	2D brick wall grids	NULXUO	[184]
181	$\{[Tb_2(pht)_2(ox)(H_2O)_4] \cdot H_2O\}_n$	μ_3-η^1:η^2:η^1:η^1 (19)	2D brick wall grids	NUYSOQ	[185]
182	$\{[Sm_2(pht)_2(mal)(H_2O)_4] \cdot 2H_2O\}_n$	μ_3-η^1:η^2:η^1:η^1 (19)	2D brick wall grids	NUYTAD	[185]
183	$\{[Eu_2(pht)_2(mal)(H_2O)_4] \cdot 2H_2O\}_n$	μ_3-η^1:η^2:η^1:η^1 (19)	2D brick wall grids	NUYTEH	[185]
184	$[Eu_2(pht)_2(ad)(H_2O)_4]_n$	μ_3-η^1:η^2:η^1:η^1 (19)	2D brick wall grids	NUYTOR	[185]
185	$[Sm_2(pht)_2(ad)(H_2O)_4]_n$	μ_3-η^1:η^2:η^1:η^1 (19)	2D brick wall grids	NUYTUX	[185]
186	$\{[Pr_3(pht)_3(mal')(H_2O)_3] \cdot H_2O\}_n$	μ_3-η^1:η^1:η^1:η^1 (16) μ_3-η^1:η^1:η^2:η^1 (18) μ_4-η^2:η^2:η^2:η^1 (34)	2D layers	NUYTIL	[185]
187	$\{[(NpO_2)_2(pht)] \cdot 4H_2O\}_n$	μ_3-η^1:η^1:η^0:η^1 (12)	2D rhombic grids	OKOCOG	[187]
188	$\{[PuO_2(pht)(H_2O)] \cdot H_2O\}_n$	μ_3-η^1:η^1:η^1:η^1 (16)	2D honeycomb grids	NAJHAI	[108]
189	$[K(UO_2)_3O(pht)_2(OH)(H_2O)]_n$	μ_4-η^1:η^2:η^2:η^1 (28)	2D layers		[109]
190	$[NH_3(UO_2)_3O(pht)_2(OH)(H_2O)]_n$	μ_4-η^1:η^2:η^2:η^1 (28)	2D layers		[109]
191	$[Co(Hpht)_2(4,4'\text{-bpy})]_n$	μ_2-η^1:η^0:η^0:η^1 (5)	3D framework	LESBEQ	[189]
192	$[Cd(Hpht)_2(4,4'\text{-bpy})]_n$	μ_2-η^1:η^0:η^0:η^1 (5)	3D framework	FEQSOK	[190]
193	$[Ni(pht)(bim\text{-}4)]_n$	μ_2-η^1:η^1:η^1:η^1 (9)	3D framework	RIWVEZ	[160]
194	$[Co(pht)(bim\text{-}b)]_n$	μ_2-η^1:η^1:η^1:η^1 (5)	3D framework	RIWVAV	[160]
195	$\{[Zn_2(pht)_2(bim\text{-}b)] \cdot 2H_2O\}_n$	μ_3-η^1:η^1:η^0:η^1 (12)	3D framework	PUPMIX	[191]
196	$[Zn(pht)(bbim)]_n$	μ_2-η^1:η^1:η^1:η^1 (9)	3D framework		[161]

N	CPs	Coordination modes displayed by the phthalate ligand (type in Scheme 1)	Structural features	Ref. code in CSD	Ref.
197	[Cu(pht)(btb)]n	μ_2-η^1:η^0:η^0:η^1 (5)	3D framework		[192]
198	$[Zn_2(pht)Cl_2(tib)]_n$	μ_3-η^1:η^1:η^0:η^1 (12)	3D framework	GUCYUZ	[192]
199	$\{[Zn_2(pht)(tib)_2](NO_3)_2\}_n$	μ_2-η^1:η^0:η^0:η^1 (5)	3D framework	GUCZEK	[192]
200	$\{[Zn_2(pht)(tib)_2]I_2\}_n$	μ_2-η^1:η^0:η^0:η^1 (5)	3D framework	GUCZIO	[192]
201	$\{[Zn_2(pht)_2(tim)] \cdot 3.5H_2O\}_n$	μ_2-η^1:η^0:η^0:η^1 (5) μ_2-η^1:η^1:η^0:η^1 (6)	3D framework	CUYTUM	[193]
202	$\{[Co_2(pht)_2(tim)] \cdot 2.5H_2O\}_n$	μ_2-η^1:η^0:η^0:η^1 (5) μ_2-η^1:η^1:η^1:η^1 (9)	3D framework	CUYVAU	[193]
203	$[Co_3(pht)_2(1,4\text{-bdc})(H_2O)_4)]_n$	μ_5-η^1:η^1:η^2:η^1 (39)	3D framework	WOCGUR	[194]
204	$[Y_2(pht)_2(1,4\text{-bdc})(H_2O)_4]_n$	μ_4-η^1:η^1:η^1:η^1 (22)	3D framework	ATIKIX	[195]
205	$[La_2(pht)_2(1,4\text{-bdc})(H_2O)_4]_n$	μ_4-η^1:η^1:η^2:η^1 (25)	3D framework	ATIJOC	[195]
206	$[Pr_2(pht)_2(1,4\text{-bdc})(H_2O)_4]_n$	μ_4-η^1:η^1:η^2:η^1 (25)	3D framework	ATIJUI	[190]
207	$[Nd_2(pht)_2(1,4\text{-bdc})(H_2O)_4]_n$	μ_4-η^1:η^1:η^2:η^1 (25)	3D framework	AVAQAP	[196]
208	$[Sm_2(pht)_2(1,4\text{-bdc})(H_2O)_4]_n$	μ_4-η^1:η^1:η^1:η^1 (22)	3D framework	AVAQET	[196]
209	$[Gd_2(pht)_2(1,4\text{-bdc})(H_2O)_4]_n$	μ_4-η^1:η^1:η^1:η^1 (22)	3D framework	ATIKAP	[195]
210	$[Dy_2(pht)_2(1,4\text{-bdc})(H_2O)_4]_n$	μ_4-η^1:η^1:η^1:η^1 (22)	3D framework	ATIKET	[195]
211	$[La_3Cu_7(pht)(ina)_8Br_6(H_2O)_4]_n$	μ_3-η^1:η^1:η^2:η^1 (18)	3D framework	HOSYUK	[200]
212	$[Ce_3Cu_7(pht)(ina)_8Br_6(H_2O)_4]_n$	μ_3-η^1:η^1:η^1:η^1 (16)	3D framework	DIVTJY	[197]
213	$[Er_4Cu_8(pht)_2(ina)_8I_7(OH)(H_2O)_5)]_n$	μ_3-η^1:η^1:η^2:η^1 (18)	3D framework	DIVTOS	[197]
214	$\{[Nd_2Cu(pht)_2(ina)_2(H_2O)_2](ClO_4)\}_n$	μ_5-η^1:η^1:η^2:η^2 (35)	3D framework	JIVSIR JIVSIR01	[198] [201]
215	$\{[Sm_2Cu(pht)_2(ina)_2(H_2O)_2](ClO_4)\}_n$	μ_5-η^1:η^1:η^2:η^2 (35)	3D framework	JIVSEN JIVSEN01	[198] [201]
216	$\{[Eu_2Cu(pht)_2(ina)_2(H_2O)_2](ClO_4)\}_n$	μ_5-η^1:η^1:η^2:η^2 (35)	3D framework	JIVSAJ JIVSAJ01	[198] [201]
217	$\{[Tb_2Cu(pht)_2(ina)_2(H_2O)_2](ClO_4)\}_n$	μ_5-η^1:η^1:η^2:η^2 (35)	3D framework	NOFRIK	[201]
218	$\{[Nd_2Cu(pht)_2(ina)_2(H_2O)_2]Cl\}_n$	μ_5-η^1:η^1:η^2:η^2 (35)	3D framework	JIVSOX	[198]
219	$\{[Er_7O(OH)_6(pht)_3(ina)_9][Cu_3Cl_4]\}_n$	μ_3-η^1:η^1:η^2:η^1 (18)	3D framework	CECDEU	[199]

Table 1. (Continued)

N	CPs	Coordination modes displayed by the phthalate ligand (type in Scheme 1)	Structural features	Ref. code in CSD	Ref.
220	$\{[Er_7O(OH)_6(pht)_3(ina)_9][Cu_3Br_4]\}_n$	μ_3-η^1:η^1:η^2:η^1 (18)	3D framework	CECDIY	[199]
221	$[NdAg(pht)(ina)_2(H_2O)]_n$	μ_4-η^1:η^2:η^2:η^1 (28)	3D framework	HETCEP	[202]

Abbreviation: py = pyridine; β-pic = 3-methylpyridine; γ-pic = 4-methylpyridine; pip = piperidine; 1-Meim = 1-methylimidazole; 2-Meim = 2-methylimidazole; 4-Meim = 4-methylimidazole; 4-Etim = 4-ethylimidazole; 1-Viim = 1-vinylimidazole; 1-Alim = allylimidazole; bzim = benzimidazole; ica = isonicotinamide; pnH_2 = propane-1,3-diammonium; tmp = trimethoprim [2,4-diamino-5-(3',4'5'-trimethoxy benzyl) pyrimidine; bpy = 2,2′-bipyridine; phen = 1,10′-phenanthroline; HL^1 = 3-(2-pyridyl)pyrazole; dpa = 2,2′-dipyridilamine; 3,4′-bpt = 1*H*-3-(3-pyridyl)-5-(4-pyridyl)-1,2,4-triazole; 4,4′-bpt = 3,5-bis(4-pyridyl)-1,2,4-triazole; 3,3′-bpt = 3,5-bis(3-pyridyl)-1,2,4-triazole; pzp = pyrazino[2,3-f][1,10]phenanthroline; bpbp = 2,2′-bis(4-pyridylmethyleneoxy)-1,1′-biphenylene; L^2 = dipyrido[3,2-*a*:2′,3′-*c*]phenazine; L^3 = 2-(furan-2-yl)-1*H*-imidazo[4,5-*f*][1,10]phenanthroline; bpp = 1,3-bis(4-pyridyl)propane; L^4 = 1,1′-(1,4-butanediyl)bis(imidazole-2-phenyl); L^5 = 1,1′-(2,2′-oxybis(ethane-2,1-diyl))bis(1*H*-imidazole); L^6 = 1,3-bis(3-methoxysalicylideneimino)propane; tacn = 1,4,7-triazacyclononane; tu = thiourea; L^7 = *N*,*N*′-dimethylthiourea; oaoH = oxamide oxime; dien = diethylenetriamine; 4,4′-dpa = 4,4′-dipyridylamine; L^8 = *N*,*N*′-bis(3-pyridyl-methyl)piperazine; azpy = 4,4′-azopyridine; phz = phthalazine; Hina = isonicotinic acid; btm = bis(1,2,4-triazole-1-yl)methane; bte = bis(1,2,4-triazole-1-yl)ethane; bim-4 = 1,1′-(1,4-butanediyl)bis(imidazole); bim-5 = 1,1′-(1,5-pentanedidyl)bis(imidazole); bim-6 = 1,1′-(1,6-hexanedidyl)bis(imidazole); bbim = 1,1′-(1,4-butanediyl)bis(2-methylbenzimidazole); bie = 2,2′-bis(1*H*-imidazolyl)ether; eda = ethylenediamine; 1,3-bdc = 1,3-benzenedicarboxylic acid; 1,4-bdc = 1,4-benzenedicarboxylic acid; L^9 = *N*,*N*′-bis(3-aminopropyl)oxamide; L^{10} = *N*,*N*′-bis(2-aminoethyl)oxamide; L^{11} = *N*,*N*'-bis(2-aminopropyl)oxamide; piz = piperazine; H_2ox = oxalic acid; H_2mal = maleic acid; H_2ad = adipic acid; ur = urea; tib = 1,3,5-tris(1-imidazolyl)benzene; btb = 1,4-bis(1,2,4-triazol-1-yl)butane.

Different from 211 and 212, the Cu(I) atoms in 213 are bridged by μ_3-I and μ_4-I to form an cationic $\{Cu_8I_7\}^+$ chain, that further links the adjacent anionic Eu-carboxylate layers, $[Er_4(pht)_2(ina)_8(OH)(H_2O)_5)]^-$, to produce a 3D framework. In 214 – 218, the Ln atoms are connected by the bridging phthalate ligands to form the cationic 2D layers, $\{Ln_2(pht)_2(H_2O)_2\}^{2+}$. The isonicotinate ligand coordinates to Cu^+ ions with the formation of linear $\{Cu(ina)_2\}^-$ units. The connection between Ln–pht layers and Cu–ina units gives rise to a pillared 3D framework having channels with dimensions of 6.2 x 15.8 Å and 7.0 x 8.1 Å. The ClO_4^- anions in 214 – 217 and Cl^- anions in 218 act as counterions and occupy the channels. It is interesting to note that small variations in the synthetic conditions, e.g. the use of only 0.385 mmol of $HClO_4$ compared to 0.5 mmol in 214–216 in the reaction with Er_2O_3, Hina, H_2Pht and $CuCl_2$ or $CuBr_2$, leads to a very unusual 3D lanthanide-transition metal framework comprising $\{Cu_3\}$ pillars and layers of $\{Er_{36}\}$ wheels with the general formula $\{[Er_7O(OH)_6(pht)_3(ina)_9][Cu_3X_4]\}_n$ (X = Cl^- (219); Br^- (220) [199]). On the one hand, the Er(III) atoms in 219 and 220 are linked by oxo and hydroxo bridges to give two types $\{Er_2\}$ and $\{Er_4\}$ cluster cores, which are further alternately connected by phthalates to form a nanosized $\{Er_{36}\}$ wheel. The $\{Er_{36}\}$ wheels are bridged to each other by sharing hydroxo bridges to generate a highly-ordered cluster layer with honeycomb arrays. On the other hand, three Cu(II) centers are bridged by Cl^- (219) or Br^- (220) ions to form a trinuclear $\{Cu_3X_4\}$ cluster. Finally, the structures of 219 and 220 are of alternating lanthanide cluster layers and the trinuclear copper clusters bridged by ina ligands form a 3D framework. A rather different 3D network was found in a 4d-4f heterometallic phthalate-based MOF with Nd^{3+} and Ag^+ ions, $[NdAg(pht)(ina)_2(H_2O)]_n$ (221), reported by Xue and Gu [202].

The CP 221 was synthesized from the similar hydrothermal reaction using Nd_2O_3, phthalic and isonicotinic acids, but heated at 160°C for 4 days. Here, the 3D motif is constructed from neodymium-phthalate chains that are interlinked by silver(I) centers and isonicotinate ligands.

ACKNOWLEDGMENTS

The authors are grateful to the Swiss National Science Foundation (SCOPES 7MDPJ065712.01/1, IB7320_110976, and IZ73Z0_127925). S.G.B. thanks the EU for supporting FP-7 International Incoming Fellowship (POLYMAG, IIF contract no. 252984).

REFERENCES

[1] Baten, S.R.; Neville, S.M.; Turner, D.R. Coordination Polymers: Design, Analysis and Application; the Royal Society of Chemistry: Cambridge, UK, 2009; pp 471.

[2] Metal-Organic Frameworks: Applications from Catalysis to Gas Storage; Farrusseng, D.; Ed.; Wiley-VCH: Weinheim, DE, 2011; pp 392.

[3] Design and Construction of Coordination Polymers; Hong, M-C.; Chen, L.; John Wiley and Sons, Inc.; Hoboken, New Jersey, 2009; pp.

[4] Kitagawa, S.; Noro, S. In Comprehensive Coordination Chemistry II; MaCleverty, J.A.; Mayer, T.J.; Elsevier: Oxford, UK, 2004; Vol. 7, pp 231–261.

[5] Sanchez, C.; Belleville, P.; Popall, M.; Nicole, L. *Chem. Soc. Rev.* 2011, 40, 696–753.
[6] Robin, A.Y.; Fromm, K.M. Coord. *Chem. Rev.* 2006, 250, 2127–2157.
[7] Kitagawa, S.; Kitaura, R., Noro, S. *Angew. Chem. Int. Ed.* 2004, 43, 2334–2375.
[8] Leong, W.L.; Vittal, J.J. *Chem. Rev.* 2011, 111, 688–764.
[9] Winpenny, R.E.P. J. *Chem. Soc., Dalton Trans.* 2002, 1–10.
[10] (a) Hoskins, B.F.; Robson, R. *J. Am. Chem. Soc.* 1989, 111, 5962–5964; (b) Hoskins, B.F.; Robson, R. *J. Am. Chem. Soc.* 1990, 112, 1546–1554.
[11] (a) Tranchemontagne, D.J.; Mendoza-Cortes, J.L.; O'Keeffe, M.; Yaghi, O.M. Chem. Soc. Rev. 2009, 38, 1257-1283; (b) O'Keeffe, M.; Peskov, M. A.; Ramsden, S. J.; Yaghi, O. M. *Acc. Chem. Res.* 2008, 41, 1782–1789.
[12] (a) Moulton, B.; Lu, J.; Mondal, A.; Zaworotko, M. Chem. Commun. 2001, 9, 863-864; (b) Perry, J.J.; Perman, J.A.; Zaworotko, M. *J. Chem. Soc. Rev.* 2009, 38, 1400–1417.
[13] Kellett, A.; O'Connor, M.; McCann, M.; McNamara, M.; Lynch, P.; Rosair, G.; McKee, V.; Creaven, B.; Walsh, M.; McClean, S.; Foltyn, A.; O'Shea, D.; Howe, O.; Devereux, M. *Dalton. Trans.* 2011, 40, 1024–1027.
[14] Ang, W.H.; Pilet, S.; Scopelliti, R.; Bussy, F.; Juillerat-Jeanneret, L.; Dyson, P.J. *J. Med. Chem.* 2005, 48, 8060–8069.
[15] Liang, X.; Weishaupl, M.; Parkinson, J.A.; Parsons, S.; McGregor, P.A.; Sadler, P.J. *Chem.-Eur. J.* 2003, 9, 4709–4717.
[16] Devereux, M.; McCann, M.; Leon, V.; Geraghty, M.; McKee, V.; Wikaira, J. *Met.-Based Drugs* 2000, 7, 275–288.
[17] Baca, S.G.; Reetz, M.T.; Goddard, R.; Filippova, I.G.; Simonov, Yu.A.; Gdaniec, M.; Gerbeleu. N. *Polyhedron* 2006, 25, 1215–1222.
[18] Porai-Koshits, M.A. *Russ. J. Struct. Chem.* 1980, 21, 146–180.
[19] Allen, F.H. *Acta Crystallogr.* 2002, B58, 380–388.
[20] Baca, S.G.; Filippova, I.G.; Gerbeleu, N.V.; Simonov, Yu.A.; Gdaniec, M.; Timco, G.A.; Gherco, O.A.; Malaestean, Yu.L. *Inorg. Chim. Acta* 2003, 344, 109–116.
[21] Baca, S.G. *International Research Journal of Pure and Applied Chemistry* 2012, 2, 1–24.
[22] Fadaei, M.R.; Morsali, A.; Mahjoub, A.R. *Z. Naturforsch.* 2005, 60b, 741–744.
[23] Rodriguez-Cuamatzi, P.; Tlahuext, H.; Hopfl, H. *Acta Crystallogr.* 2009, E65, m246.
[24] Cheng, X.-N.; Xue, W.; Zhang, W.-X.; Chen X.-M. *Chem. Mater.* 2008, 20, 5345–5350.
[25] Lindh, B. *Acta Chem. Scand.* 1967, 21, 2743–2752.
[26] Cingi, M. B.; Guastini, C.; Musati, A.; Nardelli, N. *Acta Crystallogr.* 1970, B26, 1836–1843.
[27] Cingi, M. B.; Lanfredi, A. M. M.; Tiripicchio, A.; Camellini, M. T. *Acta Crystallogr.* 1977, B33, 659–664.
[28] Cingi, M. B.; Lanfredi, A. M. M.; Tiripicchio, A.; Camellini, M. T. *Acta Crystallogr.* 1978, B34, 412–416.
[29] Adiwidjaja, G.; Rossmanith, E.; Kuppers, H. *Acta Crystallogr.* 1978, B34, 3079–3083.
[30] Cingi, M.B.; Guastini, C.; Musati, A.; Nardelli, N. *Acta Crystallogr.* 1969, B25, 1833–1840.

[31] Baca, S.G.; Simonov, Yu.A.; Gerbeleu, N.V.; Gdaniec, M.; Bourosh, P.N.; Timco, G.A. *Polyhedron* 2001, 20, 831–839.

[32] Cingi, M.B.; Lanfredi, A.M.M.; Tiripicchio, A.; Camellini, M.T. *Acta Crystallogr.* 1981, B37, 2159–2163.

[33] Bakalbassis, E.G.; Terzis, A. *Inorg. Chim. Acta* 1994, 218, 167–172.

[34] Gerbeleu, N.; Simonov, Yu.; Timco, G.; Bourosh, P.; Lipkowskii, J.; Baka, S.; Saburov, D.; Mazus, M. *Russian Journal of Inorganic Chemistry* 1999, 44, 1191–1204.

[35] Zhou, X.-G. *Acta Crystallogr.* 2007, E63, m3166.

[36] Chen, Y.; Liu, P.; Liu, Q.-Y.; Wang, J.; Cheng, W.-D. *Chin. J. Struct. Chem.* 2003, 22, 313–317.

[37] Zhao, H.-X.; Zhuang, G.-L.; Wu, S.-T.; Long, L.-S.; Guo, H.-Y.; Ye, Z.-G.; Huang, R.-B.; Zheng, L.-S. *Chem. Commun.* 2009, 1644–1646.

[38] Liu, Q.-Y.; Liu, P.; Wang, J.; Chen, Y. *Chin. J. Struct. Chem.* 2002, 21, 509–512.

[39] Li, J.; Zhou, J-H.; Li, Y-Z.; Chen. X-T. *Chin. J. Inorg. Chem.* 2004, 20, 841–844.

[40] Baca, S.G.; Filippova, I.G.; Gherco, O.A.; Gdaniec, M.; Simonov, Yu.A.; Gerbeleu, N.V.; Franz, P.; Basler, R.; Decurtins, S. *Inorg. Chim. Acta* 2004, 357, 3419–3429.

[41] Simonov, Yu.A.; Gdaniec, M.; Filippova, I.G.; Baca, S.G.; Timco, G.A.; Gerbeleu, N.V. *Russ. J. Coord. Chem.* 2001, 27, 353–359.

[42] Baca, S.G.; Malinovskii, S.T.; Franz, P.; Ambrus, C.; Stoeckli-Evans, H.; Gerbeleu, N.; Decurtins, S. *J. Solid State Chem.* 2004, 117, 8, 2841–2849.

[43] Li, R.-X.; Li, S.-X.; Liu, L.; Liu, G.-Y.; Liu, F.-Q. *Acta Crystallogr.* 2007, E63, m2875.

[44] Li, R.-X.; Li, S.-X.; Wu, Q.-Y.; Liu, G.-Y.; Liu, F.-Q. *Acta Crystallogr.* 2007, E63, m2874.

[45] Liu, F.-Q.; Li, R.-X.; Li, S.-X.; Sun, L.-S.; Liu, G.-Y. *Acta Crystallogr.* 2007, E63, m3166.

[46] Li, R.-X.; Wu, Q.-Y.; Liu, F.-Q. *Acta Crystallogr.* 2010, E66, m258.

[47] Cui, Y.-C.; Wang, J.-J.; Che, G.-B.; Li, C.-B. *Acta Crystallogr.* 2006, E62, m2761–m2763.

[48] Liu, B.-X.; Su, J.-R.; Xu, D.-J. *Acta Crystallogr.* 2004, C60, m183–m185.

[49] Xue, Y-H.; Zhang, L.; Xu, D-J. *Chin. J. Struct. Chem.* 2005, 24, 1043–1045.

[50] Bats, J.W.; Kallel, A.; Fuess, H. *Acta Crystallogr.* 1978, B34, 1705–1707.

[51] Rodrigues, B.L.; Costa, M.D.D.; Fernandes, N.G. *Acta Crystallogr.* 1999, C55, 1997–2000.

[52] Padmanabhan, M.; Joseph, K.C.; Puranik, V.G.; Huang, X.; Li, J. *Solid State Sci.* 2007, 9, 491–495.

[53] Raj, S.B.; Muthiah, P.T.; Bocelli, G.; Cantoni, A. *Inorg. Chem. Commun.* 2003, 6, 748–751.

[54] Thirumurugan, A.; Rao, C.N.R. *J. Mater. Chem.* 2005, 15, 3852–3858.

[55] Sun, Y.-G.; Wei, D.-Z.; Gao, E.-J.; Ding, F. *Acta Chim. Sinica* 2004, 62, 1362–1366.

[56] Baca, S.G.; Filippova, I.G.; Ambrus, C.; Gdaniec, M.; Simonov, Y.A.; Gerbeleu, N.; Gherco, O.A.; Decurtins, S. *Eur. J. Inorg. Chem.* 2005, 3118–3130.

[57] Baca, S.G.; Simonov, Yu.A.; Gdaniec, M.; Gerbeleu, N.; Filippova, I.G.; Timco, G.A. *Inorg. Chem. Commun.* 2003, 6, 685–689.

[58] Gao, L.; Cheng, Q.-Y.; Wang, Y.-J. *Chin. J. Struct. Chem.* 2006, 25, 1193–1196.

[59] Yao, J.; Huang, W.; Li, B.; Gou, S.; Xu, Y. *Inorg. Chem. Commun.* 2002, 5, 711–714.

[60] Krstanovic, I.; Karanovic, Lj.; Stojakovic, Dj. *Acta Crystallogr.* 1985, C41, 43–45.
[61] Bi, J.-H.; Yao, F.-H.; Huang, Z.-X.; Wang, H.-L.; Hu, N.-L. *Asian J. Chem.* 2007, 19, 5360–5364.
[62] Tian, L.; Chen, L.; Yi, L. (Chin.) *Natural Science Journal of Xiangtan University* 2003, 1, 113–116.
[63] Ye, C.-H.; Sun, H.-L.; Wang, X.-Y.; Huang, R.-L.; Li, J.-R.; Gao, S. *J. Chem. Cryst.* 2005, 35, 381–384.
[64] Tian, L.; Chen, L.; Shen, S.-H.; Wen, C.-L.; Liu, S.-L. *J. Coord. Chem.* 2007, 60, 683–689.
[65] Zhang, Y.; Li, J.; Su, Q.; Wang, Q.; Wu, X. *J. Mol. Struct.* 2000, 516, 231–236.
[66] Poleti, D.; Karanovic, Lj.; Prelesnik, B.V. *Acta Crystallogr.* 1993, C49, 1249–1251.
[67] Song, Y.-S.; Yan, B.; Chen, Z.-X. *Appl. Organometal. Chem.* 2005, 20, 44–50.
[68] Ma, Ch.; Wang, W.; Zhang, X.; Chen, Ch.; Liu, Q.; Zhu, H.; Liao, D.; Li, L. *Eur. J. Inorg. Chem.* 2004, 3522–3532.
[69] Yang, J.-M.; Zhou, Z.-H.; Zhang, H.; Wan, H.-L.; Lu, S.-J. *Inorg. Chim. Acta* 2005, 358, 1841–1849.
[70] Wang, X.-Y.; Dong, C.-C.; Deng, X.-T.; Wang, C.-G.; Hu, B. *Acta Crystallogr.* 2007, E63, m2079.
[71] Hu, T.-L.; Zou, R.-Q.; Li, J.-R.; Bu, X.-H. *Dalton Trans.* 2008, 1302–1311.
[72] Zhu, L.-G.; Zhang, J. *CrystEngComm.* 2011, 13, 553–560.
[73] Poleti, D.; Karanovic, Lj. *J. Serb. Chem. Soc.* 1997, 62, 657–660.
[74] Ou, Y.-C.; Liu, W.-T.; Li, J.-Y.; Zhang, G.-G.; Wang, J.; Tong, M.-L. *Chem. Commun.* 2011, 47, 9384–9386.
[75] Huang, F.-P.; Tian, J.-L.; Chen, G.-J.; Li, D.-D.; Gu, W.; Liu, X.; Yan, S.-P.; Liao, D.-Z.; Cheng, P. *CrystEngComm.* 2010, 12, 1269–1279.
[76] Gong, Y.; Wang, R.; Yuan, D.; Su, W.; Huang, Y.; Yue, Ch.; Jiang, F.; Hong, M. *Polyhedron* 2007, 26, 5309–5316.
[77] Wang, X.-L.; Bi, Y.-F.; Liu, G.-C.; Lin, H.-Y.; Hu, T.-L.; Bu, X.-H. *CrystEngComm.* 2008, 10, 349–356.
[78] Yang, J.; Ma, J.-F.; Liu, Y.-Y.; Ma, J.-C.; Batten, S.R. *Cryst. Growth Des.* 2009, 9, 1894–1911.
[79] Wang, X.-L.; Chen, B.-K.; Lin, H.-Y.; Bi, Y.-F.; Liu, G.-C. *J. Chem. Cryst.* 2008, 38, 339–343.
[80] Zhang, S.-C.; Sun, J. *Acta Crystallogr.* 2006, E62, m3107–m3109.
[81] Li, C.-B.; Li, Y.-W.; Fang, W.; Liu, B. *Acta Crystallogr.* 2006, E62, m3541– m3543.
[82] Chen, Y.-B.; Zhang, J.; Cheng, J.-K.; Kang, Y.; Li, Z.-J.; Yao, Y.-G. *Inorg. Chem. Commun.* 2004, 7, 1139–1141.
[83] Liu, Y.-Y.; Ma, J.-F.; Yang, J.; Ma, J.-C.; Ping, G.-J. *CrystEngComm,* 2008, 10, 565–572.
[84] Qi, Y.; Che, Y.; Luo, F.; Batten, S.R.; Liu, Y.; Zheng, J. *Cryst. Growth Des.* 2008, 8, 1654–1662.
[85] Bermejo, M.R.; Fondo, M.; Garcia-Deibe, A.; Gonzalez, A.M.; Sousa, A.; Sanmartin, J.; McAuliffe, C.A.; Pritchard, R.G.; Watkinson, M.; Lukov, V. *Inorg. Chim. Acta* 1999, 293, 210–217.

[86] Cheng, P.; Yan, S.-P.; Xie, C.-Z.; Zhao, B.; Chen, X.-Y.; Liu, X.-W.; Li, C.-H.; Liao, D.-Z.; Jiang, Z.-H.; Wang, G.-L. *Eur. J. Inorg. Chem.* 2004, 2369–2378.

[87] Burrows, A.D.; Harrington, R.W.; Mahon, M.F.; Price, C.E. *Dalton Trans.* 2000, 3845–3854.

[88] Burke, N.J.; Burrows, A.D.; Donovan, A.S.; Harrington, R.W.; Mahon, M.F.; Price, C.E. *Dalton Trans.* 2003, 3840–3849.

[89] Endres, H. Z. *Anorg. Allg. Chem.* 1984, 513, 78–88.

[90] Zhang, Y.; Jianmin, L.; Min, Z.; Wang, Q.; Wu, X. *Chem. Lett.* 1998, 1051–1052.

[91] Bakalbassis, E.G.; Paschalidis, D.G.; Raptopoulou, C.P.; Tangoulis, V. *Inorg. Chem.* 1998, 37, 4735–4737.

[92] Shakhatreh, S.K.; Bakalbassis, E.G.; Brudgam, I.; Hartl, H.; Mrozinski, J.; Tsipis, C.A. *Inorg. Chem.* 1991, 30, 2801–2806.

[93] Braverman, M.A.; Supkowski, R.M.; LaDuca, R.L. *Inorg. Chim. Acta* 2007, 360, 2353–2362.

[94] Johnston, L.L.; Martin, D.P.; Supkowski, R.M.; LaDuca, R.L. *Inorg. Chim. Acta* 2008, 361, 2887–2894.

[95] Zhu, L.-N.; Liang, M.; Wang, Q.-L.; Wang, W.-Z.; Liao, D.-Z.; Jiang, Z.-H.; Yan, S.-P.; Cheng, P. *J. Mol. Struct.* 2003, 657, 157–163.

[96] Kumar, D. K.; Das, A.; Dastidar, P. *J. Mol. Struct.* 2006, 796, 139–145.

[97] Thirumurugan, A.; Natarajan, S. *Dalton Trans.* 2004, 2923–2928.

[98] Whitcomb, D.R.; Rogers, R.D. *Inorg. Chim. Acta* 1997, 256, 263–267.

[99] Li, X.-F.; An, Y.; Yin, Y.-S. *Acta Crystallogr.* 2008, E64, m779.

[100] Thirumurugan, A.; Natarajan, S. *J. Mater. Chem.* 2005, 15, 4588–4594.

[101] Song, Y.-S.; Yan, B. *Inorg. Chim. Acta* 2005, 358, 191–195.

[102] Li, P.; Qiu, Y.; Liu, J.; Ling, Y.; Cai, Y.; Yue, S. *Inorg. Chem. Commun.* 2007, 10, 705–708.

[103] Wan, Y.-H.; Jin, L.-P.; Wang, K.-Z. *J. Mol. Struct.* 2003, 649, 85–93.

[104] Wan, Y.; Zhang, L.; Jin, L.; Gao, S.; Lu, S. *Inorg. Chem.* 2003, 42, 4985–4994.

[105] Huang, M.-H.; Liu, P.; Wang, J.; Chen, Y.; Liu, Z. *Chin. J. Struct. Chem.* 2006, 25, 453–458.

[106] Wang, G.-M.; Huang, H.-L.; Li, H.; Zheng, Q.-H. *Acta Crystallogr.* 2007, C63, m531–533.

[107] Charushnikova, I.A.; Krot, N.N.; Starikova, Z.A. *Radiochemistry* 2004, 46, 556–559.

[108] Grigoriev, M.S.; Antipin, M.Yu.; Krot, N.N.; Bessonov, A.A. *Radiochim. Acta* 2004, 92, 405–409.

[109] Mihalcea, I.; Henry, N.; Loiseau, T. *Cryst. Growth Des.* 2011, 11, 1940–1947.

[110] Shchelokov, R.N.; Mikhailov, Yu.N.; Orlova, I.M.; Sergeev, A.V.; Ashurov, Z.R.; Tashev, M.T.; Parpiev, N.A. *Koord. Khim.* (Russ.) 1985, 11, 1144–1146.

[111] Charushnikova, I.A.; Krot, N.N.; Starikova, Z.A. *Radiochemistry* 2001, 43, 496–501.

[112] Prout, C.K.; Carruthers, J.R.; Rossotti, F.J.C. *J. Chem. Soc.* (A) 1971, 3350–3354.

[113] Qiao, H.; Shen, X.; Zhang, H.; Yuan, L.; Chen, P.; Mao, H.; Hou, H. *Synth. React. Inorg. Met.-Org. Chem.* 2005, 35, 627–632.

[114] Cingi, M.B.; Lanfredi, A.M.M.; Tiripicchio, A.; Camellini, M.T. *Acta Crystallogr.* 1978, B34, 134 137.

[115] Chen, X.-Y.; Shi, W.; Cheng, P.; Chen, J.-T.; Yan, S.-P.; Liao, D.-Z.; Jiang, Z.-H. *Z. Anorg. Allg. Chem.* 2003, 629, 2034–2039.

[116] Liu, H.-L.; Mao, H.-Y.; Zhang, H.-Y.; Xu, Ch.; Wu, Q.-A.; Li, G.; Zhu, Y.; Hou, H.-W. *Polyhedron* 2004, 23, 943–948.

[117] Liu, X.; Xie, C.; Wang, X.; Shen, G.; Shen, D. *Inorg. Chem. Commun.* 2003, 6, 1433–1435.

[118] Cingi, M.B.; Lanfredi, A.M.M.; Tiripicchio, A.; Camellini, M.T. *Acta Crystallogr.* 1978, B34, 406–411.

[119] Cingi, M.B.; Lanfredi, A.M.M.; Tiripicchio, A.; Camellini, M.T. *Acta Crystallogr.* 1979, B35, 312–316.

[120] Kupers, H. *Acta Crystallogr.* 1988, C44, 2093–2095.

[121] Okaya, Y. *Acta Crystallogr.* 1965, 19, 879–882.

[122] Eremina, T.A.; Furmanova, N.G.; Malakhova, L.F.; Okhrimenko, T.M.; Kuznetsov, V.A. *Crystallogr. Rep.* 1993, 38, 554–556.

[123] Harte, S.M.; Parkin, A.; Goeta, A.; Wilson, C.C. *J. Mol. Struct.* 2005, 741, 93–96.

[124] Nelyubina, Yu.V.; Lyssenko, K.A.; Antipin, M.Yu. *Crystallogr. Rep.* 2008, 53, 192–198.

[125] Li, Y.-F.; Zhang, T.-L.; Zhang, J.-G.; Yu, K.-B. *Z. Naturforsch.* 2003, B58, 1171–1175.

[126] Stein, I.; Ruschewitz, U. *Acta Crystallogr.* 2005, E61, m141–m143.

[127] Yang, S.-Y.; Long, L.-S.; Huang, R.-B.; Zheng, L.-S.; Weng Ng, S. *Acta Crystallogr.* 2003, C59, m456±m458.

[128] Robl, C. *Z. Anorg. Allg. Chem.* 1988, 566, 144–150.

[129] Vaz, J.L.L.; Duc, G.; Petit-Ramel, M.; Faure, R.; Vittori, O. *Can. J. Chem.* 1996, 74, 359–364.

[130] Wang, S.; Hou, Y.; Wang, E.; Li, Y.; Xu, L.; Peng, J.; Liu, Sh.; Hu, Ch. *New J. Chem.* 2003, 27, 1144–1147.

[131] Luo, F.; Che, Y.-X.; Zheng, J.-M. *J. Coord. Chem.* 2008, 61, 2097–2104.

[132] Tang, Y.-Z.; Qian, Sh.-S.; Wang, X.-S.; Zhao, H.; Huang, X.-F.; Li, Y.-H.; Jiao, X.-C.; Xiong, R.-G. *Z. Anorg. Allg. Chem.* 2004, 630, 1623–1626.

[133] Wang, Y.-L.; Liu, Q.-Y.; Zhong, S.-L. *Acta Crystallogr.* 2006, C62, m395–m397.

[134] Askarinejad, A.; Torabi, A.A.; Morsali, A. *Z. Naturforsch.* 2006, B61, 565–569.

[135] Askarinejad, A.; Morsali, A. *J. Coord. Chem.* 2006, 59, 997–1005.

[136] Marandi, F.; Ghorbanloo, M.; Soudi, A.A. *J. Coord. Chem.* 2007, 60, 1557–1565.

[137] Sun, Di.; Wei, Z.-H.; Wang, D.-F.; Zhang, N.; Huang, R.-B.; Zheng, L.-S. *Cryst. Growth Des.* 2011, 11, 1427–1430.

[138] Whitcomb, D.R.; Rajeswaran, M. *J. Imaging Sci. Technol.* 2003, 47, 107–114.

[139] Ma, Ch.; Wang, W.; Zhang, X.; Chen, Ch.; Liu, Q.; Zhu, H.; Liao, D.; Li, L. *Eur. J. Inorg. Chem.* 2004, 3522–3532.

[140] Gherco, O.; Keene, T.D.; Filippova, I.G.; Baca, S.G.; Decurtins, S. *Moldavian Journal Physical Sciences* 2011, 1, 103–108.

[141] Ma, Ch.; Chen, Ch.; Liu, Q.; Liao, D.; Li, L.; Sun, L. *New J. Chem.* 2003, 27, 890–894.

[142] Köferstein, R.; Robl, Ch.; *Z. Anorg. Allg. Chem.* 2007, 633, 1127–1130.

[143] Suresh, E.; Boopalan, K.; Jasra, R.V.; Bhadbhade, M.M. *Inorg. Chem.* 2001, 40, 4078–4080.

[144] Simonov, Yu.A.; Baca, S.G.; Filippova, I.G.; Gdaniec, M.; Gherco, O.A.; Gerbeleu, N.V. *Russ. J. Coord. Chem.* 2004, 30, 727–732.
[145] Li, X.-M.; Wang, Q.-W.; Cui, Y.-Ch.; Li, Ch.-B.; Wang, R.-Z.; Liu, B. *Chin. J. Struct. Chem.* 2006, 25, 621–624.
[146] Xu, H.; Sun, R.; Li, Y.-Z.; Bai, J.-F. *Acta Crystallogr.* 2006, E62, m1156–m1158.
[147] Zhang, X.; Yi, Z.H.; Xia, W.; Yang, C.; Li, B. *Acta Crystallogr.* 2008, E64, m345–m346.
[148] Fan, L.; Qin, Ch.; Li, Y.; Wang, E.; Wang, X.; Xiao, D.; An, H.; Xu, L. *J. Coord. Chem.* 2006, 59, 883 – 890.
[149] Yang, S.-Y.; Long, L.-S.; Huang, R.-B.; Zheng, L.-S.; Ng, S.W. *Acta Crystallogr.* 2003, E59, m507–m509.
[150] Sun, C.-Y.; Li, L.-C.; Jin, L.-P. *Polyhedron*, 2006, 25, 3017–3024.
[151] Xie, X.-F.; Chen, S.-P.; Xia, Z.-Q.; Gao, S.-L. *Polyhedron* 2009, 28, 679–688.
[152] Huang, F.-P.; Tian, J.-L.; Gu, W.; Liu, X.; Yan, S.-P.; Liao, D.-Z.; Cheng, P. *Cryst. Growth Des.* 2010, 10, 1145–1154.
[153] Du, M.; Jiang, X.-J.; Zhao, X.-J. Inorg. Chem. 2007, 46, 3984–3995.
[154] Pochodylo, A.L.; Blake, K.M.; Braverman, M.A.; Farnum, G.A.; LaDuca, R.L. *Inorg. Chim. Acta* 2010, 363, 3951–3958.
[155] Zhang, L.; Ling, Y.; Li, J.; Gao, H. *Struct. Chem.* 2008, 19, 911–916.
[156] Li, B.; Li, B.; Zhu, X.; Lu, X.; Zhang, Y. *J. Coord. Chem.* 2004, 57, 1361–1367.
[157] Wang, L.-Y.; Yang, Y.; Liu, K.; Li, B.-L.; Zhang, Y. *Cryst. Growth Des.* 2008, 8, 3902–3904.
[158] Zhang, W.-L.; Liu, Y.-Y.; Ma, J.-F.; Jiang, H.; Yang, J. *Polyhedron* 2008, 27, 3351–3358.
[159] Zhang, L.-P.; Ma, J.-F.; Yang, J.; Liu, Y.-Y.; Wie, G.-H. *Cryst. Growth Des.* 2009, 9, 4660–4673.
[160] Qi, Y.; Luo, F.; Che, Y.; Zheng, J. *Cryst. Growth Des.* 2008, 8, 606–611.
[161] Liu, Y.-Y.; Jiang, Y.-Y.; Yang, J.; Liu, Y.-Y.; Ma, J.-F. *CrystEngComm.* 2011, 13, 6118–6129.
[162] Zhang, L.-P.; Yang, J.; Ma, J.-F.; Jia, Z.-F.; Xie, Y.-P.; Wie, G.-H. *CrystEngComm.* 2008, 10, 1410–1420.
[163] Liang, X.-Q.; Zhou, X.-H.; Chen, C.; Xiao, H.-P.; Li, Y.-Z.; Zuo, J.-L.; You, X.-Z. *Cryst. Growth Des.* 2009, 9, 1041–1053.
[164] Su, Z.; Xu, J.; Fan, J.; Liu, D.-J.; Chu, Q.; Chen, M.-S.; Chen, S.-S.; Liu, G.-X., Wang, X.-F.; Sun, W.-Y. *Cryst. Growth Des.* 2009, 9, 2801–2811.
[165] Escuer, A.; Vicente, R.; Mautner, F.A.; Goher, M.A.S. *Inorg. Chem.* 1997, 36, 1233–1236.
[166] Zeng, Y.-F.; Hu, X.; Zhao, J.-P.; Hu, B.-W.; Sanudo, E.C.; Li, F.-C.; Bu, X.-H. *Chem.-Eur. J.* 2008, 14, 7127–7130.
[167] Krstanovic, I.; Karanovic, L.; Stojakovic, D.; Golic, L. *Cryst. Struct. Commun.* 1982, 11, 1747–1749.
[168] Fu, X.-C.; Wang, X.-Y.; Li, M.-T.; Wang, C.-G.; Deng, X.-T. *Acta Crystallogr.* 2006, C62, m343–m345.
[169] Guo, F.; Yu, W.; Zhang, X.-L. *Chin. J. Struct. Chem.* 2008, 27, 1123–1126.

[170] Thirumurugan, A.; Natarajan, S. *Solid State Sci.* 2004, 6, 599–604.
[171] Williams, D.J.; Maginn, S.J.; Davey, R.J. *Polyhedron* 1994, 13, 1683–1688.
[172] Chen, Z.-N.; Zhang, H.-X.; Yu, K.-B.; Kang, B.-S.; Cai, H.; Su, C.-Y.; Wang, T.-W.; Lu, Z.-L. *Inorg. Chem.* 1998, 37, 4775–4781.
[173] Zhang, H.-X.; Kang, B.-S.; Xu, A.-W.; Chen, Z.-N.; Zhou, Z.-Y.; Chan, A.S.C.; Yu, K.-B.; Ren, C. J. *Chem. Soc., Dalton Trans.* 2001, 2559–2566.
[174] Chen, J.; Jiang, Z-H.; Liao, D-Z.; Yan, S-P. Chin. *J. Struct. Chem.* 2005, 24, 1017–1020.
[175] Wan, Y.; Jin, L.; Wang, K.; Zhang, L.; Zheng, X.; Lu, S. *New J. Chem.* 2002, 1590–1596.
[176] Song, Y.-S.; Yan, B.; Chen, Z.-X. *Can. J. Chem.* 2004, 82, 1745–1751.
[177] Wang, Y.; Li, X.-L.; Wang, T.W.; Song, Y. You, X.-Z. *Inorg. Chem.* 2010, 49, 969–976.
[178] Meng, Z.-R.; Zhang, Q.-Z.; Wu, X.-Y.; Chen, L.-J.; Lu, C.-Z. *Acta Crystallogr.* 2006, E62, m1033–1035.
[179] Li, X.; Zha, M.-Q.; Wang, X.-W.; Cao, R. *Inorg. Chim. Acta*. 2009, 362, 3357–3363.
[180] Wang, G.-M.; Duan, C.-S.; Liu, H.-L.; Li, H. *Acta Crystallogr.* 2008, E64, m468–m469.
[181] Wan, Y.; Zhang, L.; Jin, L.; Gao, S.; Lu, S. *Inorg.Chem.* 2003, 42, 4985–4994.
[182] Natarajan, S.; Thirumurgan, A. *Inorg. Chim. Acta* 2005, 358, 4051–4056.
[183] Si, S.; Li, C.; Wang, R.; Li, Y. *J. Coord.Chem.* 2006, 59, 215–222.
[184] Yang, Q.-F.; Yu, Y.; Song, T.-Y.; Yu, J.-H.; Zhang, X.; Xu, J.-Q.; Wang, T.-G. *CrystEngComm.* 2009, 11, 1642–1649.
[185] Wang, Z.; Xing, Y.-H.; Wang, C.-G.; Sun, L.-X.; Zhang, J.; Ge, M.-F.; Niu, S.-Y. *CrystEngComm.* 2010, 12, 762–773.
[186] Li, Y.-B.; Zha, M.-Q.; Pan, J.-G.; Li, X. *Z. Kristallogr.-New Cryst. Struct.* 2010, 225, 143–144.
[187] Charushnikova, I.A.; Krot, N.N.; Starikova, Z.A. *Radiochemistry* 2001, 43, 492–495.
[188] Yaghi, O.M.; Li, G.M.; Li, H.L. *Nature* 1995, 378, 703–706.
[189] Lightfoot, P.; Snedden, A. *J. Chem. Soc., Dalton Trans.* 1999, 3549–3551.
[190] Wang, X.; Qin, Ch.; Wang, E.; Xu, L. *J. Mol. Struct.* 2005, 737, 49–54.
[191] Chen, S.-S.; Fan, J.; Okamura, T.; Chen, M.-S.; Su, Z.; Sun, W.-Y.; Ueyama, N. *Cryst. Growth Des.* 2010, 10, 812–822.
[192] Zhu, X.; Wang, L.-X.; Liu, X.-G.; Wang, J.; Li, B.-L.; Li, H.-Y. *CrystEngComm.* 2011, 13, 6090–6098.
[193] Bai, H.-Y.; Ma, J.-F.; Yang, J.; Zhang, L.-P.; Ma, J.-C.; Liu, Y.-Y. *Cryst. Growth Des.* 2010, 10, 1946–1959.
[194] Luo, F.; Che, Y.; Zheng, J. *Cryst. Growth Des.* 2008, 8, 2006–2010.
[195] Thirumurugan, A.; Natarajan, S. *Eur. J. Inorg. Chem.* 2004, 762–770.
[196] Thirumurugan, A.; Natarajan, S. *Inorg. Chem. Commun.* 2004, 7, 395–399.
[197] Cheng, J.-W.; Zheng, S.-T.; Yang, G.-Y. *Inorg. Chem.* 2007, 46, 10261–10267.
[198] Cheng, J.-W.; Zheng, S.-T.; Ma, E.; Yang. G.-Y. *Inorg. Chem.* 2007, 6, 10534–10538.
[199] Cheng, J.-W.; Zhang, J.; Zheng, S.-T.; Zhang, M.-B.; Yang, G.-Y. *Angew. Chem., Int. Ed.* 2006, 45, 73–77.

[200] Wang, G.-M.; Li, Z.-X.; Xue, S.-Y.; Liu, H.-L. *Acta Crystallogr.* 2009, E65, m550–m551.

[201] Li, Y.-W.; Wang, Y.-H.; Li, Y.-G.; Wang, E.-B. *J. Solid State Chem.* 2008, 181, 1485–1491.

[202] Gu, X.; Xue, D. *Cryst. Growth Des.* 2006, 6, 2551–2557.

In: Phthalates
Editors: Gerardo L. Moretti and Drago Romano ISBN: 978-1-62081-994-4

Chapter 2

A CHEMICAL AND BIOLOGICAL APPROACH TO PHTHALATES CONTAMINATION IN TWO ESTUARIES FROM THE BASQUE COAST (BAY OF BISCAY): LEVELS, DISTRIBUTION AND BIOMARKER RESPONSES

Luis Bartolomé*[1]*, Nestor Etxebarria*[2]*, Mariluz Alonso*[2]*,
***Patricia Navarro*[2] *and Maren Ortiz-Zarragoitia*[3]**
[1]General Analysis Service (SGIKER),
[2]Department of Analytical Chemistry
[3]Lab. Cell Biology and Histology, Dep. Zoology and Cell Biology
School of Science and Technology
Euskal Herriko Unibertsitatea, University of the Basque Country (UPV/EHU),
Basque Country, Spain

ABSTRACT

Phthalates are esters of the phthalic acid and are a group of chemicals of high environmental relevance due to their production rates and ecotoxicological potential. Phthalates are mainly used as plasticizers in PVC plastics. Therefore, many consumer products contain specific members of this family of chemicals, including building materials, household furnishings, clothing, cosmetics, pharmaceuticals, nutritional supplements, medical devices, dentures, children's toys, glow sticks, modelling clay, food packaging, automobiles, lubricants, waxes, cleaning materials and insecticides. Di-(2-ethylhexyl) phthalate (DEHP) is one of the most widespread phthalate plasticizer, used in numerous consumer products, commodities, and building materials. Due to its classification as toxic to reproduction, the corresponding labelling (skull and crossbones) and several restrictions on marketing and use (2003/36/EC) DEHP is replaced with other phthalate plasticizers as di-isodecyl phthalate (DIDP) and di-isononyl phthalate (DINP). With the total amount of plasticizer consumption remaining unchanged in Europe, the consumption of DEHP dropped from 42% in 1999 to 21% in 2006, whereas the share of DIDP and DINP increased from 35% to 60% within the same time period (Cardogan,

2007). Also other phthalates such as di-ethyl-phthalate (DEP), di-n-butyl phthalate (DBP), butyl benzyl phthalate (BBP), and di-n-octyl phthalate (DnOP) are widely used. As the phthalate plasticizers are not chemically bound to PVC, they can leach, migrate or evaporate into indoor air and atmosphere, foodstuff, other materials, etc. Once released to the environment they tend to adsorb on particles and, in this way, have been detected in different estuarine compartments as sediment, water or biota. The effects of phthalates on aquatic wildlife have been extensively studied. Phthalate compounds are weak xenoestrogenic compounds to aquatic organisms and can alter the reproduction of exposed organism. Furthermore, they can also modulate cellular pathways such as lipid metabolism leading to oxidative stress processes. Proliferation of peroxisomes and induction of peroxisomal enzymes such as catalase and acyl-CoA oxidase have been reported on fish and molluscs.

Our work was developed in two selected estuaries from Bay of Biscay, the estuary of Bilbao and Urdaibai, showing different pollution loads. The estuary of Bilbao is one of the most industrialised and populated areas of the Bay of Biscay. Due to its geographic location, the area has been subjected to intense industrial pressure. For many years, urban and industrial effluents were discharged into the estuary without further treatment. In 1979, an Integral Plan for the Drainage of the estuary was approved and new environmental protection policies resulted in improved wastewater treatment systems. This, together with the closure of several industries, has produced a gradual recovery of the estuarine environment, with a continuing decrease in heavy metal concentrations and an increase in oxygenation. The tidal zone of the Ibaizabal River forms the estuary, although four tributaries (Kadagua, Asua, Galindo and Gobela) also discharge in the estuary.The docks and secondary channels are the most affected by urban and industrial wastes.

The estuary of Urdaibai and its surroundings have been a Biosphere Reserve by UNESCO since 1984, and are located in a rural environment about 50 km east from the Ibaizabal estuary. The Urdaibai region is characterised by the basin of the Oka River, and by some shorter rivers, including the Golako, Mape, Artike and Laga. The estuary, some 13 km long and 500 m wide, is located in relatively close proximity to several industries (metallurgic, shipyards, dyes, cutlery manufacture) located in the surroundings of Gernika (population ~20,000) upstream the Oka River.

The present chapter would report:

i. Levels and spatial distribution of phthalate esters PEs in sediments and biota samples along the both estuaries since 2002 to 2005 in different sampling campaigns. Comparison between estuaries.
ii. Bioavailability and bioaccumulation studies in both estuaries.
iii. Toxicological effects in aquatic organisms inhabiting both estuaries based on biomarker responses.

1. Introduction

Phthalates or phthalate esters (PEs) are esters of phthalic acid and they are used as functional additives of PVC. In fact, many polymers and especially PVC require the addition of many different sort of plasticizers, UV- filters, biocides or fillers to get a final product with the performance requirements in terms of stability, flexibility, durability, etc. Phthalates are the most used plasticizers and according to the ECPI (European Council of Plasticisers and Intermediates) the average production of phthalates in Europe is one million tons per year.

Phthalates are synthesized through a reaction between phthalic anhydride with alcohols (ROH and R'OH) of different sizes (from C1 to C13) and types (linear or branched). From the processing point of view, they are classified as high and low molecular weight phthalates (HWP and LWP) attending if the size of the alcohol backbone is higher or lower than 6 carbons.

The lightest PEs(C1 or C2) are used as solvents. The LWP (C4 to C6) are used as high solvating plasticisers and the HWP (C7 to C13) are general purpose plasticisers with different uses according to the weight of the PE. Though the way theses plasticisers interact with the host polymer is still under discussion, these compounds can be released from the polymer (Jahed Khaniki et al., 2006), and as a consequence of the variety of PEs and the wide range of polymer applications they are ubiquitous in the environment. As it has reported, there is a general exposition to phthalates and an increasing concern about the adverse effects that can be attributed to these compounds (Kamrin, 2012). This fact is particularly sensitive with some uses of plastics such as in food processing (Cao, 2010), indoor environments (Garcia-Jares et al., 2009), in cosmetics and personal care products (Koniecki et al., 2011), or in toys, especially when mouthing by children is highly likely (Kamrin, 2012).

Regarding the physical-chemical properties and the structures of the PEs and the abbreviations followed to name them, we have followed the values reviewed by Cousins et al. (Cousins et al., 2003). In addition to this, the environmental fate and distribution of PEs has been reviewed in the literature (Staples et al., 1997) and since their volatility and water solubility is quite low, and even lower with the increasing molecular weight, they are associated with solid particles, soils and sediments (Lu, 2009; Cousins et al., 2003), and urban wastes are the typical sinks of these types of compounds (Björklund et al., 2009, Clarke et al., 2011; Olujimi et al., 2012). A detailed summary of the most recent data regarding the distribution of PEs in sediments and biota is shown in Table 1. Additionally, PEs are not environmentally persistent since they are degraded by microorganism (Peterson and Staples, 2003).

Consequently, phthalate compounds show short half-lives and are quickly biodegraded in the environment, suggesting no significant effects on exposed wildlife (Cadogan, 1999). However, PEs are produced at high amounts and their presence in the environment is widespread, exposing wildlife to a chronic phthalate uptake scenario (WHO/IPCS, 2002). From the ecotoxicological point of view, PEs are typically included among the endocrine disruptive compounds (EDCs) and non-genotoxic carcinogens (Björklund, 2009; Heudorf, 2007). However, their biological effects are complex and in a study using DEHP exposed mice the expression of 51 different genes, such as peroxisome proliferation, xenobiotic detoxification, oxidative stress, immune response, steroid hormone metabolism, testis development and pheromone transport associated genes were modulated in the liver (Wong and Gill, 2002).

Endocrine disruptors were defined after the Weydbridge meeting (1996) as a substance or mixture that alters function(s) of the endocrine system and consequently causes adverse health effects in an intact organism or its progeny or (sub)populations. Phthalates have been described as xeno-estrogenic compounds, causing feminizing effects on exposed organisms (Latini, 2000; Matthiessen et al., 1998). PEs are classified as weak estrogens in comparison to the activity measured for the natural estrogenic hormone 17β-estradiol (Soto et al., 1995; Hewitt and Servos, 2001).

Table 1. Mean values of phthalates concentrations (mg/Kg) in sediments, along the World in the last decade

Matrix	Year	Place	BBP	DBP	DEHP	DINP	DMP	DEP	DNOP	DIBP	Total Phthalates	References
Aqueous sediment	2000	Canada	ND	ND	ND		ND		15.00		15.00	Clark et al, 2003
Aqueous sediment	2000	USA	0.19	14.60	1.40		3.60		ND		19.79	Clark et al, 2003
Aqueous sediment	2000	Europe	0.06	0.22	4.90		0.01		ND		5.19	Clark et al, 2003
Aqueous sediment	2000	Japan/Asia	ND	0.12	0.48		ND		ND		0.60	Clark et al, 2003
North Sea sediment	2000	Netherlands	ND	ND	1.41		ND	ND	0.05		1.46	Klamer et al, 2005
River sediment	2002	Taiwan (China)	0.20	6.30	4.60			0.2			11.30	Yuan et al,2002
Estuary sediment	2003	Biscay (Spain)	0.68	0.40	12.80		ND	0.06	ND		13.93	Cortazar et al, 2005
Tidal salt marsh sediment	2003	California (USA)			1.63						1.63	Hwang et al, 2006
Estuary sediment	2003	Biscay (Spain)	0.68	0.40	12.75		ND	0.25	ND		14.07	Bartolomé et al, 2005
Soil of costal line	2004	Greece		0.07	0.32		0.02	0.01		0.05	0.48	Grigoriadou et al, 2011
Aqueous sediment	2004	Greece		0.01	0.08		0.03	0.01		0.27	0.40	Grigoriadou et al, 2011
Marine sediment	2004	Greece		0.15	1.21		0.01	0.07		0.28	1.72	Grigoriadou et al, 2011
Dam sediment	2004-05	South Africa		0.40	0.20		ND	0.17			0.77	Fatoki et al, 2010
Channel sediment	2004-05	South Africa		0.33-1.83	0.04-0.17		ND	0.19			0.19	Fatoki et al, 2010
River sediment	2004-05	South Africa		0.35	0.04		ND	0.16			0.55	Fatoki et al, 2010
River sediment	2004-05	South Africa		0.64	0.41		ND	0.22			1.27	Fatoki et al, 2010
River sediment	2004-05	Taiwan (China)			1.00						1.00	Huang et al, 2008
Lake sediment	2005	China	0.03	0.28	1.30	0.067	0.04	0.13	0.02	5.80	7.67	Zeng et al, 2008

Matrix	Year	Place	BBP	DBP	DEHP	DINP	DMP	DEP	DNOP	DIBP	Total Phthalates	References
River sediment	2005	Nigeria (Africa)		0.59	0.17		0.21	0.20			1.18	Adeniyi et al, 2011
Fresh water sediment	2006	Netherlands	0.44	0.09	4.30						4.39	Peijnenburg et al, 2006
Marine sediment	2007	Santander (Spain)		ND	0.70		ND	ND			0.70	Antizar-Ladislao et al, 2009
Estuary sediment 1	2007-08	Biscay (Spain)		ND	0.62-ND		ND	ND			0.62	Puy-Azurmendi et al, 2010
Estuary sediment 2	2007-08	Biscay (Spain)		0.975-ND	2.94-1.12		ND	2.95-ND			6.87-1.12	Puy-Azurmendi et al, 2010
Estuary sediment 3	2007-08	Biscay (Spain)		0.66-ND	1.04-2.48		ND	4.34-ND			6.00-2.48	Puy-Azurmendi et al, 2010
Estuary sediment 4	2007-08	Biscay (Spain)		0.89-ND	1.78-ND		ND	3.32-ND			5.99-ND	Puy-Azurmendi et al, 2010
Estuary sediment 5	2007-08	Biscay (Spain)		0.78-ND	1.53-ND		ND	0.97-ND			3.29-ND	Puy-Azurmendi et al, 2010
Gulf sediment	2008	Suez (Africa)		13200.00	15600.00			2000.00			30800.00	Rushdi et al, 2009
Marine sediments	2011	NE Spain	0.71		0.04		0.02	2.44			3.19	Sánchez-Ávila et al, 2011

1-5. Estuary places ordered from smallest to largest distance to the sea.

Recombinant gene expression assays also demonstrated that selected PEs such as DBP, BBP and DEHP can compete with 17β-estradiol activating estrogen receptor dependent genes in vitro (Zacharewski et al., 1998).

As a consequence of this concern, regulatory developments in Europe have resulted in a change in PE consumption, with the higher PEs (DINP and DIDP) replacing DEHP as the plasticiser of choice owing to the fact that DIDP and DIP are not classified as hazardous. Although most applications are shown to pose no risk when studied using recognised methods of risk assessment DEHP has been classified as a Category 1A reprotoxin and is now on the Annex XIV of the EU REACH (Registration, Evaluation, Authorisation and Restriction of Chemicals) legislation.

In this chapter we have reviewed the distribution of PEs and the ecotoxicogical status derived from the presence of PEs in two estuaries of the Basque Coast: the estuaries of Bilbao and Urdaibai. These two estuaries represent the two outmost examples of environmental scenarios of the Cantabrian Coast, being the estuary of Bilbao the polluted and heavily altered one, and the estuary of Urdaibai a natural reserve of the Biosphere since 1984. Broadly speaking, these two estuaries have been deeply studied in many previous works (Azkona et al., 1986; Cearreta et al. 2000, Leorri et al. 2008, Cortazar et al. 2008; Monge-Ganuza et al 2008; Liria et al. 2009; Puy-Azurmendi et al., 2010a; Bartolomé et al. 2011) and this review is focused on the integrative view of the chemical distribution and the ecotoxicological effects of PEs.

2. Analytical Methods

The analysis of PEs in environmental samples has to deal with two important drawbacks: first of all, the low concentrations that are typically found, especially in waters, required an efficient extraction and pre-concentration factor as well as a selective clean up method, and, in addition to this, the problems coming from the ubiquity of PEs, including the laboratory plastic ware, makes fairly easy the contamination of the samples.

Regarding the contamination of samples PEs were found in many of the common laboratory materials and reagents, such as tubing, cork, glass wool, filter paper, alumina, sodium sulphate, septum, and so on (Petrovic et al. 2001). Sources of PEs in the laboratory environment were investigated by Fankhauser-Noti and Grob (2007), finding, for instance, DBP and DEHP in commercially available hexane at levels of 100 μg/L, and contamination of glassware, very common due to the presence of phthalates in laboratory air. Therefore, every possible measure should be taken to reduce PEs contamination in the materials and reagents used in the laboratory (Fankhauser-Noti and Grob 2007) and efforts should also be made to develop methods that have the least number of steps and use the minimum amounts of solvents, adsorbents, and other laboratory supplies in order to minimize blank levels of PEs and to improve the detection limits.During the determination of these compounds, the separation of PEs from the sample matrix is a very important stage. For simple sample matrices, such as water, clean-up procedure is not required and the extraction and pre-concentration procedures used are often simple, i.e. the traditional liquid-liquid extraction (LLE), solid-phase extraction (SPE) and solid-phase microextraction (SPME), and thus detection limits at ppt levels are commonly achievable (Holadová and Hajslová 1995,

Farahani et al. 2008; Jara et al. 2000; Prapatpong and Kanchanamayoon 2010; Peñalver et al. 2000, Luks-Betlej et al. 2001, Polo et al. 2005).

In this sense, the use of SPME provides extraction and pre-concentration in one step, thus eliminating a significant number of the undesirable steps associated with conventional sample preparation techniques, due to its simplicity, effectiveness, no use of solvents or complicated apparatus, sensitivity and portability (Kotowska et al. 2006). Several studies have developed the SPME in water samples using fibers coated with a thin film of different sorbents (Luks-Betlej et al. 2001; Li et al. 2006; Prokůpková et al. 2002). Water is a simple matrix, which allows the direct immersion of the fiber. Nevertheless, SPME was performed immersing directly the fiber into the aqueous sample (Luks-Betlej et al. 2001; Li et al. 2006) or into the previously extracted sample (Asadollahzadeh et al. 2010). Although SPME is normally used when the separation is carried out by gas chromatography (GC), on-line in-tube SPME has also been coupled to liquid chromatography LC-UV for the determination of PEs in water samples (Mitani et al. 2004; Cháfer-Pericás et al. 2008) or to liquid chromatography-mass spectrometry (LC-MS) (Möder et al. 1998). Limits of detection (LODs) obtained using SPME were in the 5-50 ng/L range, the recoveries were typically in the 90-113% and the maximum relative standard deviation was around 10%.

He et al. (2010) have applied a novel molecularly imprinted polymer (MIP) to a SPME device, which could be coupled directly to gas chromatography-mass spectrometry (GC-MS). Dibutyl phthalate (DBP) was used as the template molecule, achieving detection limits for the PE samples within the range of 2.2–20.8 ng/L and recoveries of up to 95–105%.

A novel sample preparation method based on the same principles of SPME, the stir bar sorptive extraction (SBSE), has been also developed to determine PEs in water samples (Serôdio and Nogueira 2000; Brossa et al. 2005; Prieto et al. 2008). In the SBSE, the amount of polydimethylsiloxane typically coated is substantially higher than on a SPME fiber and thus, much higher recoveries and lower detection limits can be reached.

On the other hand, concerning the liquid–liquid microextraction (LLME), several alternatives have been optimized and applied to the determination of PEs in water samples. The first one, it is a method called dispersive liquid–liquid microextraction (DLLME), based on dispersion of tiny droplets of the extraction liquid within the aqueous solution. Its advantages are its rapidity, low cost and high recoveries of the analytes (Liang et al. 2008; Xu et al. 2007). The limits of detection achieved using this technique coupled to GC-MS have been in the low ppt level. Relative standard deviations and recoveries of the method were less than 12% and around 80%, respectively (Farahani et al. 2007; Panagiotou et al. 2009). Another approach of the LLME is a novel method, termed ionic liquid cold-induced aggregation dispersive liquid–liquid microextraction (IL-CIA-DLLME), which improves the extraction efficiency and reduces the extraction time (Zhang et al. 2011).

Other alternative of the LLME is the use of membranes as selective barrier between the extraction solvent and the water sample. The membranes used could be porous (Bergström et al. 2007) or non-porous (Prieto et al. 2008), depending of the complexity of the sample matrix.

For the determination of PEs in sediments and tissues, a solid-liquid extraction is necessary and, normally, also a clean-up step before the instrumental analysis. Among the techniques employed for the extraction of the PEs, we can cite the microwave-assisted extraction (Cortazar et al. 2005), the focused ultrasound-assisted extraction (Errekatxo et al. 2008), the pressurized solvent extraction (Cortazar et al. 2005) and the classical ultrasonic bath extraction (Sánchez-Avila et al. 2011) and Soxhlet (Wang et al. 2006). After the extraction, although the clean-up is usually carried out by SPE using Florisil (Cortazar et al. 2005; Errekatxo et al. 2008; Sánchez-Avila et al. 2011) or alumina followed by Florisil (Wang et al. 2006), other authors preferred a reverse-phase HPLC before GC-MS analysis (Cho et al. 2006) or no clean-up step for seaweed analysis in GC-FID (Chi et al. 2006). In the case of sediments, the sulphur interference was removed by adding activated copper (Cortazar et al. 2005; Errekatxo et al. 2008) or mercury (Wang et al. 2006).

Last but not least, we should notice that there is no certified reference material available for PEs, adding other difficulty to the determination of these pollutants.

3. First Case Study: Urdaibai Estuary

The estuary of Urdaibai is a shallow, macro-tidal estuary located in the Basque Country (Bay of Biscay, 43° 22'0N, 2° 40'0W), northern Spain (see Figure 1). Physically it is a rather shallow estuary (average 2.6 m) and it extends roughly 12 km and shows a maximum width of 1.2 km.

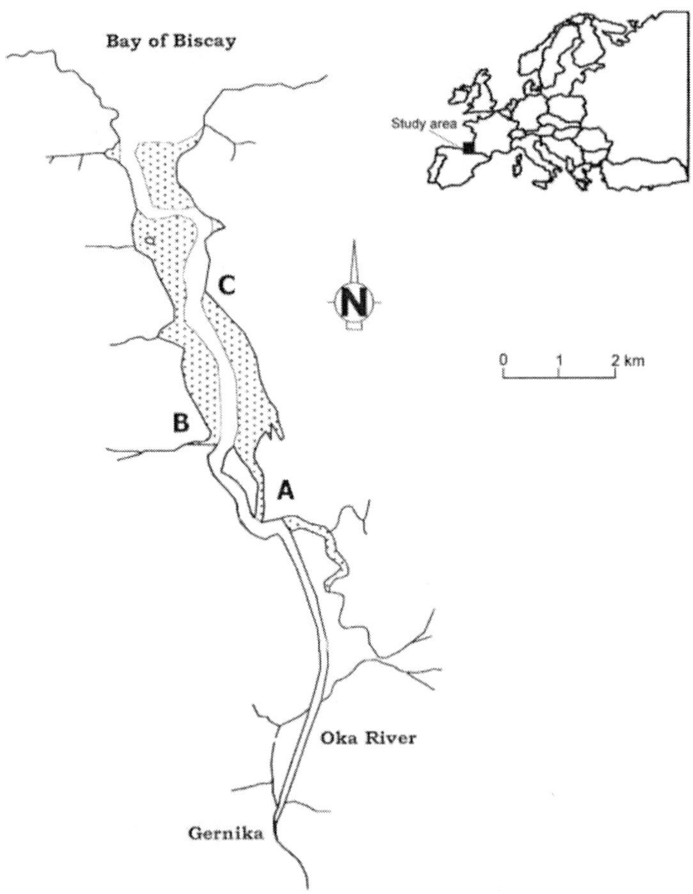

Figure 1. The Urdaibai estuary and the location of the sampling points: Arteaga (A), Murueta (B), Kanala (C), and Gernika (D).

From June 2002 to September 2004 and every three months (10 sampling campaigns) both sediments and oysters were collected manually from three sampling points at low tide (~1 m). On the right bank, Kanala and Arteaga are two wild sites with sandy and muddy sediments, respectively.

On the left bank, the station of Murueta is enclosed in a shipyard, and the sediments are also sandy. From the inner station located in Gernika, only sediments were taken since no oysters were found. Sediment samples were collected manually from the superficial sediment (<10 cm depth), stored in pre-cleaned flasks (10% (v/v)) nitric acid bath for 24 h and dried at 110 °C overnight, and 1 ml of formaldehyde 4% was added to avoid bacterial growth. Oysters (*Crassostrea sp.*) of 90–120 mm shell length were collected close to the water line, rinsed with natural water and introduced into plastic bags. Water samples were collected manually from the centre in the bed of the river, stored in pre-cleaned flasks (10% (v/v)) nitric acid bath for 24 h and dried at 110 °C overnight, and 1 ml of formaldehyde 4% was added to avoid bacterial growth. In the case of water and sediment samples, flasks were filled to maximum capacity to avoid loss of the volatile fraction. All the samples were transported in cooled boxes to the laboratory. Sediment samples were frozen (-20 °C) and freeze-dried at low temperatures (-46/-52 °C) and pressures (0.17/0.22 mbar). Sediment samples were sieved and two particle-size fractions (<63 µm and between 63 µm and 250 µm) were analysed. Samples were kept in the fridge at 4 °C until analysis. The individual oysters were dissected with a clean scalpel blade to separate the soft tissues from the shells. Around twenty individuals were dissected out, frozen, homogenised and freeze-dried as explained above. Freeze-dried samples were ground in a ball mill and kept in the fridge at 4 °C until analysis.

The protocol for the determination of PEs is thoroughly described in the works of Bartolomé et al. (2005) and Cortazar (2009). Other micro-contaminants were simultaneously determinated at the same time (Polycyclic aromatic hydrocarbons PAHs, polychlorinated biphenyl PCBs and nonylphenol ethoxilates NPEOs). Briefly, 1.0 g of dry sediment or 2.0 g of dry oyster homogenante were accurately weighed, and 15 ml of acetone as extracting solvent, were added before submitting to microwave assisted extraction. In the case of sediment samples, 1 g of activated copper was added in order to eliminate sulphur from the extract. The extract was filtered and evaporated to ca. 1 ml of iso-octane using nitrogen blow-down evaporation. The iso-octane extract was loaded onto 2 g (sediments) and 5 g (oysters) Florisil SPE cartridge that had been previously conditioned with 5 ml of n-hexane. The elution solvent was 5 ml or 9 ml of ethyl acetate to elute the PEs from the sediments and oyster tissues, respectively. In the case of water samples, 300 mL with 60 g of NaCl previously added were passed through a SPE Lichrolut EN cartridge (200 mg) that had been previously conditioned with 3 ml of ethyl acetate, 3 mL of methanol and 3mL of Milli-Q reagent water. The elution solvent was 9 mL of ethyl acetate.

The eluates were concentrated to dryness, re-dissolved in 500 µl (sediments), 250 µl (oysters) or 200 µl (waters) of iso-octane and kept in the dark at -18 °C until analysis. The extracts were analysed in a gas chromatograph coupled to a mass spectrometer.

3.1. Phthalates Levels and Distribution in Water, Sediment and Biota

Figure 2 shows the total PEs concentrations detected in Urdaibai sediments within this sampling period (2002-2004). This total concentration includes concentration for dimethyl phthalate (DMP), diethyl phthalate (DEP), dibutyl phthalate (DBP), butyl benzyl phthalate (BBP), bis(2-ethyl-hexyl) phthalate (DEHP), di-n-octylphthalate (DOP). In general, unless in the October 2003 sampling campaign, the concentrations ranged between 1 to 15 mg kg^{-1} with 4.3 mg kg^{-1} of mean concentration along the estuary. The most contaminated sampling point was Gernika with a mean of 5 mg kg^{-1} although the highest value was detected in Murueta (top value of 25 mg kg^{-1}). DMP, DOP and BBP levels were almost all below their detection limits. Attending to the concentration levels in Urdaibai sediments, DEHP and DBP were placed in the high category, DEP at the second category and the rest below the detection limit. This finding is in agreement with the fact that DEHP and DBP are the most commonly produced PEs. In addition, we found that concentrations of all eight PEs were up to 1000 times higher in the sediment samples than in the surface water samples. This finding agrees with the physical-chemical properties of PEs. As a matter of fact, Al-Omran and Preston (1987), reported that PEs are soluble by fulvic and humic acids, and are capable of being adsorbed onto particulate matter, with sediments serving as a final sink. Nevertheless, no correlations were found between PEs concentrations and TOC (R^2:0,00) or particle size (R^2:0,09). The range of PEs concentrations found in sediments is comparable to other estuaries distributed through the world (see table 1). It is noteworthy that the concentration found in Urdaibai Biosphere Reserve estuary is comparable to the concentrations found in polluted estuaries such as in Taiwan sediments (Yuan et al, 2002) or Canada (Clarck et al, 2003). This fact shows us the ubiquity of these compounds in presume pristine areas due to its excessive production and use. Resource managers often use Sediment Quality Guidelines (SQGs) to interpret contamination data with respect to impacts on sediment dwelling organisms. The most widely used SQGs are empirical guidelines based on the statistical analysis of matched sediment chemistry and biological effects data.

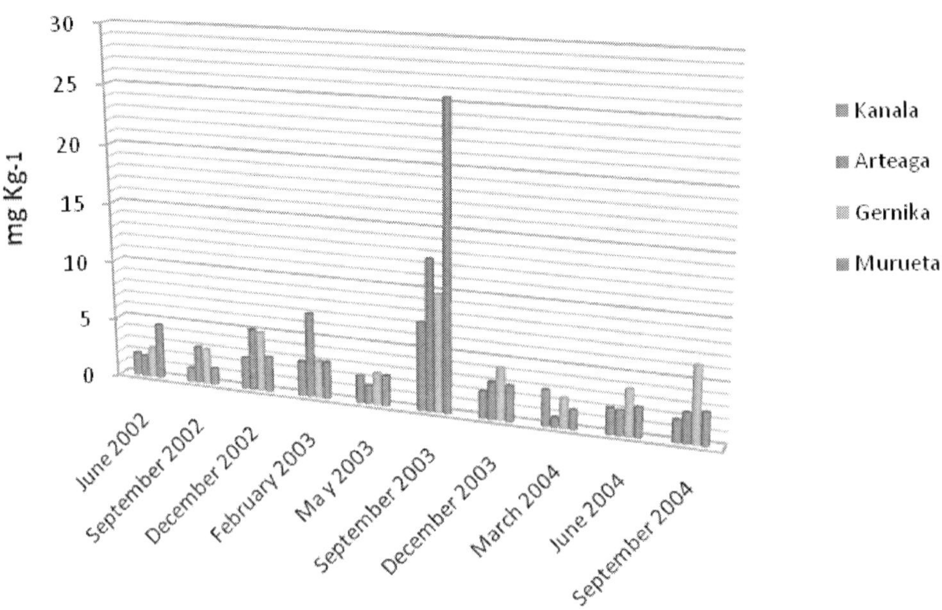

Figure 2. PEs mean concentrations (mg Kg^{-1}) in four sampling sites from Urdaibai estuary in 8 sampling campaigns (2002-2004).

Although the benthic community is often the focus of concern in sediment quality assessments, few SQGs based on benthic community effects are available; those that are available are typically based on data sets with limited geographic scope and may not be applicable to other regions. Various statistical approaches have been used to develop empirical SQGs. In the case of PEs, commonly used approaches include apparent effects threshold (AET; Barrick et al. 1988). According to these AET values, Figure 3a shows that three of the 6 compounds analysed (DEP, DBP and DEHP) exceeded the AET limit in some sampling campaigns. In the case of DEHP, AET limit was exceeded in 60% of the samples analysed (see Figure 3b). This is an expected result considering that 90% of the PEs total concentration was due to DEHP. Finally, we can see that Gernika is the place where the AET limit is exceeded more often (85%) which suggests that these PEs are close to the primary sources.

In Figure 4 the PE concentrations in oysters are shown. In the three sample sites (Kanala, Murueta and Kanala) no seasonal variation was observed.

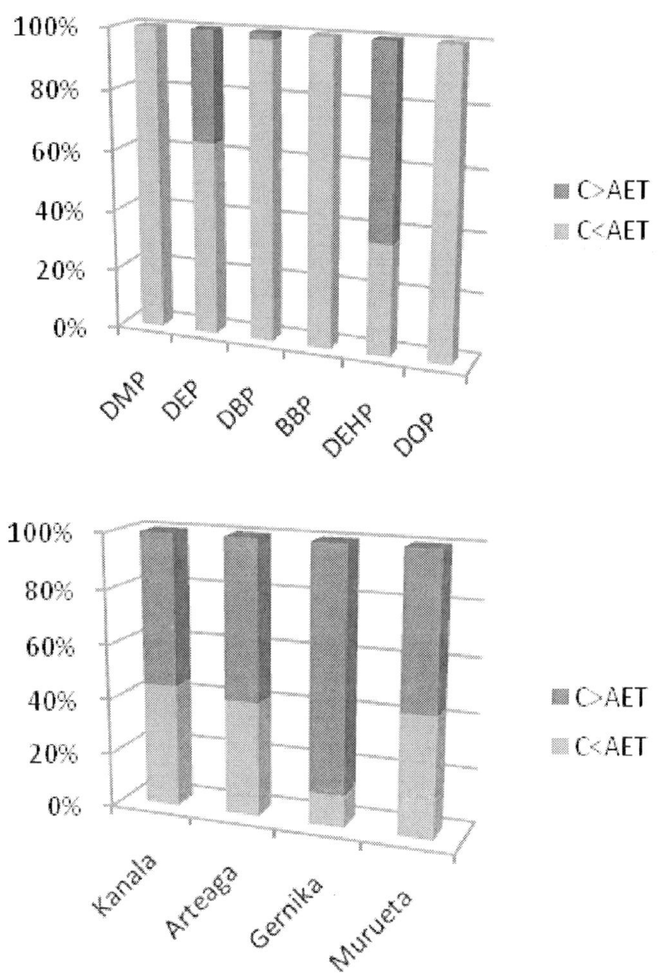

Figure 3. Frequency of PEs reach the value of AET in sediments sampled between 2002 and 2004 (a) and frequency at which DEHP concentration exceeds the value of AET in the different sampling stations between 2002 and 2004 (b).

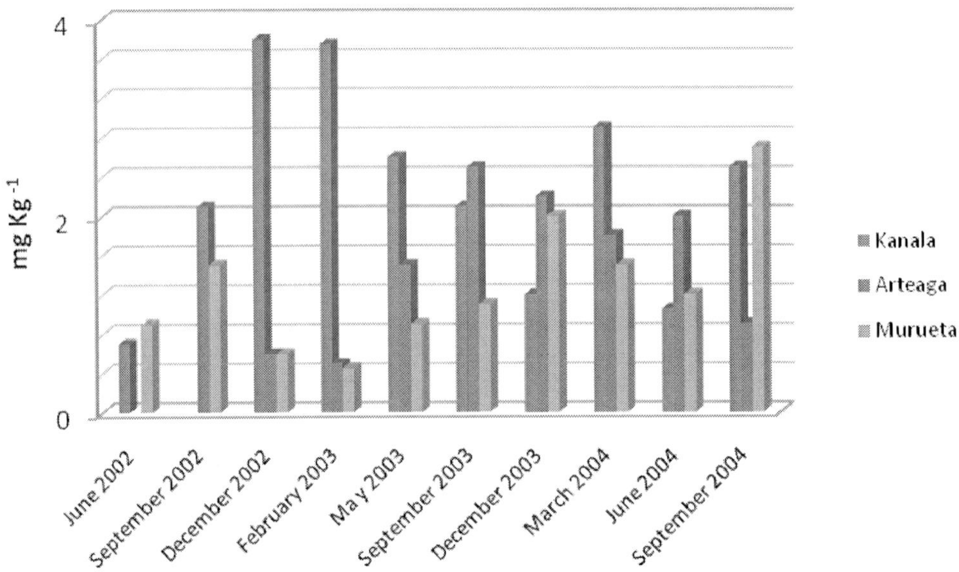

Figure 4. Total concentrations of PEs (mg kg^{-1}) analyzed in Urdaibai oysters from 2002 to 2004.

The seasonal behaviour of these compounds could be related with oyster developmental stages. Normally seasonal cycle of gamete development is often related with variations in the lipid percentage of the organisms. In this case it were not found significant correlations between the concentration of PE and the percentage of lipid (R^2:0.11), therefore we cannot link the seasonal pattern found in PEs concentrations with o the seasonal cycle of gamete development.

DMP, BBP and DBP were below their respective detection limits; DEP and DOP were found in only a few samples. DEHP was the most abundant with concentrations reaching up to 3.2 mg kg^{-1} at some locations and clearly exceeding the acceptable threshold concentration for DEHP. There was no apparent relationship between concentrations and closeness of likely sources (e.g., shipyard, urban areas) although a slight downward trend in concentrations could be observed as we move downstream. Possible contamination sources (e.g., the polyethylene sample container) were examined and discarded because leached PEs had a compound profile that was different from the profiles in the oyster extracts.

In the water samples DEP, DBP and DEHP were found in all samples, whereas detection frequencies for DMP, BBP and DOP reached up 5%, 2% and 8%, respectively. The concentrations of the various PEs varied strongly. The most abundant compound in the influent was DEHP (4.7–20.4 ng L^{-1}), followed by DEP (0.7–5.2 ng L^{-1}) and DBP (n.d.-4.2 ng L^{-1}). The remaining phthalates BBP, DMP and DOP were of minor importance with concentrations around or below 1 ng L^{-1}. Surprisingly, comparable distributions and comparable concentrations were reported by Dargnat et al. (2009) in wastewaters. It should be considered that downstream Gernika there is a wastewater treatment plant and this can contribute, together with the urban area, to the observed values of PEs in the estuary.

According to Baumard and co-workers (1999 a,b), it is possible to estimate the source of the PAHs based on the relation between the bioaccumulation factor (BAF) and the octanol/water partition coefficient ($logK_{ow}$) of each compound. BAF is defined as the ratio between the concentration found in the oyster and the concentration found in the sediment, as shown in Eq. 1.

$$BAF = \frac{C_{biota}}{C_{sed}} \qquad \text{Eq.1}$$

Applying this approach to the case of PEs, from the values of the regression coefficients (R^2) and the slopes (m) of the BAF vs. $\log K_{ow}$ plot, four different uptake pathways have been suggested: (i) the water soluble fraction ($R^2 \sim 0.5$, m ~ 30); (ii) the water soluble ($R^2 \sim 0.3$, m ~ 19); (iii) the water soluble and the particulate matter ($R^2 \sim 0.1$, m ~ 14); and (iv) the particulate matter ($R^2 \sim 0.0$, m ~ 6). The regression coefficients (R^2) and the slopes obtained in the different campaigns of this study are shown in Figure 5. In the first campaigns, both parameters were almost zero ($R^2 < 0.1$ and m ~0), as those obtained by Baumard et al. (1999a) for bivalves near sediments and with a high content of particulate matter.

This is likely the expected behaviour due to the tidal regime and the high suspended particulate matter of this estuary. However, in the last campaigns (February 2003 to September 2004) a sharp increase is observed in both the regression coefficient (0.6–0.8) and slopes (5 to 20). This change is interpreted as a change in the uptake pathway from the above mentioned particulate matter to mixed water soluble and particulate matter. The same conclusion was found for the PAHs study in these mussel samples (Cortazar et al, 2008).

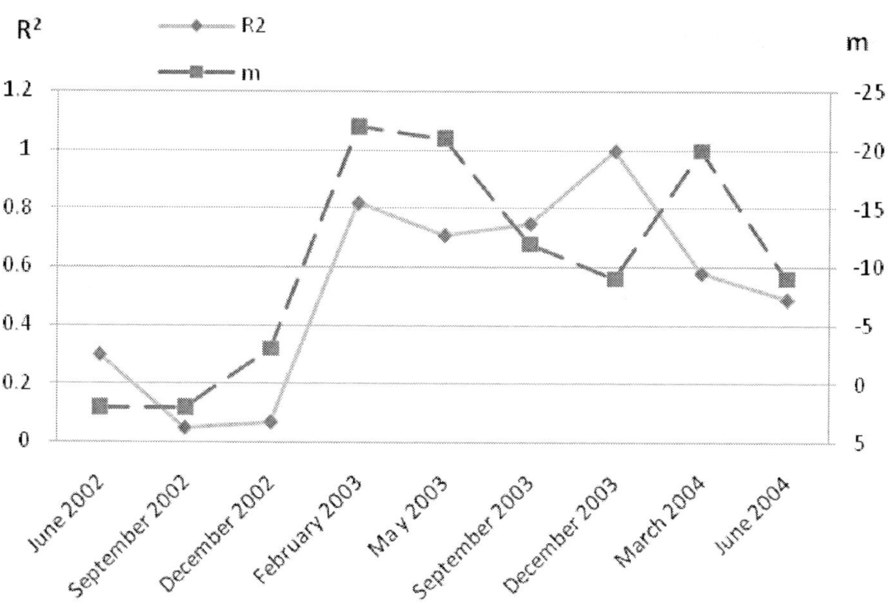

Figura 5. Regression coefficients (R^2) and slopes (inverted order) obtained for the regressions between BAF and $\log K_{ow}$ in the three sites and along the sampling campaigns.

3.2. Biological Approach to PEs Contamination in Urdaibai

In this section we have reviewed biological responses based on the biomarker approach on bivalve and fish populations from Urdaibai in the period 1991-2007. We have focused on responses associated with organic xenobiotic exposure, some of them already mentioned in the introduction section, that could be related to but not exclusive of PE exposure.

In general terms biological responses measured in sentinel organisms in the Urdaibai estuary followed a seasonal dependent cycle, that were only disturbed by an acute contamination exposure scenario. Structure of the digestive gland and digestion activity of bivalves from Urdaibai is higher in summer months than in winter months. Lysosomes in digestive cells are larger in summer than in winter. These trends have been associated with more food bioavailability during summer period (Marigomez et al., 2004; Soto et al., 1995).

The activity of antioxidant enzymes in mussels and oysters was higher in winter months than in summer months. Peroxisomal activities, however, were higher in summer than in winter months (Orbea et al., 2002). Accordingly, contaminant body burdens in mussels and oysters followed a seasonal pattern in bivalves from Urdaibai. Accumulation of organic contaminants was higher in winter than in summer. This trend could be explained by higher lipid content of bivalves during winter months than in summer months, probably due to the presence of lipid reserves required for gametogenesis. In mussels from Urdaibai, gametogenic cycle starts on November and progresses during winter reaching gonad ripening in March-April. Spawning occurs during May-July. At the end of summer a new gametogenic cycle starts with the accumulation of lipid reserves (Ortiz-Zarragoitia and Cajaraville, 2010).

During the period covering 1991-2002 no significant alterations on biomarker responses were measured in mussels, oysters and fish populations from Urdaibai, except changes associated with seasonal trends (Orbea et al,. 2002; Marigómez et al., 2004; Soto et al., 1995). Thus, these data indicated that Urdaibai could be considered a low contaminated environment (Orbea and Cajaraville, 2006; Orbea et al., 2002). However, after the arrival of the fuel oil spilled by the tanker Prestige on the North West Iberian Peninsula in November 2002, the ecological situation of Urdaibai changed. Furthermore, in the last two decades new pressures such as an increase in local population and the develpment of new industries have lead to a worsening in the health status of aquatic wildlife populations from Urdaibai. During a study covering 2003-2006 period mussels from Urdaibai showed enlarged lysosomes in the digestive gland that did not follow the above described seasonal trend (Garmendia et al., 2011a). Accordingly, higher prevalence of histopathological alterations such as haemocyte infiltrations of digestive tubules, and brown cell aggregations were described in the same mussel populations (Garmendia et al., 2011b). These results were accompanied by higher expression levels of the antioxidant enzyme catalase in the digestive gland of mussels in July 2004 (Bilbao, 2007). However, no changes on catalase protein levels not in acyl-CoA enzyme gene expression levels were detected at the same period (Apraiz et al., 2009; Bilbao, 2007). Thus, a worsening on mussel health in Urdaibai was observed (Garmendia et al., 2011c).

Ortiz-Zarragoitia and Cajaraville (2010) described histopathological alterations in the gonad tissue of mussels from Urdaibai in January March and April 2003. Female mussels showed higher incidence of oocyte atresia and gametogenic follicle infiltrations by haemocytes (phagocytic cells). In male mussels high levels of vitellogenin-like proteins were detected in March 2003. These responses could be associated with the arrival of the fuel oil to the coast line of Urdaibai. However, on March 2004 an abnormal high prevalence of intersex mussels was detected in Urdaibai. These intersex mussels showed mainly male follicles containing oocytes. Thus, feminisation of male mussels was detected in mussel population from Urdaibai (Ortiz-Zarragoitia and Cajaraville, 2010). These effects could not be associated with the impact of the fuel oil, since no estrogenic effects for the *Prestige* fuel oil have been described (Ortiz-Zarragoitia et al., 2011). Therefore, new sources of xeno-estrogenic contaminants were present in Urdaibai. Together with the presence of intersex mussels, gametogenic follicles of small size were detected in April 2004 (Ortiz-Zarragoitia et al., 2011). The biological meaning of this histopathological alteration in mussels is not well known, and has been proposed as a consequence of imperfect gamete ripeness caused by stress conditions (poor food availability, temperature changes, contamination) (Villalba, 1995).

Nevertheless, similar prevalences of small sized follicles, showing mature gametes surrounded by high amount of reserve material in the mantle, was described for the same season in other 21 mussel populations affected by the *Prestige* oil spill (Ortiz-Zarragoitia et al., 2011). Although alterations associated to fuel oil impact tend to disappear by April 2006, in a recent study covering April 2007-April 2008 in Urdaibai,we showed high prevalence of oocyte atresia and gamete necrosis in mussels, and delayed gametogenesis in oysters (Puy-Azurmendi et al., 2010). These data demonstrates the presence of endocrine disruptor compounds in Urdaibai. Moreover, high levels of PEs were measured in sedmients analysed during the same period. Total PE levels in sediments reached values as high as 8000 ng/g, which were more than 100 times higher than those reported previously by Chaler et al. (2004) but similar to those presented in above. Finally, it is worth mentioning the presence of high prevalence of intersex fish detected in Urdaibai. Thicklip grey mullets collected downstream the town of Gernika, showed elevated plasma levels of vitellogenin (marker of xeno-estrogenicity) and intersex gonads, indicating feminisation of male fish (Puy-Azurmendi et al, submitted).

4. Second Case Study: Bilbao Estuary

Arriluze is located in the mouth of the Bilbao estuary on the Bay of Biscay (43°82`00 N, 38° W) (Figure. 6). As described by Cearreta et al. (2000), this estuary is a short (approximately 15 km long) tideway with a moderate freshwater flow rate (25 m^3 s^{-1} on average) from the Nervión and Ibaizabal rivers, which are the principal tributaries of the estuary.

The Bilbao estuary was originally the most extensive estuarine area in northern Spain. However, its natural features have been dramatically modified by urban, industrial, and port developments.

The management of abundant local iron ore led to early industrial development in the mid-nineteenth century. The original estuary was rapidly decreased in size through land reclamation to form a tidal channel as a navigable watercourse from the city to the open sea, which was completed by 1885.

During the last 150 years, urban and industrial effluents have been discharged into the estuary without further treatment. The system degenerated in such way that, in the 1970s, water samples showed extremely low oxygenation rates together with high organic matter contents and metal concentrations.

Mussels, 45–65 mm in length, were collected manually every 2 months from November 2002 to March 2004. Samples were collected close to the water line during low tide from the same sampling point, rinsed with natural water, and introduced into plastic bags.

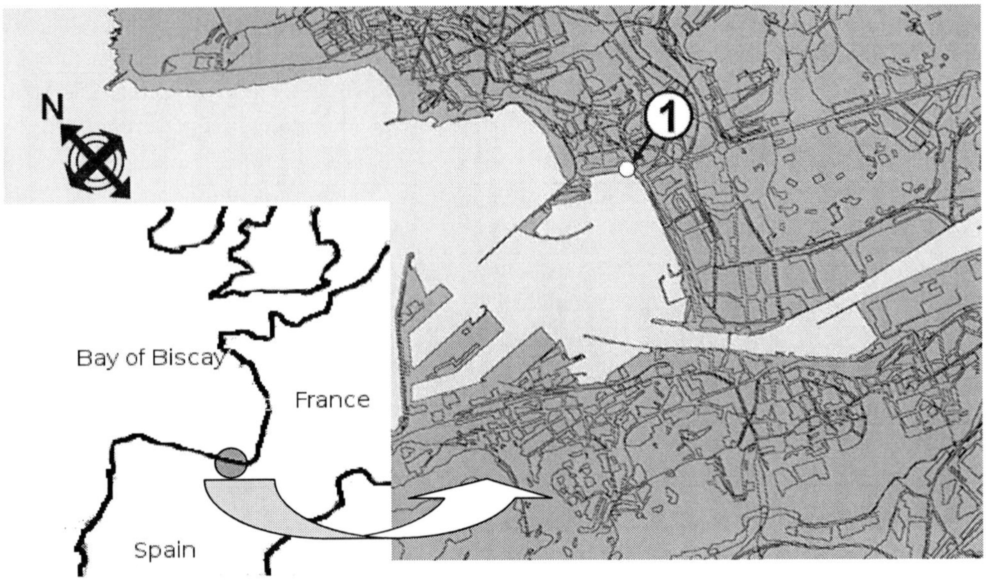

Figure 6. Estuary of Bilbao and location of Arriluze (1).

Surface water temperature, salinity, pH, and dissolved oxygen were measured at each collection. Those parameters were fairly constant for all collections e.g., pH ranged from 8.05 to 8.47 and salinity ranged from 24 to 33%. Samples were then transported on ice to the laboratory.

Each individual was dissected with a clean scalpel blade to separate the soft tissues from the shells. From each collection, 20 mussels were used for condition index measurement. For the PEs analysis 20 additional mussels were dissected out, pooled, homogenised, and freeze-dried. The protocol for simultaneous determination of PAHs, PCBs, and PEs was based on the previous works of Bartolomé et al. (2005) and Navarro et al. (2006).

As can be seen in Table 2, PEs concentrations found in Bilbao estuary are much higher than concentrations recorded in the Urdaibai estuary.

Table 2. Concentration of PEs in mussel tissues (mg kg^{-1} dw) from November 2002 to March 2004

	Nov02	Jan03	Mar03	May03	Jul03	Nov03	Jan04	Mar04
DMP	0.64	<LOD	0.28	<LOD	<LOD	<LOD	<LOD	<LOD
DEP	0.26	0.19	0.54	0.39	0.75	0.70	0.52	0.08
DBP	0.50	0.40	0.48	1.31	8.05	<LOD	0.10	0.83
BBP	<LOD	1.76	0.66	<LOD	0.73	6.43	0.89	0.09
DEHP	24.1	25.28	2.12	2.86	2.38	1.34	1.45	0.29
DOP	<LOD	0.02	<LOD	0.13	<LOD	0.07	0.11	<LOD
ΣPEs	25.5	27.65	4.08	4.69	11.91	8.54	3.07	1.29

<LOD≡ Under limit of detection. DMP dimethyl phthalate, DEP diethyl phthalate, DBP dibutyl phthalate, BBP butyl benzyl phthalate, DEHP bis(2-ethyl-hexyl) phthalate, DOP di-n-octylphthalate.

Not only that but also the total PEs concentration in this area was the highest concentration relative to other contaminants such as PAHs or PCBs. PEs concentration range was from 1.3 to 27 mg kg^{-1}, essentially because of contributions of DEHP and DBP, whereas PCBs showed the lowest concentrations (from 0.05 to 0.20 mg Kg^{-1}).

In this study, the level of alkali-labile phosphate (ALP), an indirect measurement for vitellogenin-like proteins in gonads and a potential endocrine disruption biomarker, was measured in mussels. Peroxisome proliferation, measured in terms of acyl-CoA oxidase (AOX) activity induction and/or increased peroxisomal density, has been demonstrated to respond to changes in the bioavailability of PAHs and PCBs in mussels both in vitro and in vivo (Cancio et al. 1998; Orbea and Cajaraville 2006). Sex and gamete development stage (gonadal index [IG)]) were also determined histologically with the aim of assessing possible disturbances in reproduction. According to the above some environmental pollutants can cause alterations in gametogenesis, sex determination, differentiation, and development in aquatic animals (WHO/IPCS 2002). To link the observed distribution of the trace contaminants and the variation of the three biomarkers, a regression of both sets of data was performed.

Because the chemical set of data is highly redundant, linear regression methods are not suitable because high collinearities are present. In these particular cases, bilinear regression models are the most appropriate because they can extract main variation patterns from the chemical set of data (X) that have relevance for the biologic set of data (Y). In this work, we used partial least squares (PLS) models and, instead of making the regression for all the biomarkers, chose a single model (PLS1) for each biomarker. Nevertheless, we must keep in mind that the data set is quite short (eight different samples) and that this fact restricts the ruggedness of the regression models; therefore, the interpretation can only be tentatively considered.

The X matrix was logarithmically transformed to assure normality of the variables. In addition, because the concentration ranges of the chemicals were significantly different, those variables were standardised so they had the same weight in the regression models. From those regression models, two principal components were required to explain B95% of the Y variance and B50% of the X variance. The models we built explained the variability of the three biomarkers reasonably well. In fact, although the limited amount of experimental data did not allow for good behaviour of the models and thus the subsequent calculations, prediction of those variables (within an uncertainty of 10%) could be performed based on the chemical data.

From the regression models, we obtained a deeper knowledge about the most significant variables. Because those relations were obtained from the loading weights of each model, the scatter plots of the first two loading weights are given in Figure. 7. As can be seen, the projection of each biomarker is also included to facilitate interpretation of each plot. In the case of ALP, we can see a net positive effect with some PAHs and PCBs, and just the opposite effect with DEHP. In the case of AOX, we can discriminate the net positive effect of some metals (Pb, and Zn), whereas the DEHP show, in this case also, a net negative effect. Finally, in the case of the IG, we can see a different picture because some PAHs, mostly the methylated derivatives and As, showed a positive effect, whereas only DBT showed a net negative effect.

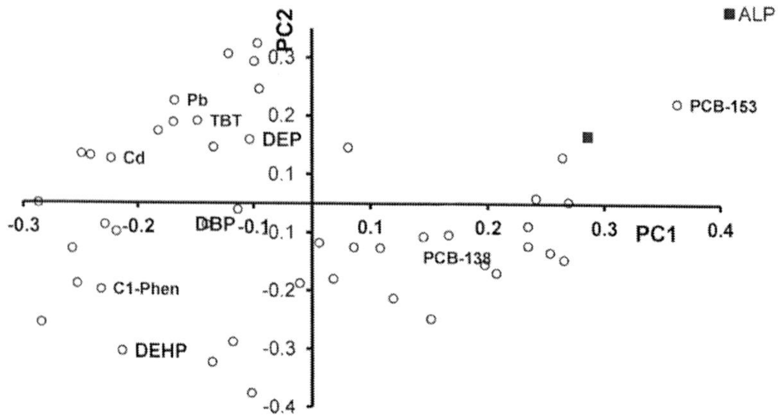

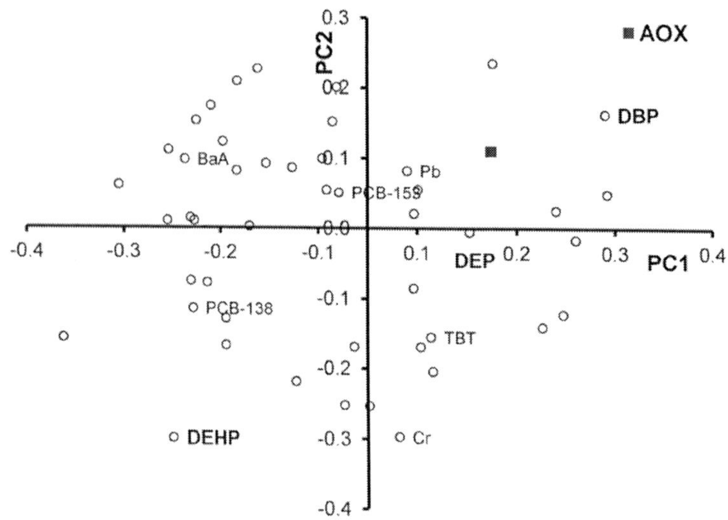

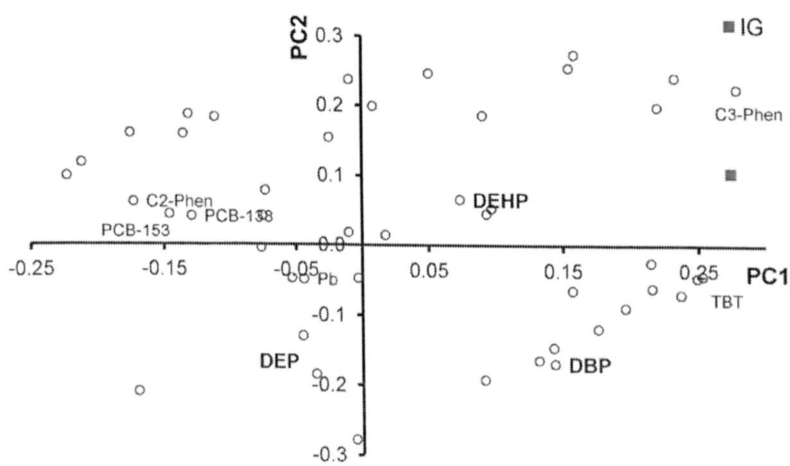

Figure 7. Scattered plots of the loading weights of the first two PCs: a) ALP, b) AOX; c) IG. Some of the most significant chemical variables (empty circles) are labeled, while the projection of the biomarker on this plane is the black square.

ACKNOWLEDGMENTS

This work was supported by the Spanish Ministry of Science and Technology through the project PRESTEPSE(VEM2003- 20082-CO6-01) and by the Basque Government (IMPRES,ETORTEK/SPRI). SGIker human support (UPV/EHU, MICINN, GV/EJ, ERDF and ESF) is gratefully acknowledged.

REFERENCES

Adeniyi, A.A., Okedeyi, O.O., Yusuf, K.A., (2011). Flame ionization gas chromatographic determination of phthalate esters in water, surface sediments and fish species in the Ogun river catchments. Ketu. Lagos. Nigeria. *Environ. Monit. Assess.* 172, 561–569.

Al-Omran, L.A., Preston, M.R., (1987) The interactions of phthalate esters with suspended particulate material in fresh and marine waters *Environ. Pollut.,* 46 (1987), 177–186.

Antizar-Ladislao B., (2009). Polycyclic aromatic hydrocarbons. polychlorinated biphenyls. phthalates and organotins in northern Atlantic Spain's coastal marine sediments. *J. Environ. Monit.* 11, 85-91.

Arias, A.H., Spetter, C.V., Freije, R.H. and Marchoveccio, J.E., (2010). Polycyclic aromatic hydrocarbons in water, mussels (Brachidontes sp., Tagelus sp.) and fish (Odontesthes sp.) from Bahía Blanca Estuary, Argentina. *Estuar Coast Shelf* S. 85, 1 67-81.

Apraiz, I., Cajaraville, M.P., Cristobal, S., (2009). Peroxisomal proteomics: biomonitoring in mussels after the Prestige's oil spill. *Mar. Pollut. Bull.*58, 1815-1826.

Azkona A., Henkins S.H., Roberts H.M.G. 1984. Sources of contamination of the estuary of the River Nervión, Spain- a case study. *Water Science Technology*, 16, 95-125.

Asadollahzadeh, H., Noroozian, E., Maghsoudi, Sh. (2010) Solid-phase microextraction of phthalate esters from aqueous media by electrochemically deposited carbon nanotube/polypyrrole composite on a stainless steel fiber. *Anal. Chim. Acta* 669, 32–38.

Barrick, R., S. Becker, R. Pastorok, L. Brown and H. Beller. (1988). Sediment quality values refinement: 1988 update evaluation of Puget Sound AET. U.S. *Environmental Protection Agency*. Seattle, WA.

Bartolomé, L., Cortazar, E., Raposo, J.C., Usobiaga, A., Zuloaga, O., Etxebarria, N. and Fernández, L.A. (2005). Simultaneous microwave-assisted extraction of polycyclic aromatic hydrocarbons, polychlorinated biphenyls, phthalate esters and nonylphenols in sediments. *J. Chromatogr* A. 1068, 229–236.

Bartolomé L., Etxebarria N., Raposo J.C. 2011. Contaminants in estuaries in relation to human activities, In (XX, ed) Nova.

Baumard, P., Budzinski, H., Garrigues, P., Dizer, H., Hansen, P.D., 1999a. Polycyclic aromatic hydrocarbons in recent sediments and mussels (Mytilus edulis) from the western Baltic Sea: occurrence, bioavailability and temporal variations. *Mar. Environ. Res.* 47, 17–47.

Baumard, P., Budzinski, H., Garrigues, P., Narbone, J.F., Burgeot, T., Michel, X., Bellocq, J., 1999b. Polycylic aromatic hydrocarbon (PAH) burden of mussels (Mytilus sp.) in different marine environments in relation with sediment PAH contamination, and bioavailability. *Mar. Environ. Res.* 47, 415–439.

Bergström, S., Barri, T., Norberg, J., Jönsson, J.A., Mathiasson, L. (2007) Extracting syringe for extraction of phthalate esters in aqueous environmental samples. *Anal. Chim.* Acta 594, 240–247.

Bilbao, E. (2007) Cloning and expression analysis of peroxisome proliferation marker genes in aquatic organisms, effects of organic xenobiotics. PhD Thesis, UPV/EHU, pp 374.

Björklund K., Cousins A.P., Strömvall A.M., Malmqvist P.A. (2009) Phthalates and nonylphenols in urban runoff: Ocurrence, distribution and area emission factors. *Sci. Tot. Environ.* 407: 4665-4672.

Brossa, L., Marcé, R. M., Borrull, F., Pocurull, E. (2005) Determination of Endocrine Disruptors in Environmental Water Samples by Stir Bar Sorptive Extraction-Liquid Desorption - Large Volume Injection-Gas Chromatography. *Chromatographia* 61, 61–65.

Cadogan, D.F. (1999) Health and environmental effects of phthalate plasticisers for poly(vinilchloride)- an update. *Plast. Rubber Compos.* 28: 476-481.

Cancio I., Orbea A., Volkl A., Fahimi H.D., Cajaraville M.P. (1998) Induction of peroxisomal oxidases in mussels: Comparison of effects of lubricant oil and benzo(a)pyrene with two typical peroxisome proliferators on peroxisome structure and function in peroxisomal enzyme activities and peroxisomal structure in mussels Mytilus galloprovincialis. *Toxicol. Appl. Pharmacol.* 149:64–72.

Cao, X.L. (2010). Phthalate Esters in Foods: Sources, Occurrence, and Analytical Methods; *Comprehensive Reviews in Food Science and Food Safety* 9, 21-43.

Cearreta, A., Irabien, M.J., Leorri, E., Yusta, I., Croudace, I.W. and Cundy, A.B. (2000). Recent anthropogenic impacts on the Bilbao Estuary, Northern Spain: Geochemical and microfaunal evidence. *Estuar. Coast. Shelf S.* 50, 571-592.

Cháfer-Pericás, C., Campíns-Falcó, P., Prieto-Blanco, M.C. (2008) Automatic in-tube SPME and fast liquid chromatography: A cost-effective method for the estimation of dibuthyl and di-2-ethylhexyl phthalates in environmental water samples. *Anal. Chim. Acta* 610, 268–273.

Chaler, R., Cantón, L., Vaquero, M., Grimalt, J.O. (2004) Identification and quantification of n-octyl esters of alkanoic and hexanedioic acids and phthalates as urban wastewater markers in biota and sediments from estuarine areas. *J. Chromatogr.* A 1046, 203-210.

Chi, J., Liu, H., Li, B., Huang, G.-L. (2006) Accumulation and Biodegradation of Dibutyl Phthalate in Chlorella vulgaris. *Bull. Environ. Contam. Toxicol.* 77, 21–29.

Cho, J.Y., Choi, J.S., Park, N.G., Shin, H.W., Hong, Y.K. (2006) Antifouling properties of plasticizers BPBG and DNOP against algal spore and mussel attachment. *Botanica Marina* 49, 263–265.

Clark, K., Cousins, I.T., Mackay, D.,·Yamada, K., (2003). Observed Concentrations in the Environment. *The Handbook of Environmental Chemistry.* 3, 125–177.

Clarke B.O., Smith S.R. (2011). Review of ‘emerging’ organic contaminants in biosolids and assessment of international research priorities for the agriculture use of biosolids. *Environ. Int.* 37, 226-247.

Cortazar, E., Bartolomé, L., Delgado, A., Etxebarria, N., Fernández, L.A., Usobiaga, A., Zuloaga, O. (2005) Optimisation of microwave-assisted extraction for the determination of nonylphenols and phthalate esters in sediment samples and comparison with pressurised solvent extraction. *Anal. Chim. Acta* 534, 247–254.

Cortazar E., Bartolomé L., Arrasate S., Usobiaga A. Raposo J.C., Zuloaga O., Etxebarria N. 2008. Distribution and bioaccumulation of PAHs in the UNESCO protected natural reserve of Urdaibai, Bay of Biscay. *Chemosphere*, 72, 1467-1474.

Cortazar E., (2009). Desarrollo de métodos para el análisis de micro-contaminantes orgánicos en muestras medioambientales. Ph D Thesis. University of the Basque Country (UPV/EHU).

Cousins I.T., Mackay D., Parkeon T.F. (2003). Physical-chemical properties and evaluative fate modeling of phthalate esters. In: Staples CA (ed.) The handbook of environmental chemistry. *Phthalate esters*, 3Q, Berlin, Springer, 57-84.

Dargnat, C., Teil, M.J., Chevreuil, M., Blanchard, M., (2009). Phthalate removal throughout wastewater treatment plant: case study of Marne Aval station (France). *Sci. Total Environ.* 407, 1235–1244.

Errekatxo, A., Prieto, A., Zuloaga, O., Usobiaga, A., Etxebarria, N., Fernández, L.A. (2008) Simultaneous extraction of several persistent organic pollutants in sediment using focused ultrasonic solid-liquid extraction. *Anal. Bioanal. Chem.* 392, 1471–1478.

Fankhauser-Noti, A, Grob, K. (2007) Blank problems in trace analysis of diethylhexyl and dibutyl phthalate: investigation of the sources, tips and tricks. *Anal. Chim. Acta* 582, 353–60.

Farahani, H., Ganjali, M.R., Dinarvand, R., Norouzi, P. (2008) Screening method for phthalate esters in water using liquid-phase microextraction based on the solidification of a floating organic microdrop combined with gas chromatography–mass spectrometry. *Talanta* 76, 718–723.

Farahani, H., Norouzi, P., Dinarvand, R., Ganjali, M.R. (2007) Development of dispersive liquid–liquid microextraction combined with gas chromatography–mass spectrometry as a simple, rapid and highly sensitive method for the determination of phthalate esters in water samples. *J. Chromatogr.* A 1172 ,105–112.

Fatoki, O.S., Bornman, M., Ravandhalala, L., Chimuka, L., Genthe, B., Adeniyi, A., (2010). Phthalate ester plasticizers in freshwater systems of Venda. South Africa and potential health effects. *From Water SA*. 36, 1 117-126.

Garcia-Jares, C., Regueiro J., Barro R., Dagnac Th., Llomparet M. (2009) Analysis of industrial contaminants in indoord air. Part 2. Emerging contaminants and pesticides. *J. Chromatogr.* A 1216, 567-597.

Garmendia, L., Izagirre, U., Cajaraville, M.P., Marigomez, I. (2011a) Application of a battery of biomarkers in mussel digestive gland to assess long-term effects of the Prestige oil spill in Galicia and the Bay of Biscay: lysosomal responses. *J. Environ. Monit.* 13, 901-914.

Garmendia, L., Soto, M., Vicario, U., Kim, Y., Cajaraville, M.P., Marigomez, I. (2011b) Application of a battery of biomarkers in mussel digestive gland to assess long-term effects of the Prestige oil spill in Galicia and Bay of Biscay: tissue-level biomarkers and histopathology. *J. Environ. Monit.* 13, 915-932.

Garmendia, L., Soto, M., Ortiz-Zarragoitia, M., Orbea, A., Cajaraville, M.P., Marigomez, I. (2011c) Application of a battery of biomarkers in mussel digestive gland to assess long-term effects of the Prestige oil spill in Galicia and Bay of Biscay: correlation and multivariate analysis. J. *Environ. Monit.* 13, 933-942.

Grigoriadou, A., Schwarzbauer, J., (2011). Non-target Screening of Organic Contaminants in Sediments from the Industrial Coastal Area of Kavala City (NE Greece). *Water, Air and Soil Pollution*. 214, 1-4 623-643.

He, J., Lv, R., Zhan, H., Wang, H., Cheng, J., Lu, K., Wang, F. (2010) Preparation and evaluation of molecularly imprinted solid-phase microextraction fibers for selective extraction of phthalates in an aqueous sample. *Anal. Chim.* Acta 674, 53–58.

Heudorf U., Mersch-Sundermann V., Angerer J. (2007). Phthalates: toxicology and exposure. *Int. J. Hyg. Environ. Health*, 210, 623-634.

Hewitt, M., Servos, M. (2001) An overview of substances present in Canadian aquatic environments associated with endocrine disruption. *Water Qual. Res. J. Canada* 36, 191-213.

Holadová, K., Hajslová, J. (1995) A comparison of different ways of sample preparation for the determination of phthalic acid esters in water and plant matrices. *Int. J. Environ. Anal. Chem*. 59, 43-57.

Huang, P., Tien, C., Sun, Y., Hsieh, C., Le, C., (2008). Occurrence of phthalates in sediment and biota: Relationship to aquatic factors and the biota-sediment accumulation factor. *Chemosphere*. 73, 4 539–544.

Hwang, H., Green, P.G., Young, T.M., (2006). Tidal salt marsh sediment in California. USA. Part 1: Occurrence and sources of organic contaminants. *Chemosphere*. 64, 1383–1392.

Jara, S., Lysebo, C., Greibrokk, T., Lundanes, E. (2000) Determination of phthalates in water samples using polystyrene solid-phase extraction and liquid chromatography quantification. *Anal. Chim. Acta* 407, 165–171.

Jahed Khaniki, G.H.R., Yunesian, M., Naddafi, K., Nouri, J., Ali Mohammadi, M. (2006) Release of the phthalate esters into water stored in plastic tumblers, *Journal of Applied Sciences* 6, 2666-2669.

Kamrin M.A. (2012). Phthalate risks, phthalate regulation, and public health: A review. *J. Toxicol. Environ. Health*, Part B. 12, 157-174.

Klamer, H.J.C., Leonards, P.E.G., Lamoree, M.H., Villerius, L.A., Åkerman, J.E., Bakker, J.F., (2005). A chemical and toxicological profile of Dutch North Sea surface sediments. *Chemosphere*. 58, 11 1579–1587.

Koniecki D., Wang R., Moody R.P., Zhu J. (2011) Phthalates in cosmetic and personal care products: concentrations and possible dermal exposure. *Environ. Res*. 111, 329-336.

Kotowska, U., Garbowska, K., Isidorov, V.A. (2006) Distribution coefficients of phthalates between absorption fiber and water and its using in quantitative analysis. *Anal. Chim. Acta* 560, 110–117.

Latini, G. (2000) Potential hazards of exposure to di-(2-ethylhexyl)-phthalate in babies. *Biol. Neonate* 78, 269-276.

Leorri E., Cearreta A., Irabien M.J., Yusta I. 2008. Geochemical and microfaunal proxies to assess environmental quality conditions during the recovery process of a heavily polluted estuary: The Bilbao estuary case (N. Spain). *The Science of the Total Environment*, 396, 12-27.

Li, X., Zhong, M., Xu, Sh., Sun, C. (2006) Determination of phthalates in water samples using polyaniline-based solid-phase microextraction coupled with gas chromatography. *J. Chromatogr.* A 1135, 101–108.

Liang, P., Xu, J., Li, Q. (2008) Application of dispersive liquid–liquid microextraction and high-performance liquid chromatography for the determination of three phthalate esters in water samples. *Anal. Chim. Acta* 609, 53–58.

Liria P., Garel E., Uriarte A. 2009. The effects of dredging operations on the hydrodynamics of an ebb tidal delta: Oka estuary, northern Spain. *Continental Shelf Research*, 29, 1983-1994.

Lu, C. (2009). Prediction of Environmental Properties in Water–Soil–Air Systems for Phthalates, *Bull Environ. Contam.* Toxicol) 83:168–173.

Luks-Betlej, K., Popp, P., Janoszka, B., Paschke, H. (2001) Solid-phase microextraction of phthalates from water. *J. Chromatogr.* A 938, 93–101.

Marigomez, I., Soto, M., Orbea, A., Cancio, I., Cajaraville, M.P. (2004) Biomonitoring of environmental pollution in the Basque coast using molecular, cellular and tissue-level biomarkers: an integrative approach, Chap 14. In: Borja A, Collins M (eds) Oceanography and marine environment in the Basque Country. Elsevier Oceanography Ser 70. Elsevier, Amsterdam, p 335–364.

Matthiessen, P. (1998) Effects on fish of estrogenic substances in English rivers. In Principles and processes for evaluating endocrine disruption in wildlife (Kendall, R.J., Dickerson, R.L., Giesy, J.P. et al. Eds.), SETAC Publication Series, pp 239-247.

Monge-Ganuzas M., Cearreta A., Iriarte E. 2008. Consequences of estuarine sand dredging and dumping on the Urdaibai Reserve of the Biosphere (Bay of Biscay): the case of the "Mundaka left wave". *J. Iberian Geol.* 34, 215-234.

Mackintosh, C.E., Maldonado, J., Hongwu, J., Hoover, N., Chong, A., Ikonomou, M.G., Gobas, F.A.P.C., (2004). Distribution of Phthalate Esters in a Marine Aquatic Food Web. Comparison to Polychlorinated Biphenyls. *Environ. Sci. Technol.* 377, 2011–2020.

Mitani, K., Izushi, F., Kataoka, H. (2004) Analysis of Phthalate Contamination in Infusion Solutions by Automated On-Line In-Tube Solid-Phase Microextraction Coupled with High-Performance Liquid Chromatography. *J. Anal. Toxicol.* 28, 575-580.

Möder, M., Popp, P., Pawliszyn, J. (1998) Characterization of water-soluble components of slurries using solid-phase microextraction coupled to liquid chromatography–mass spectrometry. *J. Microcol. Sep.* 10, 225-234.

Olujimi O.O., Fatoki O.S., Odendaal J.P., Daso A.P. (2012). Chemical monitoring and temporal variation in levels of endocrine disrupting chemicals (priority phenols and phthalate esters) from selected wastewater treatment plant and freshwater systems in Republic of South Africa. *Microchem. J.* 101, 11-23.

Orbea, A., Ortiz-Zarragoitia, M., Sole, M., Porte, C., Cajaraville, M.P. (2002) Antioxidant enzymes and peroxisome proliferation in relation to contaminant body burdens of PAHs and PCBs in bivalve molluscs, crabs and fish from the Urdaibai and Plentzia estuaries (Bay of Biscay). *Aquat. Toxicol.* 58, 75-98.

Ortiz-Zarragoitia, M. Cajaraville, M.P. (2010) Intersex and oocyte atresia in a mussel population from the Biosphere's Reserve of Urdaibai (Bay of Biscay).*Ecotoxicol. Environ. Safe.* 73, 693-701.

Ortiz-Zarragoitia, M., Garmendia, L., Barbero, M.C., Serrano, T., Marigomez, I., Cajaraville, M.P. (2011) Effects of the fuel oil spilled by the Prestige tanker on reproduction parameters of wild mussel populations. *J. Environ. Monit.* 13, 84-94.

Navarro P., Cortazar E., Bartolomé L., Deusto M., Raposo J.C., Zuloaga O. et al (2006) Comparison of solid phase extraction, saponification and gel permeation chromatography for the clean-up of microwave-assisted biological extracts in the analysis of polycyclic aromatic hydrocarbons. *J. Chromatogr. A.* 1128:10–16.

Panagiotou, A.N., Sakkas, V.A., Albanis, T.A. (2009) Application of chemometric assisted dispersive liquid–liquid microextraction to the determination of personal care products in natural waters. *Anal. Chim. Acta* 649, 135–140.

Peijnenburg, W.J.G.M., Struijs J., (2006). Occurrence of phthalate esters in the environment of the Netherlands. *Ecotoxicol. Environ. Saf.* 63, 204–215.

Peñalver, A., Pocurull, E., Borrull, F., Marcé, R.M. (2000) Determination of phthalate esters in water samples by solid-phase microextraction and gas chromatography with mass spectrometric detection. *J. Chromatogr. A* 872, 191-201.

Peterson D.R., Staples CA. (2003) Degradation of phthalate esters in the environment. In: Staples CA (ed.) The handbook of environmental chemistry. Phthalate esters, 3Q, Berlin, Springer, 57-84.

Petrovic, M., Eljarrat, E., López de Alda, M.J., Barceló, D. (2001) Analysis and environmental levels of endocrine disrupting compounds in freshwater sediments. *TrAC Trends Anal. Chem.* 20, 637-648.

Polo, M, Llompart, M, García-Jares, C, Cela, R. (2005) Multivariate optimization of a solid-phase microextraction method for the analysis of phthalate esters in environmental waters. *J. Chromatogr. A* 1072, 63–72.

Prapatpong, P., Kanchanamayoon, W. (2010) Determination of Phthalate Esters in drinking Water using Solid-phase Extraction and Gas Chromatography. *J. Appl. Sci.* 10, 1987-1990.

Prieto, A., Telleria, O., Etxebarria, N., Fernández, L.A., Usobiaga, A., Zuloaga, O. (2008) Simultaneous pre-concentration of a wide variety of organic pollutants in water samples. Comparison of stir bar sorptive extraction and membrane-assisted solvent extraction. *J. Chromatogr. A* 1214, 1–10.

Prokůpková, G., Holadová, K., Poustka, J., Hajšlová, J. (2002) Development of a solid-phase microextraction method for the determination of phthalic acid esters in water. *Anal. Chim. Acta* 457, 211–223.

Puy-Azurmendi, E., Ortiz-Zarragoitia, M., Kuster, M., Martínez, E., Guillamón, M., Domínguez, C., Serrano, T., Barbero, M.C., López de Alda, M., Bayona, J.M., Barceló, D., Cajaraville, M.P., (2010). An integrated study of endocrine disruptors in sediments and reproduction-related parameters in bivalve molluscs from the Biosphere's Reserve of Urdaibai (Bay of Biscay). *Marine Environmental Research.* 69, S63–S66.

Puy-Azurmendi, E., Ortiz-Zarragoitia, M., Kuster, M., Martínez, E., Guillamón, M., Domínguez, C., Serrano, T., Barbero, M.C., López de Alda, M., Bayona, J.M., Barceló, D., Cajaraville, M.P., (2010). An integrated study of endocrine disruptors in sediments and reproduction-related parameters in bivalve molluscs from the Biosphere's Reserve of Urdaibai (Bay of Biscay). *Marine Environmental Research.* 69, S63–S66.

Puy-Azurmendi, E., Ortiz-Zarragoitia, M., Kuster, M., Martinez, E., Guillamon, M., Pilar, A., Atienza, J., Puchades, R., Maquieira, A., Dominguez, C., Lopez de Alda, M., Fernandes, D., Porte, C., Bayona, J.M., Barcelo, D., Cajaraville, M.P. Exposure and effects of endocrine disruptors in thicklip grey mullet (Chelon labrosus) from the Biosphere Reserve of Urdaibai (Bay of Biscay). *Sci. Total Environ.* Submitted.

Rushdi, A.I., Kassim, T.A.T.A., Simoneit B.R.T., (2009). Organic tracers in sediments from the coastal zone of Ras Abu el-Darag. *Gulf of Suez. Environ. Geol.* 58, 1675–1687.

Sánchez-Avila, J., Fernández-Sanjuan, M., Vicente ,J., Lacorte, S. (2011) Development of a multi-residue method for the determination of organic micropollutants in water, sediment and mussels using gas chromatography–tandem mass spectrometry. *J. Chromatogr.* A 1218, 6799– 6811.

Serôdio, P., Nogueira, J.M.F (2004) Multi-residue screening of endocrinc disruptcrs chemicals in water samples by stir bar sorptive extraction-liquid desorption-capillary gas chromatography–mass spectrometry detection. *Anal. Chim. Acta* 517, 21–32.

Soto, A.M., Sonnenschein, C., Chung, K.L., Fernandez, M.F., Olea, N., Serrano, F.O. (1995) The E-SCREEN assay as a tool to identify estrogens - an update on estrogenic environmental-pollutants. *Environ. Health Perspec.* 103, 113-122.

Soto, M., Kortabitarte, M., Marigomez, I. (1995) Bioavailable heavy-metals in estuarine waters as assessed by metal/shell-weight indexes in sentinel mussels Mytilus-galloprovincialis. *Mar. Ecol. Progress Ser.* 125, 127-136.

Orbea A., Cajaraville MP. (2006) Peroxisome proliferation and antioxidant enzymes in transplanted mussels of tour Basque estuaries with different levels of polycyclic aromatic hydrocarbon and polychlorinated biphenyl pollution. *Environ. Toxicol. Chem.* 25:1616–1626.

Staples CA, Peterson DR, Parkenton TF, Adams W.J. (1997). The environmental fate of phthalate esters: A literature review. *Chemosphere* 35, 667-749.

Villalba, A. (1995) Gametogenic cycle of cultured mussel, Mytilus-galloprovincialis, in the Bays of Galicia (NW Spain) *Aquaculture* 130, 269-277.

Wang, X. T., Ma, L. L., Sun, Y. Z., Xu, X. B. (2006) Phthalate Esters in Sediments from Guanting Reservoir and the Yongding River, Beijing, People's Republic of China. Bull. *Environ. Contam. Toxicol.* 76, 799–806.

WHO/IPCS (2002) Global assessment of the state-of-the-science of endocrine disruptors. In: Damstra T, Barlow S, Bergman A, Kavlock R, Van der Kraak G (eds) *World Health Organization* (WHO/PCS/EDC/02.2).

Wong, J.S., Gill, S:S. (2002) Gene expression changes induced in mouse liver by di(2-ethylhexyl) phthalate. Toxicol. Appl. Pharmacol. 185, 180-196.

Wong C.S., Capel P.D., Nowell L.H. (2001) National-scale, field-based evaluation of the biota-sediment accumulation factor model. *Environ. Sci. Technol.* 35:1709–1715.

Xu, J., Liang, P., Zhang, T. (2007) Dynamic liquid-phase microextraction of three phthalate esters from water samples and determination by gas chromatography. *Anal. Chim. Acta* 597, 1–5.

Yuan, S.Y., Liu, C., Liao, C.S., Chang, B.V, (2002). Occurrence and microbial degradation of phthalate esters in Taiwan river sediments. *Chemosphere* 49, 1295–1299.

Zacharewski, T.R., Meek, M.D., Clemons, J.H., Wu, Z.F., Fielden, M.R., Matthews, J.B. (1998) Examination of the in vitro and in vivo estrogenic activities of eight commercial phthalate esters. *Toxicol. Sci.* 46, 282-293.

Zhang, H., Chen, X., Jiang, X. (2011) Determination of phthalate esters in water samples by ionic liquid cold-induced aggregation dispersive liquid–liquid microextraction coupled with high-performance liquid chromatography. *Anal. Chim. Acta* 689, 137–142.

Zeng, F., Cui, K., Xie, Z., Liu, M., Li, Y., Lin, Y., Zeng, Z., Li, F., (2008). Occurrence of phthalate esters in water and sediment of urban lakes in a subtropical city. Guangzhou. South China. *Environment International*. 34, 372–380.

In: Phthalates
Editors: Gerardo L. Moretti and Drago Romano

ISBN: 978-1-62081-994-4

Chapter 3

PHTHALATE ESTERS: BIOACCUMULATION AND INTRACELLULAR SIGNAL MODIFICATIONS IN *IN VIVO* AND *IN VITRO* MODELS

L. Marchetti, M. G. Sabbieti and D. Agas
School of Biosciences and Biotechnology,
University of Camerino, Camerino, MC, Italy

ABSTRACT

Phthalates, a class of chemicals used in a whole range of industrial applications, are considered to be ubiquitous global contaminants and endocrine disruptors (EDs). These hormone-mimicking compounds bioaccumulate within organ tissues in several animals, including fishes, amphibians and mammals with different and selective patterns of distribution.

The endocrine disrupting effects are probably related to the absorption of these chemicals via the alimentary canal. In particular, phthalates have been seen to be stored in the oxyntic cells of the gastric tubular glands in fishes and amphibians.

A relation between bioconcentration and organ functions was also observed. Moreover, *in vitro* studies on rat osteoblastic cultures demonstrated that benzyl butyl phthalate and di-n-butyl phthalate bioaccumulate, modify actin cytoarchitecture and exert mitogenic effects involving microfilament disruption and nuclear actin/lamin A regulation.

On the other hand, primary mouse calvarial osteoblasts, treated with the above chemicals showed DNA base lesions with an increase of apoptotic markers induced by the p53-related pathway.

These data strongly suggest that chronic exposure to phthalates could probably affect new bone formation and matrix deposition with clinical implications on bone homeostasis and mineral density.

Understanding the way of action of phthalates through the study of autochthon or laboratory animals could reveal the mechanism by which these compounds, as EDs, regulate the fate of the organism.

MAIN DOCUMENT

Many compounds, called endocrine disrupting chemicals (EDCs) or endocrine disruptors (EDs), introduced into the environment by human activities are capable of disrupting the endocrine system of animals including invertebrates (annelids, mollusks), fishes, birds and mammals. Endocrine disruption could be profound because of the crucial role that hormones play in controlling development and tissues homeostasis. Phthalic acid diesters, commonly known as phthalates, are a group of organic chemicals affecting the body's hormones, that have a variety of industrial uses, including use as plasticisers in a wide range of house-hold and consumer goods, uses in lubricating oils and uses as carriers for perfumes in cosmetics.

The release of phthalates into the environment may occur during the production and distribution of these chemicals throughout the manufacture or the use of phthalate-based products, or after the waste of these products. Phthalate esters could bioaccumulate, bioconcentrate and biomagnify then exerting toxic effects in a wide scale of organisms. The bioaccumulation criterion identifies chemicals as "bioaccumulative" if they exhibit bioaccumulation (BAF) or bioconcentration factors (BCF) greater than 5000 referred to aquatic organisms. In absence of BAF or BCF data "bioaccumulative" substances are defined as compounds with octanol-water partition coefficients (KOWs) greater than 10^5 (Gobas et al., 2003). In particular, phthalate esters, as global contaminants, are characterized by a moderate resistance to degradation, widely distributed in the environment, and often occurring at low levels in food (Group, 1986; Sharman et al., 1994). Their toxic potential was referred to many years ago (Mayer et al., 1972) and it has been found that they can also be regarded as EDs with estrogenic activity (Jobling et al., 1995; Harris et al., 1997). It has been demonstrated that specific phthalates, including di-ethylhexyl phthalate, diethyl phthalate (DEP), benzyl butyl phthalate (BBP) and di-n-butyl phthalate (DBP), could play potential activities as EDs. Indeed, EDs exert their action by binding to the estrogen receptor and regulating the activity of estrogen responsive genes (Jobling et al., 1995; Guillette et al., 1996; Harris et al., 1997; Longnecker et al., 1997; Safe, 1998). Phthalates bioaccumulate in plants, invertebrates, fishes, and higher animals albeit mammals efficiently metabolize and excrete phthalates when stored in low concentrations in tissues; in particular, some phthalates could be metabolized and eliminated from the body within 48 h of exposure in mammals (Kluwe, 1982). Bioaccumulation models can be useful tools in the investigation of the mechanism of phthalate ester bioaccumulation. The merit of such models is in investigating the role on uptake and elimination processes contribute to the observed bioaccumulation factors (Gobas et al., 2003). For example, these models could be used to estimate whether a chemical substance is predominantly absorbed by an organism from the water via the respiratory surface or through the dietary route. Furthermore, molecular biology studies regarding the intracellular pathways implicated could elucidate the way of action of the phthalates in all studied models.

Invertebrate Model

Due to their hydrophobicity, phthalate esters are often believed to have a high potential to bioconcentrate and bioaccumulate in aquatic organisms. Oehlmann and collaborators (2009) stated that early investigations on annelids and mollusks focused on bioaccumulation and acute toxicity of phthalates, provoke wider biological effects, including mitotic inhibition, induction of chromosomal aberrations and effects on larval development.

Considering soil-dwelling organisms, studies on the uptake and toxicity of the phthalates dimethyl phthalate (DMP), DEP, DBP and Di-(2-ethylhexyl)-phthalate DEHP (at a concentration of 5 mg kg^{-1} for 25 days) in the worm *Eisenia fetida* have shown that high molecular weight phthalates (DBP and DEHP) were incorporated but with relatively low accumulation factors, and DMP and DEP were apparently not taken up (Xiao-yu et al. 2005). It is possible that the lack of any uptake of DMP and DEP may be as a consequence of their higher microbial degradation rates and a faster metabolic degradation (Albro et al. 1993). Furthermore, Neuhauser et al. (1986) investigated the acute toxicity of phthalates and found similar LC_{50} values for DMP in four soil-dwelling annelid species (*Allolobophora tuberculata*, *E. fetida*, *Eudrilus eugeniae*, *Perionyx excavatus*). Together, these studies demonstrated that phthalates bioaccumulate in terrestrial worms, but they have a low acute toxicity. Moreover, comparisons of the acute toxicity of several phthalate esters [DMP, DEP, DBP, BBP, di-n-hexylphthalate (DHP) and DEHP] in various aquatic organisms including the annelid *Lumbriculus variegatus* have shown that the higher molecular weight phthalates DHP and DEHP were not acutely toxic for the concentration range tested (<2.3–47.8 μg DHP l^{-1}, 10.2–69.1 μg DEHP l^{-1}) (Call et al. 2001). The toxicity of lower molecular weight phthalates was positively correlated with their K_{ow}, indicating non-polar narcosis as the underlying mode of action.

Studies performed in adult marine tubeworms *Pomatoceros lamarckii*, demonstrated that exposure to DMP decreased fertilization success at a threshold concentration of $1x10^{-5}$ M DMP and induced a significant increase in the number of aberrations in chromosome separations in oocytes at anaphase at concentrations $\geq 1x10^{-7}$ M (Dixon et al. 1999; Wilson et al. 2002). Furthermore, DMP has been shown to affect larval development in the polychaete worm at a threshold concentration $\geq 1x10^{-5}$M.

The effects of phthalates have also been studied on various biochemical pathways such as antioxidant and peroxisomal marker enzymes in marine mussels *Mytilus galloprovincialis* and *M. edulis* (Orbea et al. 2002; Cajaraville and Ortiz-Zarragoitia 2006). Exposure to 500 μg DEHP l^{-1} for 21 days was shown to result in a significant increase in the catalase and acyl-CoA oxidase (AOX) activity and an inhibition of the superoxide and manganate superoxide dismutase in *M. galloprovincialis*. The peroxisomal volume density in the digestive gland of *M. edulis* was significantly enhanced at 50 mg diallylphthalate (DAP) per liter after three weeks of exposure, whereas the chemical had no significant effect on AOX activity in this species (Cajaraville and Ortiz-Zarragoitia 2006). DAP was able to decrease the phospho-protein level (a mussel Vitellogenin-like protein) but did not impair ovarian follicles, oocytes and spermatogenesis in *M. edulis* when exposed to 50 μg DAP l^{-1} for three weeks (Aarab et al. 2006). In another 21 day study, DAP was also shown to increase micronuclei frequency, a marker of genotoxicity, and fragmented-apoptotic cells in the gills of *M. edulis* at 50 μg DAP l^{-1} (Barsiene et al. 2006).

These studies could be correlated with the observations of Tagatz and collaborators (1986) on annelid or mollusk populations, affirming that exposure to DBP altered the community structure and colonization profile of the zoomacrobenthos under laboratory and field conditions.

Acute toxicity for phthalates differs significantly for different species of crustaceans. Call and collaborators (2001) stated that *Hyalella azteca* is 10–20-fold more sensitive than *D. magna*, to the higher molecular weight phthalates DHP and DEHP, although both above phthalates did not exhibit toxic effects in *H. azteca*. Moreover, studies performed in *Artemia salina* larvae by Acey and collaborators (2002) showed that the maximal accumulation of DBP within the larvae of this brine shrimp occurred prior to the development of the early acute toxic response, which was, in turn, correlated with the expression of a DNBP-hydrolyzing enzyme. The concomitant induction of toxicity by DBP with the expression of DNBPase, suggests that the enzyme is a mediator of the adverse response. Furthermore Thuren and Woin (1991) reported that in crustaceans DBP decreased locomotor activity in *Gammarus pulex* at an exposure concentration of 500 $\mu g\ l^{-1}$.

Fish Model

There is large information about endocrine disruptor effects in fish regarding their sexual maturation, puberty and sex behaviour (Group, 1986; Sumpter and Jobling, 1995; Arukwe, 2001) and bioaccumulation in various tissues (Menghi et al., 2002; Capacchietti et al., 2007).

Indeed, it was found a relevant bioaccumulation of phthalates in the gastric glands and enterocytes of the anterior intestine of the cichlid fish *Tilapia rendalli* Boulenger (Menghi et al., 2002). Besides on these observation the authors hypothesized that phthalic esters present in enterocytes and gastric glands can interfere with the reproductive capability of fishes in addition to influence the correct digestion and assimilation of nutrients. In addition, the authors evidenced that *Tilapia*, as well as other aquatic organisms, appear concurrently or sequentially to be exposed to a wide array of chemicals from various sources such as BBP and DBP (Menghi et al., 2002).

Furthermore it was demonstrated that phthalate esters localized in the gastrointestinal tract of two aquaculture species–as the freshwater rainbow trout *Oncorhynchus mykiss* (W.) and the marine shi drum *Umbrina cirrosa* (L.). The endocrine disrupting chemicals enter via the alimentary canal and then not only reach the target cells but also persist differently in specialized cells (Capacchietti et al., 2007). Qualitative evaluation of results revealed that, as far as the stomach is concerned, the *Umbrina cirrosa* exhibited a very intense immunoreactivity at gland level in contrast to a modest staining in *Oncorhynchus mykiss*. These differences reflect a diverse occurrence of glycoconjugates involved in transport processes associated with gastric secretion and with dilution of ingested seawater (Pedini et al., 2001). Conversely, the trout revealed the strongest reactivity at connective tissue level in both the anterior and the posterior region of the stomach. In the intestinal tract of both species, goblet cells failed to stain, whereas enterocytes showed the highest binding of phthalates only restricted to the apical cytoplasm. However, strong cytoplasmic immunohistochemical staining was only observed in enterocytes of trout intestinal cecae and proximal intestine. The authors emphasized the fact that the bioaccumulation of phthalates inside cells specialized in secretions as well as absorption functions could make hypothesized that they can interfere with the digestive process carrying to implications for the physiology of the digestive system.

The peculiar micro-environment responsible for the maintenance of adequate pH probably allows the phthalate concentration with effects on pepsinogen activation and consequential reduction of peptic activity and loss of digestive efficiency. Even if the two teleost fishes investigated showed several differences linked to species, habitat, alimentation and metabolism, it was found that the target organ tissues were the same (Capacchietti et al., 2007).

In this view Gobas and collaborators (2003) argued that the structural similarity between lipids and phthalate esters may favour such a process as pH and enzymatic conditions in the gastro-intestinal tract are tailor-made for the hydrolysis of lipids and perhaps phthalate esters. The uptake that would still occur is directly from the water. Model calculations illustrated that in the absence of dietary uptake the bioaccumulation factors can be expected to drop with increasing organism size (which correlates well with trophic level) as it has been observed in the field study. This is due to the fact that with increasing organism size (and reducing area-to-volume ratio), the gill elimination and fecal egestion rates drop and become negligible compared to growth rates or even small metabolic transformation rates. This results in smaller bioaccumulation factors for larger organisms. The above hypothesis does not require the occurrence of a high rate of metabolism in the fish. The bioaccumulation behaviour of the lower molecular weight phthalate esters is not consistent with a high rate of metabolism. It is therefore possible that phthalate esters are fairly slowly metabolized after they have been absorbed (Gobas et al.,2003).

The above findings are associated with the fact that the most of the oestrogenic chemicals are lipophilic and hydrophobic and, hence, have a strong tendency to bioconcentrate in aquatic organisms, both plants and animals (Ahel et al., 1993). Investigations of oestrogen-mimicking substances (Jobling et al., 1995; Harris et al., 1997) in different organs and species, and widely used in aquaculture, have a particular importance. This is in view of the pathologies or abnormalities correlated with massive presence of phthalates and altered biomarkers, such as vitellogenin (Sumpter and Jobling, 1995).

It is widely verified that EDs can interact with estrogen or androgen receptors and thus act as agonists or antagonists of endogenous hormones. Increasing evidence shows that EDs may also modulate the activities/expressions of steroidogenic enzymes (Kortner and Arukwe 2007). Carnevali and collaborators (2010) stated that exposure of female zebrafish to DEHP led to a significant increase of circulating levels of vitellogenin. Since estrogen is well known for its effect on promoting vitellogenesis, this observation supports the hypothesis that DEHP has estrogenic effects in zebrafish. However, this species exposed to the highest dose of DEHP showed a significant decrease in the number of post-vitellogenic oocytes and the number of ovulated eggs was dramatically decreased by all treatments. Moreover, the authors argued that all environmental relevant doses of DEHP affect vitellogenesis, indicating its estrogenic potency. Different dose-related effects have been observed in relation to maturation and ovulation process. The lowest doses have a stronger negative effects on signals inducing maturation (LHR and mPRb), while the highest doses have a greater impact on the inhibition of ovulation (ptgs2). Also this study revealed that all doses of DEHP strongly impair oocyte maturation and ovulation by influencing the expression of factors involved in these processes. These results were in line with previous findings reporting that BBP was oestrogenic and induces VTG synthesis in male fathead minnow (*Pimephales promelas*) (Harries et al. 2000), and, similarly, to induce VTG in carp (*Cyprinus carpio*) exposed to 1–20 mg DEP over a period of four weeks (Barse et al. 2007). In salmon (*Salmo salar*), exposure to 1500 mg DEHP in the food for a prolonged period (four months) during early life resulted in a small incidence of intersex (Norman et al. 2007).

These *in vivo* effects are consistent with the mode of action of BBP and DEHP established *in vitro* where they have been shown to bind to the oestrogen receptor (ER) and elicit a weakly oestrogenic response in fish cell cultures (Jobling et al. 1995) and in the yeast oestrogen screen (YES; containing the human ER) (Harris et al. 1997). In female Japanese medaka (*Oryzias latipes*), exposure to concentrations in excess of 1 mg DEHP caused a marked reduction in both VTG and in the percentage of females containing mature oocytes, suggesting an anti-oestrogenic mode of action (Kim et al. 2002).

This anti-oestrogenic effect presumably arises through competition with the endogenous oestrogens for the receptor. *In vitro* receptor interaction studies have also shown that BBP and DEHP possess antiandrogenic activity (Sohoni and Sumpter 1998).

Other *in vivo* properties of phthalates in fish include effects on the metabolic pathways involved in steroid biosynthesis and metabolism. In a study by Thibaut and Porte (2004), DEHP and DBP altered the activity of enzymes involved in the synthesis of endogenous steroid hormones and their metabolism in carp, effects that can lead to the alteration of the balance between endogenous oestrogens and androgens, and thus alterations in their reproductive physiology.

These findings are supported by studies in mammals in which phthalates have also been shown to alter the expression and enzymatic activity of enzymes involved in the synthesis and metabolism of sex steroids in the testis (Liu et al. 2005). Considering all above findings, it is interesting to underlined that possible differences in bioaccumulation patterns will in parallel affects on key biotransformation enzyme genes and circulating levels of sex steroid hormones with concomitant differences in tissue toxicological effects.

Several phthalates sub lethal effects have been reported in freshwater species that may reduce the survival of fish in the wild. In a study performed by Ghorpade and collaborators (2001), acetylcholinesterase activity was reduced in fish exposed to DEP resulting in sluggish, non motile behaviour. Wibe and collaborators (2002, 2004) showed that shoaling and bottom-dwelling behaviour in threespine stickle-back, *Gasterosteus aculeatus*, were altered as a result of exposure to BBP. In addition, exposure to 5 μg DEP l^{-1} caused alterations in the general behaviour of common carp (Barse et al. 2007). Other behavioural effects may include loss of equilibrium, depressed locomotor activity, loss of startle response and darkened colouration (DeFoe et al., 1990).

DBP and DEHP have been shown to modulate the function of carp phagocytic cells suggesting that phthalates are also immune modulators (Watanuki et al., 2003). It has to be noticed that experimental exposures are often expressed in mg/L whereas measured concentrations of phthalates in aquatic environments are usually in μg/L. (Giam 1984; Tan, 1995; Huang et al., 1999).

Many experiments exposing DEHP via the water at concentrations >3μg l^{-1} have, nevertheless, clearly established concentration-related biological effects (Forget-Leray et al. 2005).

Experimentally defined toxic concentrations generally exceed environmental concentrations, therefore the acute toxicity of phthalates appears of slight concern. However, exposure to water concentrations below toxic levels may pose a problem because high internal concentrations may be achieved via bioaccumulation of phthalates.

Amphibian Model

It is generally accepted that many species of amphibians are on the decline, and some face extinction (Houlahan et al., 2000; Stuart et al., 2004). A number of factors including habitat loss, increased susceptibility to disease and parasites, climate changes, acid precipitation, introduced invasive predators, increased exposure to ultraviolet radiation, and environmental contaminants, including EDCs, potentially contribute to these trends (Blaustein et al., 2002). Several studies indicate that EDCs cause reproductive abnormalities and limb malformations in amphibians (Burkhart et al., 1998; Hayes et al., 2003; Kloas et al., 1999; Levy et al., 2004; Ohtani et al., 2000). According to the 12th Annual National Listing of Fish Advisories released by the EPA in 2004, 35% of total lake acres and 24% of river miles in the United States are contaminated with one or more of 40 different pollutants including mercury, PCBs, DDT, chlordane, and polyvinyl chloride (contains phthalates). In this view, Lee and Veeramachaneni (2005) sustained that loss of reproductive competence as a result of exposure to EDCs, such as DBP, is a fundamental aspect to be considered in preserving environmental health. Impairment of the spermatogenic process in a species could significantly reduce the number of offspring and may render the species unsustainable.

The studies of the effects of phthalates in amphibians have been focused on their toxicity, their potential to induce teratogenic effects and their endocrine-disrupting activities, most notably effects on reproduction, sexual differentiation and the thyroid system.

Information about the incidence and localization of phthalate esters in amphibians were provided by Menghi and collaborators (2003). Accordingly, it was observed a bioaccumulation of o-phthalate esters in the gastrointestinal tract of the most common frog, *Rana esculenta Linnaeus*. In particular, phthalates were immunodetected in stomach and intestine (proximal and distal tracts). In addition, it was noted the lack of phthalates accumulation inside the nucleus of reactive cells suggesting that specific phthalates may influence nuclear activities without entering inside the nucleus but by modulating growth factors and signalling transduction pathways.

The massive bioaccumulation of these chemicals in distinct compartments of untreated subjects, found in green frogs examined, may be correlated with the exposition to both aquatic and terrestrial environment (Menghi et al., 2003). These findings could be of particular concern since these environmental contaminants may be relevant factors in global amphibian declines. The bioconcentration of o-phthalate esters, regarded as global contaminants *in vivo* and *in vitro*, which after digestion, absorption and accumulation in the enterocytes could perturb the endocrine system have to be noticed; in the case of the amphibians too, it has to be underlined that phthalate esters can affect reproduction by estrogenic modes of action that produce severe effects including abnormal sexual differentiation (Jobling et al., 1995; Harris et al., 1997).

In addition, the potential of autonomous hormonal steroidogenesis in liver and small intestine of frog, *Rana esculenta Linnaeus*, has been investigated and it has been found that both liver and intestine can be independent sources of hormonally active steroids (Belvedere et al., 2001). Furthermore, knowledge of amphibian biology and endocrinology indicated that aquatic vertebrates such as amphibians are very suitable models for the study of endocrine disruptors mainly of anthropogenic origin and distributed in surface water (Kloas, 2002); probably, among other endocrine disruptors, phthalates could contribute to changes of amphibian populations via adverse effects on reproduction by abnormal sexual differentiation and on the thyroid system by acceleration or retardation of metamorphosis. Indeed, the effects of dibutyl phthalate on gonadal sex differentiation of genetically male tadpoles of *Rana rugosa* have been examined and it has been found that the histological examination of the gonads showed the typical structure of testes in the control tadpoles; conversely, treatments with diluted solutions of DBP and 17-β-estradiol caused the undifferentiated gonads to develop into gonads of complete or partial ovarian structure. DBP, at concentrations of 0.1, 1 or 10 μm, was estimated to be 1.000-fold less potent than 17-β-estradiol; nevertheless, this phthalate ester is an environmentally dangerous hormone that disrupts the pathways of testicular differentiation in genetically male animals (Ohtani et al., 2000).

Focusing on toxicity studies, Larsson and Thuren (1987) found that exposure of *Rana arvalis* eggs to DEHP via sediment decreased successful hatchings with increasing concentrations. Studies with DBP using the frog embryo teratogenesis assay *Xenopus* (FETAX) have shown developmental malformations at 500 μg and enhanced mortality rates at concentrations of 10 and 15 mg (Lee et al. 2005). Moreover, Lee and Veeramachaneni (2005) have demonstrated that DBP interfered with maintenance of testicular integrity in *Xenopus laevis*.

They also affirmed that both testis length and width were reduced in DBP-treated frogs and seminiferous tubular diameter was significantly reduced in all DBP treatment groups, along with a decrease in the average number of cell nests per tubule. Moreover, after DBP treatment have been observed disorganization of seminiferous tubules and poor differentiation of the germinal epithelium, as hallmarks of testicular dysgenesis, even at low concentrations of phthalates administration (Lee and Veeramachaneni, 2005).

Under the above observations it is clear that phthalates are continuously being released into the environment, where it may act synergistically or additively with many other contaminants, leading to significant changes in behaviour, physiology, and eventually population dynamics through bioaccumulation and bioconcentration in amphibians organisms.

Avian Model

In the ecosystem, some hydrophobic EDs are introduced into the fatty tissues of organisms, where they bioaccumulate over time and, thereby, infiltrate the food chain. The risk of EDs exposure can vary according to the position in the food chain; therefore, some ground-dwelling, fish eating and predatory birds that live or feed around areas contaminated with EDs are often exposed to highly concentrated levels of EDs.

From the distant 80's it was observed that low levels of DBP and DEHP have been shown to be present in chicken eggs collected from retail stores in Japan. The concentration of DBP and DEHP in the egg white were in the range of trace to 0.15 ppm and 0.05 to 0.40 ppm, respectively, the phthalates were not detected in the egg yolk. After DEHP administration to laying hens, it was observed that the eggs were contaminated with the DEHP (Ishida et al., 1981).

Ishihara and collaborators (2003) have performed *in vitro* studies regarding the effects of various EDCs on thyroid of Japanese quail (*Coturnix japonica*). They showed specific and short-term interactions of BBP such a potent inhibitor of triiodothyronine [^{125}I]T3 binding to transthyretin receptor (qTTR) with BBP IC_{50} values of 10^2–10^3 higher than that of T3. They stated that there is a possibility that EDCs bound to qTTR and may bioaccumulate in tissues with transthyretin (TTR) receptors. Receptor mediated uptake of TTR has been reported in mammalian hepatomas, adenocarcinomas and neuroblastomas (Divino and Schussler, 1990a), astrocytes (Divino and Schussler, 1990b), ependymomas (Kuchler-Bopp et al., 2000), and chicken oocytes (Vieira et al., 1995). Interestingly, in chicken and Japanese quail the T3/T4 ratio is higher in yolk than in plasma (Sechman and Bobek, 1988; Wilson and McNabb, 1997) suggesting the presence of a specific T3-uptake system on the plasma membrane of these oocytes. Therefore, the authors concluded that the developing embryo would be exposed to potentially hazardous chemicals, such as BBP, but also claimed that the experimental conditions do not reflect the *in vivo* situation where the bioaccumulation and the metabolism of EDs may greatly influence their intracellular concentrations (Ishihara et al., 2003).

Abdul-Ghani and collaborators (2011) have considered the phthalate effects in chick model claiming that DEHP and DBP exerts teratogenicity, congenital malformations, defects in major biochemical markers, DNA damages and an abolishment of the behavioural functions. The authors argued that the phthalates provoke congenital malformations such as gastroschisis and omphalocele, and only DBP was associated with motor dysfunction. They concluded that other defects, particularly the behavioural ones, are related to the effect of the teratogen during later stages of embryonic development and hatching (Abdul-Ghani et al., 2011; Hutchings, 1978).

Mammalian Model

Phthalate effects on mammalian species have been largely investigated. Studies performed in rats and rabbits demonstrated that DBP induces histopathological changes within seminiferous epithelium, including widespread germ cell loss, vacuolization of Sertoli cell cytoplasm, Leydig cell hyperplasia, presence of giant cells, dysgenetic seminiferous tubules, and atypical germ cells resembling carcinoma *in situ* cells (Fisher et al., 2003; Higuchi et al., 2003). It is well known that bone remodeling is controlled by various hormones and cytokines (Rodan, 1992). An imbalance between bone formation and bone resorption leads to metabolic bone diseases (Parfitt, 1987). Since estrogens deficiency is also liable for important bone diseases, as osteoporosis, the role of estrogens as well as estrogen-mimicking compounds, like phthalates, in regulating activities of bone cells is crucial to understand some bone metabolism pathologies. In this contest, some of these chemicals have been found to cause a significant increase in the number of skeletal malformations as deformity of the thoracic vertebrae and fusion of the vertebral arches in rat fetuses (Ema et al., 1993). Also, teratogenicity and embryo lethality was observed in 11-20 day rat fetuses of dams fed with a diet containing 2% of BBP when cleft palate was observed, whereas the administration of BBP during the first half of pregnancy produced embriolethality (Ema et al., 1992). It has been investigated the *in vitro* effects of BBP and DBP on rat osteoblasts claiming interesting results on the intracellular localization of fibroblast growth factor-2 (FGF-2), one of the most important regulator of bone remodeling (Hurley et al., 1994; Hurley et al., 1998). In particular, the above estrogen-mimicking substances induced a perinuclear FGF-2 accumulation and subsequent translocation into the nucleus (Menghi et al., 2001). It has to be considered that fibroblast growth factor acts directly on nuclear events such as transcription in different ways; indeed, gene transcription can be inhibited, unchanged or enhanced (Nakanishi et al., 1992). Based on these data it should be claimed the possible role of BBP and DBP during differentiation and bone formation, suggesting that the embryotoxicity and skeletal malformations in rats (Ema et al., 1992, 1993) could be attributed to the fact that phthalates interfere with the intracellular signaling and functions of FGF-2. It has been demonstrated that susceptibility to the teratogenicity of BBP varies with the developmental stage at the time of administration (Ema et al., 1992). Accordingly, it was observed that pregnant rats given DBP by gastric intubation at a dose of 750, 1000, or 1500 mg/kg/day on gestation days 7–9, 10–12, or 13–15 resulted in 100% post-implantation loss in the 1500 mg/ kg/day dose group, an increase in the number of skeletal malformations (deformity of vertebral column) when treated with 750 and 1000 mg DBP/kg/day on gestation days 7–9, and a dose-dependent increase in the number of external and internal malformations such as cleft palate and fusion of the sternebrae when exposed on gestation days 13–15 (Ema et al., 1994).

The *in vitro* studies performed by Menghi and collaborators (2001) on clonal cell line Py1a, derived from fetal (21 days of gestation) rat calvariae, provided significant findings regarding the developmental toxicity of BBP that can exert the maximum effect during mesenchymal differentiation and bone formation and resorption; indeed, an altered FGF-2 modulation can lead to embryolethality and teratogenicity in relation with the critical periods of gestation in which the administration of dietary BBP occurs as previous illustrated (Ema et al., 1992, 1993). Moreover, Marchetti and collaborators (2002) described the effects of two phthalate esters, BBP and DBP on actin cytoskeleton in rat osteoblastic Py1a cells. They further argued that after phthalate stimulation these endocrine disruptors act rapidly, transiently, and in a dose- and time-related manner on cell morphology involving changes in microfilament organization. Since it is well known that cytoskeleton plays a central role in the expression of phenotype (Folkman and Moscona, 1978; Ben-Ze'Ev and Amsterdam, 1987), the phthalates induced changes on actin cytoskeleton rearrangement; for that reason these chemicals are, also, considered strong actin modulators. Indeed, actin filaments are crucial for the maintenance of cell shape and an altered assembly of actin fibers can be responsible for the perturbation of focal contacts and compromise the correct adhesion of osteocytes (Tanaka-Kamioka et al., 1998).

In this contest, it has been demonstrated a decreased actin synthesis after BBP administration but an increased amount of actin after removal of the effector has been found (Agas et al., 2007). These molecular data suggested that BBP effects on cytoskeleton could involve translational and/or posttranslational steps. In addition, the re-establishment of the cytoskeletal original features involves transcriptional steps. Regarding the non filamentous and filamentous actin distribution after BBP administration, cells showed a peculiar labeling at nuclear envelope level. In addition, a close spatial distribution of lamin A and nuclear actin was observed in the nucleoplasmic side of the inner nuclear membrane after BBP treatement (Agas et al., 2007). Nuclear actin may play a critical role in mRNA processing and transport and, recently, some authors showed an involvement of actin and a subset of hnRNP proteins for nuclear mRNA (Percipalle et al., 2001, 2002). Moreover, it has been proposed that lamins and nuclear actin could be structurally and functionally linked and that they are involved in facilitating essential nuclear functions, including DNA replication and transcription (Shumaker et al., 2003).

The actin microfilament disruption is reported to be of importance as a potentiator for cell growth (Gordon, 2002). Agas and collaborators (2007) stated that treatment with phthalates (BBP) increased cell viability in rat osteoblasts due to increased cyclin D3 expression which specifically requires extracellular mitogenic stimuli for its activation. BBP as xenoestrogen, mimicking 17 β-estradiol (Zacharewski et al., 1998; Yu et al., 2003) which is reported to induce the expression of both cyclin D2 and D3, lead to the promotion of cell growth in osteoblasts (Fujita et al., 2002). It is also known that by reducing the amount of p27, a cyclin-dependent kinase inhibitor, a decreased amount of F-actin was observed (Kawauchi et al., 2006). A D-cyclin is reported to play a central role in the regulation of the actin cytoskeleton organization, consequently controlling also cell shape and cytokinesis (Cuomo et al., 2005). Accordingly, the endocrine disruptor BBP affecting actin distribution causes an increase in cell proliferation that may be due in part to modulation of cyclin D3 (Agas et al., 2007).

Regarding the *in vitro* studies on mouse model, Sabbieti and collaborators (2009) demonstrated that both BBP and DBP provoke an increase of DNA damage and a related phosphorylation of specific check points molecules as ATM (ser-1981) and phospho-p53 (ser-15 and ser-20) in MC3T3-E1 osteoblasts and mouse primary calvarial osteoblasts. These findings were concordant with recent studies employing comet assay to measure the DNA damage induced by DEHP demonstrating that DEHP induced DNA damage in the nervous system of mice and oxidative damage in the internal organs of mice in an apparently dose-dependent manner (Martino-Andrade and Chahoud, 2010). Other studies demonstrated DNA damage induced by DEHP, as indicated by long term induction of 8-OH-dG, in the liver of male rats (Cattley and Glover, 1993; Takagi et al., 1992). In humans, urinary monoethyl phthalate at environmental levels is associated with increased DNA damage in sperms (Duty et al., 2003). Futhermore, it has been determined the effects of phthalate administration on critical apoptosis regulators and it was found a decrease of mitochondrial potential, cytochrome C release and caspases activation (Sabbieti et al., 2009).

Interestingly, it has been verified that BBP and DBP daily administration amplified all the apoptotic markers suggesting that phthalates could exert cumulative effects. The complex of these data indicated that the exposure of mouse osteoblasts to BBP or DBP could caused bone perturbance; indeed, induction of DNA damage and activation of apoptotic pathways were observed by multiple approaches. The chronic exposure to phthalates at low concentrations could probably affect new bone formation and matrix deposition with clinical implications on bone homeostasis and mineral density (Sabbieti et al., 2009). In the same study, the authors debated that different models have dissimilar responses to EDs depending on the expression of cell cycle key regulation proteins, such as p53. Indeed, in mouse osteoblasts knocked-down for the p53, phthalates induced cell proliferation consistent with increased c-myc and cell cycle regulators levels. A similar response was, also, observed in rat Py1a osteoblasts, possibly for the reason that phospho-p53 levels were not substantially modified after BBP administration. Accordingly, the above findings showed that in p53 silenced mouse osteoblasts, phthalates increased proliferation by involvement of cyclins D. Taken together the above *in vitro* findings confirmed that phthalates exert their proliferative/apoptotic effects through p53 signaling regulation in the osteoblastic cell models (Sabbieti et al., 2009).

It has been stated that in mature animals each phthalate get rise a different toxicity profile (Shea, 2003). The liver, kidneys, thyroid, bone and testes are common targets for general toxicity from oral exposures. Much of the concern about phthalates arises from reports beginning in the 1980s showing several of them to be carcinogens in rodents. DEHP causes liver cancers (National Toxicology Program, 1982) and di-isononyl phthalate (DINP) causes kidney and liver cancers in rodents (National Toxicology Program, 2000). The mechanism of liver neoplasia caused by DEHP is believed to be attributable to peroxisome proliferation and a cascade of cellular events that do not occur in the human liver (Youssef and Badr, 1998) but this theory remains to be confirmed (Melnick, 2001). The mechanism of carcinogenesis of DINP in rodent liver is not fully understood but may also involve peroxisome proliferation. The development of kidney neoplasm in rodents caused by DINP may also be mediated through a mechanism that is not relevant in humans (Caldwell, 1999) probably involving a different signalling cascade of checkpoint molecules like p53, and Bcl-2 family proteins as mentioned by Sabbieti and collaborators (2009). Nonetheless, research indicates that carcinogenic risk to humans from at least some of the phthalates may be lower than that to laboratory animals, and focus has shifted to other toxic endpoints. A case control study conducted in Northern Mexico found significantly higher levels of monoethyl phthalate (MEP), a metabolite of diethyl phthalate (DEP), in breast cancer cases compared to controls, while evidence for phthalates' association with premature puberty and breast development is conflicting (Lopez-Carrillo et al., 2010).

CONCLUSION

Phthalates can produce fetal death, malformations, and reproductive toxicity with different profiles for each chemical (National Toxicology Program, 2003). The different phthalates can also have quite different potencies. The extent of these toxicities and their applicability to humans remain incompletely characterized and controversial.

Developmental and reproductive toxicity, and tissue bioaccumulation have been studied in laboratory and wild animal systems with different conclusions depending on several conditions such as environmental concentrations, laboratory doses tested, route of administration, timing of exposure, and endpoints studied. Some studies use only high doses of a chemical to determine if any hazard exists. Other studies designs use finer dose increments to establish the dose-response relationship to determine if there is a dose that is not associated with any adverse effect. Some other ones look at endpoints, such as gross malformations and fetal demise, or examine tissues (such as bones, liver and thyroid) for histological or biochemical abnormalities.

While bioavailable environmental levels of phthalates may be low, due to their physical and chemical properties, small amounts may accumulate within organisms to much higher levels through bioaccumulation. Some studies stated the fact that albeit DBP does not extensively bioaccumulate in most environmental strata, it is continuously being released into the environment, where it may act synergistically or additively with many other contaminants, leading to significant changes in behavior, physiology, and eventually population dynamics in all organisms. Other studies affirmed a moderate bioconcentration in wild autochthon species (fish and amphibians) depending on the environmental distribution of phthalate esters.

Levels of phthalates found in sediment and biota highlighted a stronger contamination, which is consistent with the particular production and regulation of these classes of chemicals.

Conclusions about human toxicity must be extrapolated from animal studies after considering the extent and strength of existing data sets, assessing the uncertainties remaining, and making judgments about the similarity of animal systems to human systems. Thus, a more meticulous comprehension of the modes of action of these chemicals will help in developing a broader understanding of their potential on tissues damage and in the identification of the most vulnerable species to their biological effects. The bioaccumulation, bioconcentration, biomagnification levels and the synergistic properties of phthalates in fishes, amphibians and lower mammalians could be a good support for indicate how they could exert their effects and interfere on human health.

REFERENCES

Agas D, Sabbieti MG, Capacchietti M, Materazzi S, Menghi G, Materazzi G, Hurley MM, Marchetti L. (2007) Benzyl butyl phthalate influences actin distribution and cell proliferation in rat Py1a osteoblasts. *J. Cell. Biochem.* 101:543-51.

Aarab N, Lemaire-Gony S, Unruh E, Hansen PD, Larsen BK, Andersen OK, Narbonne JF. (2006) Preliminary study of responses in mussel (Mytilus edulis) exposed to bisphenol A, diallyl phthalate and tetrabromodiphenyl ether. *Aquat. Toxicol.* 78: 86–92.

Acey RA, Bailey S, Healy P, Jo C, Unger TF, Hudson RA. (2002) A butyrylcholinesterase in the early development of the brine shrimp (Artemia salina) larvae: a target for phthalate ester embryotoxicity? *Biochem. Biophys. Res. Commun.* 299: 659-662.

Ahel M, McEvoy J, Giger W. (1993). Bioaccumulation of the lipophilic metabolites of nonionic surfactants in freshwater organisms. *Environ. Pollut.* 79: 243-248.

Albro PW, Corbett JT, Schroeder JL. (1993) The metabolism of di-(2-ethylhexyl)phthalate in the earthworm Lumbricus terrestris. *Comp. Biochem. Physiol.* C104: 335–344.

Barse AV, Chakrabarti T, Ghosh TK, Pal AK, Jadhao SB. (2007) Endocrine disruption and metabolic changes following exposure of Cyprinus carpio to diethylphthalate. *Pest. Biochem. Physiol.* 88: 36–42.

Barsiene J, Syvokiene J, Bjornstad A. (2006) Induction of micronuclei and other nuclear abnormalities in mussels exposed to bisphenol A, diallyl phthalate and tetrabromodiphenyl ether-47. *Aquat. Toxicol.* 78: 105–108.

Belvedere P, Dalla Valle L, Vinello S, Carnevali O, Colombo L. (2001) Hormonal steroidogenesis in liver and small intestine of the green frog, *Rana esculenta* L. *Life Sci.* 69: 2921-2930.

Ben-Ze'Ev A, Amsterdam A. (1987) In vitro regulation of granulosa cell differentiation. Involvement of cytoskeletal protein expression. *J. Biol. Chem.* 262: 5366-5376.

Blaustein AR, Kiesecker JM. (2002) Complexity in conservation: Lessons from the global decline of amphibian populations. *Ecol. Lett.* 5: 597–608.

Burkhart JG, Helgen JC, Fort DJ, Gallagher K, Bowers D, Propst TL, Gernes M, Magner J, Shelby MD, Lucier G. (1998) Induction of mortality and malformations in Xenopus laevis embryos by water sources associated with field frog deformities. *Environ. Health Perspect.* 106: 841–848.

Cajaraville MP, Ortiz-Zarragoitia M. (2006) Specificity of the peroxisome proliferation response in mussels exposed to environmental pollutants. *Aquat. Toxicol.* 78: 117–123.

Caldwell DJ. (1999)Review of mononuclear cell leukemia in F-344 rat bioassays and its significance to human cancer risk: a case study using alkyl phthalates. *Regul. Toxicol. Pharmacol.* 30: 45– 53.

Call DJ, Markee TP, Geiger DL, Brooke LT, VandeVenter FA, Cox DA, Genisot KI, Robillard KA, Gorsuch JW, Parkerton TF, Reiley MC, Ankley GT, Mount DR. (2001) An assessment of the toxicity of phthalate esters to freshwater benthos. 1. Aqueous exposures. *Environ. Toxicol. Chem.* 20: 1798–1804.

Capacchietti M, Sabbieti MG, Materazzi S, Materazzi G, Menghi G, Marchetti L. (2007) Phthalate esters immunolocalized in the gastrointestinal tract of shi drum Umbrina cirrosa (L.) and rainbow trout, Oncorhynchus mykiss (W.). *Histol. Histopathol.* 22: 15-21.

Carnevali O, Tosti L, Speciale C, Peng C, Zhu Y, et al. (2010) DEHP Impairs Zebrafish Reproduction by Affecting Critical Factors in Oogenesis. *PLoS ONE* 5(4): e10201

Cattley RC, Glover SE. (1993) Elevated 8-hydroxydeoxyguanosine in hepatic DNA of rats following exposure to peroxisome proliferators: relationship to carcinogenesis and nuclear localization. *Carcinogenesis* 14: 2495–2499.

Crisp TM, Clegg ED, Cooper RL, WoodWP, Anderson DG, Baetcke KP, et al. (1998) Environmental endocrine disruption: an effects assessment and analysis. *Environ. Health Perspect.* 106 (Suppl. 1): 11–56.

Cuomo ME, Knebel A, Platt G, Morrice N, Cohen P, Mittnacht S. (2005) Regulation of microfilament organization by Kaposi sarcoma-associated herpes virus-cyclin. CDK6 phosphorylation of caldesmon. *J. Biol. Chem.* 280: 35844–35858.

David RM, Moore MR, Finney DC, Guest D. (2000) Chronic toxicity of di(2-ethylhexyl)phthalate in rats. *Toxicol. Sci.* 55: 433–443.

DeFoe DL, Holcombe GW, Hammermeister DE. (1990) Solubility and toxicity of eight phthalate esters to four aquatic organisms. *Environmental Toxicology and Chemistry* 9: 623-636.

Divino CM, Schussler GC, (1990a) Receptor-mediated uptake and internalization of transthyretin. *J. Biol. Chem.* 265: 1425–1429.

Divino, C.M., Schussler, G.C., 1990b. Transthyretin receptors on human astrocytoma cells. *J. Clin. Endocrinol. Metab.* 71: 1265–1268.

Dixon DR, Wilson JT, Pascoe PL, Parry JM. (1999) Anaphase aberrations in the embryos of the marine tubeworm Pomatoceros lamarckii (Polychaeta: Serpulidae): a new in vivo test assay for detecting aneugens and clastogens in the marine environment. *Mutagenesis* 14: 375–383.

Duty SM, Singh NP, Silva MJ, Barr DB, Brock JW, Ryan L, et al. (2003) The relationship between environmental exposures to phthalates and DNA damage in human sperm usingthe neutral comet assay. Environ. Health Perspect. 111: 1164–1169.

Ema M, Itami T, Kawasaki H. (1992) Embryolethality andnteratogenicity of butyl benzyl phthalate in rats. *J. Appl.Toxicol.* 12: 179–183.

Ema M, Itami T, Kawasaki H. (1993) Teratogenic phase specificity of butyl benzyl phthalate in rats. *Toxicology* 79: 11-19.

Ema M, Amano H, Ogawa Y. (1994) Characterization of the developmental toxicity of di-n-butyl phthalate in rats. *Toxicology* 86: 163–174.

Fisher JS, Macpherson S, Marchetti N, Sharpe RM. (2003) Human testicular dysgenesis syndrome: A possible model using in-utero exposure of the rat to dibutyl phthalate. *Human Reprod.* 18: 1383–1394.

Folkman J, Moscona A. (1978). Role of cell shape in growth control. *Nature* 273: 345-349.

Forget-Leray J, Landriau I, Minier C, Leboulenger F. (2005) Impact of endocrine toxicants on survival, development, and reproduction of the estuarine copepod Eurytemora affinis (Poppe). *Ecotoxicol. Environ. Saf.* 60: 288–294.

Fujita M, Urano T, Horie K, Ikeda K, Tsukui T, Fukuoka H, Tsutsumi O, Ouchi Y, Inoue S. (2002) Estrogen activates cyclin-dependent kinases 4 and 6 through induction of cyclin D in rat primary osteoblasts. *Biochem. Biophys. Res. Commun.* 299: 222–228.

Ghorpade NV, Mehta M, Khare P, Sinkar S, Rao CV. (2001) Toxicity study of diethyl phthalate on freshwater fish *Cirrhina mrigala. Exotoxicology and Environmental Safety* 53: 255-258.

Giam CS, Atlas E, Powers MA, Leonard JE. (1984). Phthalic acid esters. In: The Handbook of Environmental Chemistry; Anthropogenic Substances, Vol. 3 Springer-Verlag, Berlin, Germany pp.67-142.

Gobas F, Mackintosh CE, Webster G, Ikonomou M, Parkerton TF, Robillard K. (2003). Bioaccumulation of Phthalate Esters in Aquatic Food-Webs. *The Handbook of Environmental Chemistry* Vol. 3, Part Q: 201–225.

Gordon SR. (2002) Microfilament disruption in a noncycling organized tissue, the corneal endothelium, initiatesmitosis. *Exp. Cell. Res.* 272: 127–134.

Gray TJ, Butterworth KR. (1980) Testicular atrophy produced by phthalate esters. Arch Toxicol Suppl 4: 452–455.

Group EF Jr. (1986) Environmental fate and aquatic toxicology studies on phthalate esters. *Environ. Health Perspect.* 65: 337-340.

Guillette LJ Jr, Guillette EA. (1996) Environmental contaminants and reproductive abnormalities in wildlife: implications for public health? *Toxicol. Ind. Health* 12: 537-550.

Harries JE, Runnalls T, Hill E, Harris CA, Maddix S, Sumpter JP, Tyler CR. (2000) Development of a reproductive performance test for endocrine disrupting chemicals using pair-breeding fathead minnows (Pimephales promelas). *Environ. Sci. Technol.* 34: 3003–3011.

Harris CA, Henttu P, Parker MG, Sumpter JP. (1997) The estrogenic activity of phthalate esters. *In Vitro Environ. Health Perspect.* 105: 802-811.

Hayes T, Haston K, Tsui M, Hoang A, Haeffele C, Vonk A. (2003) Atrazine-induced hermaphroditism at 0.1 ppb in American leopard frogs (Rana pipiens): Laboratory and field evidence. *Environ. Health Perspect.* 111: 568–575.

Higuchi T, Palmer T, Gray JS, Veeramachaneni DNR. (2003) Effects of dibutyl phthalate in male rabbits following in utero, adolescent, or postpubertal exposure. *Toxicol. Sci.* 72: 301–313.

Houlahan JE, Findlay CS, Schmidt BR, Meyer AH, Kuzmin SL. (2000). Quantitative evidence for global amphibian declines. *Nature* 404: 752–755.

Huang GL, Sun HW, Song ZH. (1999) Interactions between dibutyl phthalate and aquatic organisms. *Bulletin of Environmental Contamination and Toxicology* 63: 759-765.

Hurley MM, Lee SK, Raisz LG, Bernecker P, Lorenzo J. (1998) Basic fibroblast growth factor induces osteoclast formation in murine bone marrow cultures. *Bone* 22: 309-16.

Hurley MM, Abreu C, Gronowicz G, Kawaguchi H, Lorenzo J. (1994) Expression and regulation of basic fibroblast growth factor mRNA levels in mouse osteoblastic MC3T3-E1 cells. *J. Biol. Chem.* 269: 9392-9396.

Ishida M, Suyama K, Adachi S. (1981) Occurrence of dibutyl and di(2-ethylhexyl) phthalate in chicken eggs. *J. Agric. Food Chem.* 1: 72-74.

Ishihara A, Nishiyama N, Sugiyama S, Yamauchi K. (2003) The effect of endocrine disrupting chemicals on thyroid hormone binding to Japanese quail transthyretin and thyroid hormone receptor. *Gen. Comp. Endocrinol.* 134: 36-43.

Jobling S, Reynolds T, White R, Parker MG, Sumpter JP. (1995) A variety of environmentally persistent chemicals, including some phthalate plasticizers, are weakly estrogenic. Environ. Health Perspect. 103: 582-587.

Kawauchi T, Chihama K, Nabeshima Y, Hoshino M. (2006) Cdk5 phosphorylates and stabilizes p27kip1 contributing to actin organization and cortical neuronal migration. *Nat. Cell. Biol.* 8: 17–26.

Kim EJ, Kim JW, Lee SK. (2002) Inhibition of oocyte development in Japanese medaka (Oryzias latipes) exposed to di-2-ethylhexyl phthalate. *Environ. Int.* 28: 359–365.

Kloas W. (2002). Amphibians as a model for the study of endocrine disruptors. *Int. Rev. Cytol.* 216: 1-57.

Kloas W, Lutz I, Einspainier R. (1999). Amphibians as a model to study endocrine disruptors: II. Estrogen activity of environmental chemicals in vitro and in vivo. *Sci. Total environ.* 225: 59–68.

Kluwe WM. (1982) Overview of phthalate ester pharmacokinetics in mammalian species. *Environ. Health Perspect.* 45: 3–10.

Kortner TM, Arukwe A (2007) The xenoestrogen, 4-nonylphenol, impaired steroidogenesis in previtellogenic oocyte culture of Atlantic cod (Gadus morhua) by targeting the StAR protein and P450scc expressions. *Gen. Comp. Endocrinol.* 150: 419–529.

Kuchler-Bopp S, Dietrich JB, Zaepfel M, Delaunoy JP, (2000) Receptor-mediated endocytosis of transthyretin by ependymoma cells. *Brain Res.* 870: 185–194.

Lamb JCt, Chapin RE, Teague J, Lawton AD, Reel JR. Reproductive effects of four phthalic acid esters in the mouse. *Toxicol. Appl. Pharmacol.* (1987) 88: 255–269.

Larsson P, Thure´n A. (1987) Di-2-ethylhexylphthalate inhibits the hatching of frog eggs and is bioaccumulates by tadpoles. *Environ. Toxicol. Chem.* 6: 417–422.

Lee SK, Veeramachaneni DNR. (2005) Subchronic Exposure to Low Concentrations of Di-n-Butyl Phthalate Disrupts Spermatogenesis in Xenopus laevis Frogs. *Toxicol. Sciences* 84: 394–407.

Lee SK, Owens GA, Veeramachaneni DN. (2005) Exposure to low concentrations of di-n-butyl phthalate during embryogenesis reduces survivability and impairs development of Xenopus laevis frogs. *J. Toxicol. Environ. Health A* 68: 763–772.

Levy G, Lutz I, Kruger A, Kloas W. (2004) Bisphenol A induces feminization in Xenopus laevis tadpoles. *Environ. Res.* 94: 102–111.

Liu KJ, Lehmann KP, Sar M, Young SS, Gaido KW. (2005) Gene expression profiling following in utero exposure to phthalate esters reveals new gene targets in the etiology of testicular dysgenesis. *Biol. Reprod.* 73: 180–192.

Longnecker MP, Rogan WJ, Lucier G. (1997) The human health effects of DDT (dichlorodiphenyltrichloroethane) and PCBS (polychlorinated biphenyls) and an overview of organochlorines in public health. Annu. Rev. Public Health 18: 211-244.

Lopez-Carrillo L, Hernandez-Ramirez RU, Calafat AM, Torres-Sanchez L, Galvan-Portillo M, Needham LL, Ruiz-Ramos R, Cebrian ME. (2010) Exposure to phthalates and breast cancer risk in northern Mexico. *Environ. Health Perspect* 118: 539-544.

Marchetti L, Sabbieti MG, Menghi M, Materazzi S, Hurley MM, Menghi G. (2002) Effects of phthalate esters on actin cytoskeleton of Py1a rat osteoblasts. *Histol. Histopathol.* 17: 1061-1066.

Martino-Andrade AJ, Chahoud I. (2010) Reproductive toxicity of phthalate esters. Mol. Nutr. Food Res. 54: 148–157.

Mayer FL, Stalling DL, Johnson JL. (1972) Phthalate esters as environmental contaminants. *Nature* 238: 411-413.

Melnick RL. (2001) Is peroxisome proliferation an obligatory precursor step in the carcinogenicity of di(2-ethylhexyl)phthalate (DEHP)? *Environ. Health Perspect.* 109: 437– 442.

Menghi G, Marchetti L, Sabbieti MG, Menghi M, Materazzi S. (2003) In situ visualization of o-phthalate esters in gastrointestinal tract of the frog Rana esculenta. *Histol. Histopathol.* 18: 371-377.

Menghi G, Sabbieti MG, Marchetti L, Menghi M, Materazzi S, Hurley MM. (2001) Phthalate esters influence FGF-2 translocation in Py1a rat osteoblasts. *Eur. J. Morphol.* 39: 155-62.

National Toxicology Program, Center for the Evaluation of Risks to Human Reproduction. NTP-CERHR Expert Panel Report on Di-isononyl Phthalate. Alexandria, VA: Center for the Evaluation of Risks to Human Reproduction, US Department of Health and Human Services; 2000.

National Toxicology Program, Center for the Evaluation of Risks to Human Reproduction. Final CERHR expert panel reports on 7 phthalate esters. 2003.

National Toxicology Program. Carcinogenesis Bioassay of Di-(2-ethylhexyl) Phthalate (CAS No. 117–81-7) in F344 Rats and B6C3F1 Mice (Feed Study). Research Triangle Park, NC: National Toxicology Program; 1982.

Neuhauser, E. F., Durkin, P. R., Malecki, M. R. and Anatra, M. 1986 Comparative toxicity of ten organic chemicals to four earthworm species. Comp. Biochem. Physiol. 83:;197–200.

Norman, A., Borjeson, H., David, F., Tienpont, B. and Norrgren, L. 2007 Studies of uptake, elimination, and late effects in Atlantic salmon (Salmo salar) dietary exposed to di-2-ethylhexyl phthalate (DEHP) during early life. Arch. Environ. Contam. Toxicol. 52:; 235–242.

Ohtani H., Miura I. and Ichikawa Y. (2000). Effects of dibutyl phthalate as an environmental endocrine disruptor on gonadal sex differentiation of genetic males of the frog *Rana rugosa*. Environ. Health Perspect. 108:; 1189-1193.

Orbea A, Ortiz-Zarragoitia M, Cajaraville MP. (2002) Interactive effects of benzo(a)pyrene and cadmium and effects of di(2-ethylhexyl) phthalate on antioxidant and peroxisomal enzymes and peroxisomal volume density in the digestive gland of mussel *Mytilus galloprovincialis* Lmk. Biomarkers 7: 33–48.

Parfitt AM. (1987) Bone remodelling and bone loss-understanding the pathophysiology of osteoporosis. *Clin. Obstet. Gynecol.* 30: 789-811.

Percipalle P, Jonsson A, Nashchekin D, Karlsson C, Bergman T, Guialis A, Daneholt B. (2002) Nuclear actin is associated with a specific subset of hnRNP A/B-type proteins. *Nucleic Acids Res.* 30: 1725–1734.

Rodan GA. (1992) Introduction to bone biology. *Bone* 13: S3-S6.

Sabbieti MG, Agas D, Santoni G, Materazzi S, Menghi G, Marchetti L. (2009) Involvement of p53 in phthalate effects on mouse and rat osteoblasts. *J. Cell. Biochem.* 107: 316-27.

Safe SH. (1998) Interactions between hormones and chemicals in breast cancer. Annu. Rev. Pharmacol. Toxicol. 38: 121-158.

Sechman A, Bobek S, (1988) Presence of iodothyronines in the yolk of the hen's egg. *Gen. Comp. Endocrinol.* 69: 99–105.

Sharman M, Read WA, Castle L, Gilbert J. (1994) Levels of di-(2-ethylhexyl)phthalate and total phthalate esters in milk, cream, butter and cheese. *Food Addit. Contam.* 11: 375-385.

Shea K. (2003) Pediatric Exposure and Potential Toxicity of Phthalate Plasticizers *Pediatrics* 111: 1467 -1474.

Shumaker DK, Kuczmarski ER, Goldman RD. (2003) The nucleoskeleton: Lamins and actin are major players in essential nuclear functions. *Curr. Opin. Cell. Biol.* 15: 358– 366.

Sohoni P, Sumpter JP (1998) Several environmental oestrogens are also anti-androgens. *J. Endocrinol.* 158: 327–339.

Stuart SN, Chanson JS, Cox NA, Young BE, Rodrigues ASL, Fischman DL, Waller RW. (2004) Status and trends of amphibian declines and extinctions worldwide. *Science* 306: 1783–1786.

Sumpter JP, Jobling S. (1995) Vitellogenesis as biomarker for estrogenic contamination of the aquatic environment. *Environ. Health Perspect.* 103: 173-178.

Tagatz, ME, Plaia GR, Deans CH. (1986) Toxicity of dibutyl phthalate-contaminated sediment to laboratoryand field-colonized estuarine benthic communities. Bull. Environ. *Contam. Toxicol.* 37: 141–150.

Takagi A, Sai K, Umemura T, Hasegawa R, Kurokawa Y. (1992) Hepatomegaly is an early biomarker for hepatocarcinogenesis induced by peroxisome proliferators. *J. Environ. Pathol. Toxicol. Oncol.* 11: 145–149.

Tan G.H. (1995) Residue levels of phthalate esters in water and sediment samples from the Klang River Basin. *Bull. Environ. Contam. Toxicol.* 54: 171-176.

Tanaka-Kamioka K, Kamioka H, Ris H, Lim SS. (1998) Osteocyte shape is dependent on actin filaments and osteocyte processes are unique actin-rich projections. *J. Bone Miner. Res.* 13: 1555-1568.

Thibaut R, Porte C. (2004) Effects of endocrine disrupters on sex steroid synthesis and metabolism pathways in fish. *J. Steroid Biochem. Mol. Biol.* 92: 485–494.

Thuren A, Woin P. (1991) Effects of phthalate-esters onthe locomotor-activity of the fresh water amphipod *Gammarus pulex*. *Bull. Environ. Contam. Toxicol.* 46: 159–166.

Vieira AV, Sanders EJ, Schneider WJ, (1995) Transport of serum transthyretin into chicken oocytes. A receptor-mediated mechanism. *J. Biol. Chem.* 270: 2952–2956.

Watanuki H, Gushiken Y, Sakai M. (2003) *In vitro* modulation of common carp (*Cyprinus carpio L.*) phagocytic cells by Di-n-butyl phthalate and Di-2-ethylhexyl phthalate. *Aquatic Toxicology* 63: 119-126.

Wibe AE, Billing A, Rosenqvist G, Jenssen BM. (2002) Butyl benzyl phthalate affects shoaling behaviour and bottom-dwelling behaviour in three spine stickleback. *Environmental Research Section A* 89: 180-187.

Wibe AE, Fjeld E, Rosenqvist G, Jenssen BM. (2004) Postexposure effects of DDE and butylbenzylphthalate on feeding behavior in threespine stickleback. *Ecotoxicol. Environ. Saf.* 57: 213–219.

Wilson CM, McNabb FMA. (1997) Maternal thyroid hormones in Japanese quail eggs and their influence on embryonic development. *Gen. Comp. Endocrinol.* 107: 153–165.

Wilson JT, Dixon DR, Dixon LRJ. (2002) Numerical chromosomal aberrations in the early life-history stages of a marine tubeworm, Pomatoceros lamarckii (Polychaeta: Serpulidae). *Aquat. Toxicol.* 59: 163–175.

Xiao-yu H, Bei W, Shuzhen Z, Xiao-quan S. (2005) Bioavailability of phthalate congeners to earthworms (Eisenia fetida) in artificially contaminated soils. *Ecotoxicol. Environ. Saf.* 62: 26–34.

Youssef J, Badr M. (1998) Extraperoxisomal targets of peroxisome proliferators: mitochondrial, microsomal, and cytosolic effects. Implications for health and disease. *Crit. Rev. Toxicol.* 28: 1– 33.

Yu Z, Zhang L, Wu D. (2003) Estrogenic activity of some environmental chemicals. *Wei Sheng Yan Jiu* 32: 10–12.

Zacharewski TR, Meek MD, Clemons JH, Wu ZF, Fielden MR, Matthews JB. (1998) Examination of the in vitro and in vivo estrogenic activities of eight commercial phthalate esters. *Toxicol. Sci.* 46: 282–293.

In: Phthalates
Editors: Gerardo L. Moretti and Drago Romano

ISBN: 978-1-62081-994-4

Chapter 4

INVESTIGATION OF DI-2 ETHYL HEXYL PHTHALATE (DEHP) BEHAVIOUR IN CONTACT WITH LIQUID FOOD SIMULANTS, BLOOD, SALIVA AND SOIL

N. Belhaneche-Bensemra, N. Lardjane, D. Ikermoud and M. O. Boussoum

Laboratoire des Sciences et Techniques de l'Environnement,
Ecole Nationale Polytechnique, El- Harrach, Alger, Algeria

ABSTRACT

Polyvinyl chloride (PVC) is a versatile material that can be formulated for various applications, ranging from every day products to highly specialized ones. In addition to the polymeric component, low molecular compounds such as stabilizers, lubricants, plasticizers... are added for processing and achieving the desired chemical and mechanical properties. The one used in the highest proportion is the plasticizer. Among the most commonly used plasticizers are phthalates which have been used in the production of flexible PVC for over 50 years (cable and wire covers, foodstuffs packaging- fresh meat, vegetables, cheese...-, blood bags, toys and children's items, ...). However, for the past 20 years, phthalates have come under considerable attention from media, legislative and environmental concerns.

This chapter is intended to focus on some recent scientific research in the field. Investigations on the extraction and determination of released phthalates were carried out in various media:

- Liquid fatty and aqueous simulants
- Blood
- Saliva
- Soil

The biodegradation of di-2 ethyl hexyl phthalate in the soil was evaluated by using a respirometric test based on the determination of CO_2 production.

INTRODUCTION

Poly (vinyl chloride) (PVC) is among the most widely produced and used polymeric materials in the plastics industry. Its use is widespread and diverse, ranging from every day products to highly specialized applications [1]. These commercial products, in addition to the polymeric component, contain a number of additives. PVC is still the main polymer of additive consumption as most of the plasticizers and heat stabilizers are used there. The one used in the highest - proportion is the plasticizer. Plasticizers are low- melting solids or high-boiling organic liquids added to plastic polymers in order to aid flow and processing, to extend and modify the natural properties of the resin and to develop new, technologically important properties not present in the resin itself [2]. They are used to impart flexibility to an otherwise rigid polymer. Among the most commonly used plasticizers are di-2 ethyl hexyl phtalate (DEHP) and di-2-ethyl hexyl adipate (DEHA) [3].

Plasticizers and phthalates in particular have been used in the production of flexible PVC for over 50 years for applications ranging from cable and wire covers, children toys to medical device and consumers products.

PVC films have found wide applications in the packaging of large variety of foodstuffs, such as fresh meat, fruits and vegetables, and cheeses. However plasticizers frequently possess a high mobility and, in contrast to the polymeric chains, are capable of migrating from the packaging material into the packed food [4-31]. According to current legislation (directive 90/128/EEC and its amendments; EEC1990), the overall migration to a food stuff from food contact plastics must be less than 10 mg of plastics compounds per dm^2 of surface area.

Some of the migrants may affect quality of the packed product as exhibited by sensorily determinable changes (odour and / or taste) or by toxicological symptoms from ingestion [32, 33].

For the past 20 years, phthalates have come under considerable attention from media, legislative and environmental concerns [34- 40]. Although no direct evidence has been found on the toxic effect of phthalates to human beings, it has been proved that high dosage and long term exposure of phthalates to rodents resulted in liver concern and adverse effect on the reproductive development for young male rats [41]. In addition phthalates were suspected to increase asthma and bronchial obstruction in children [42].

Therefore, the detection and quantification of contaminants migrating from the polymers are essential for the safety assessment of food contact plastic packaging materials.

For applications in the packaging of food stuffs, migration testing must be performed. Natural migration is simulated in model tests to determine the migrated or extracted additives in food simulants which can more easily be analysed.

The amount of package components that may be leached by food or food simulating solvents depends on the original concentration of the particular component or migrant in the polymer, its solubility in the solvent and / or the partition coefficient between the polymer and solvent as well as temperature and time [43].

On the other hand, the use of plastic for the fabrication of disposable medical devices has increased markedly in recent years. The development of such devices has unquestionably represented a definite advance in medical safety and storage, handling and delivery of biologic products. Phthalates are also commonly found in blood bags, dialysis tubing, paints, lacquers, cosmetics, coatings of capsules etc [21, 27]. One of the oldest medical uses of plasticized PVC is the blood pocket. PVC remains the material of choice for blood and its products storage. The blood transfusions had become current during the Second World War but it's the glass which was used as container. The plastic blood pocket appeared in 1949, since PVC became the most popular material, with increasingly varied applications. It is only in 1970 that Jaeger emitted for the first time the possibility of contamination of the blood stored in PVC pockets [11]. Since, other studies were carried out in order to confirm this possibility of migration as well as the factors influencing this phenomenon and to identify the different toxic metabolites which can be formed by reaction of plasticizer with blood or its derivatives [45]. Studies highlighted the dependence of this phenomenon of migration with the temperature of conditioning, the time of contact, the agitation and the rate of lipids in the fluid stored in these pockets [46], Thus, it was proved that the DEHP can migrate from the PVC bags and form toxic metabolites in blood and its derivatives [27].

Furthermore, almost 10% [14] children's toys are made of plasticized PVC. Phthalates are the most used plasticizers. Babies and young children bite and lick toys available to them.

They thus undergo direct contact and are subject to greater exposure at least extended to the various constituents of plastic. The exhibition is therefore more significant and chronic in children compared to adults because they are at an important phase of their growth [38].

After their service life, the greater part of post-consumer PVC products is assigned to demolition waste or municipal solid waste. Currently, the most common option is to dispose end-of- life PVC products in sanitary landfills [1]. Landfills and their long-term behavior, however, constitute a source of uncertainty and risk. The loss of plasticizers and other additives may occur when PVC products are placed as wastes in landfill sites, and in this case, it could be a hazard to the environment [47, 48]. The loss of additives from PVC is a complex process, involving three physical processes:

- Diffusion of additives up to the interface,
- Transport through the interface,
- Loss of additives from the surface into the surrounding medium.

The loss of plasticizers from the surface of PVC into the surrounding medium can take place through three specific processes [19]:

- Evaporation at which the plasticizer goes into the surrounding gas medium (most often the air).
- Extraction: at which the plasticizer goes into the surrounding liquid medium.
- Migration: at which the plasticizer in direct contact with the surface of the other material, migrates into that material.

In 1994, the Nordic environmental authorities raised questions regarding the possible environmental impact originating from PVC products and their additives in landfills [1]. Today, all the phthalates used in great quantities in the PVC products are omnipresent in the environment. This is because phthalates are not chemically bound in plastics, and few are slowly released from them to the external environment (water, air, soil, food, etc.). People as well as animals can be minimally exposed to these compounds through ingestion, inhalation or dermal exposure.

This chapter is intended to focus on some recent scientific research in the field. Investigations on the extraction and determination of released phthalates were carried out in various media such as:

- Liquid fatty and aqueous simulants

Kinetic studies of the specific migration of DEHP were carried out by using different analytical methods such as gas chromatography (GC), Fourier transform infrared (FTIR) spectroscopy and differential scanning calorimetric (DSC) analysis.

The influence of various parameters such as temperature, agitation, nature of food simulant and initial concentration of plasticizer was considered

- Blood

The interactions between plasticized PVC bags which are used to store blood and plasma, red globules and platelets were investigated. The tests were carried out in real conditions of storage by respecting the temperatures and the time of conservation.

Samples from control bag and plasma, red globules and platelets bags in extreme conditions of conservation time were analysed by using gas chromatography coupled with mass spectrometry (GC/MS).

- Saliva

The specific migration of DEHP from plasticized PVC into synthetic saliva was investigated. Various analytical methods such as the rate of mass variation, FTIR spectroscopy and scanning electron microscopy (SEM) were used.

- Soil

The aim of this work was to gain knowledge of the release of DEHP from plasticized PVC in landfills. The ageing of the PVC samples was investigated in a landfill soil for 24 months. The variations of the mass of the samples, of density, glass transition temperature (Tg), and mechanical properties (tensile and shore D hardness) were considered. Furthermre, FTIR spectroscopy and microscopic visual investigations were carried out.

- DEHP biodegradation

The biodegradation of DEHP in the soil was evaluated by using a respirometric test based on the determination of CO_2 production.

Experimental

1. Determination of Phthalates in Liquid Fatty and Aqueous Simulants

1.1. Materials and Preparation of PVC Samples

Commercial grades of PVC (Shintech from Itochu, France) and additives were used as received. Several formulations containing 4.5 % of the stabilizer (Irgastab 17 MOK from Ciba-Geigy, Switzerland), 4.5 % of the lubricant (Loxiol GH4 from Henkel, Germany) and, respectively, 30, 40, 50 and 60 % of DEHP (Palatinol AH-L from BASF, Germany) were realized. PVC and additives were mixed in a two-roll mill at 180°C. The specimens for migration tests were melt compressed at 150°C under a pressure of 12 bars during 15 minutes. Then, circular samples of 2 mm in thickness and 19 mm in diameter were cut.

1.2. Migration Testing

Migration tests were conducted using four food simulants (sunflower oil, distilled water, 3% (w/v) aqueous acetic acid, 15% (v/v) aqueous ethanol). These food simulants represent all fatty, liquid and moist foods and beverages, except for beverages with a high alcoholic strength.

The test conditions used were 10 days at 25 and 45°C with and without agitation. Ten (10) circular samples were immersed in 100 ml of each food simulant. A circular sample and 10 ml of food simulant were taken off every day to be analyzed. Each sample was wiped and weighed. The rate of variation of the mass (τ) was determined as a function of time following the relation:

$$\tau(\%) = \frac{m_t - m_0}{m_0} \times 100 \tag{1}$$

where: m_0 = initial mass before immersion, m_t = mass of the sample at the time t.

1.3. Gas Chromatography

A Philips PYE Unicam 304 GC with a flame ionization detector was used. Chromatographic conditions were as follows: column OVI of dimensions 1.8 m × 4mm o.d; column temperature: 280°C; injection temperature: 300°C; detector temperature: 300°C; nitrogen flow: 20.5 ml /min.

1.4. FTIR Analysis

A Philips type PU 9800 FTIR spectrophotometer was used. The food simulants were placed between two KBr pellets and analyzed directly. On the other hand, the PVC circular samples were dissolved in tetrahydrofuran (THF). After evaporation of the solvent, a polymeric film was recuperated and analyzed.

1.5. DSC Analysis

Glass transition temperatures were measured with a Perkin Elmer DSC- 7 apparatus. Each sample was heated from 5 to 110°C under nitrogen with a heating rate of 10°C/ min.

2. Determination of Phthalates in Blood

2.1. Materials

Commercial triple blood bag samples made from P-PVC were used. The bags containing red globules were kept at temperature of 4 °C for 42 days.

The bags containing platelets were kept at temperature of 18 °C under agitation for five days while the bags containing plasma were kept at temperature of -30 °C for 6 months.

2.2. GC Analysis

The preparation of the samples to be analyzed was carried out according to the protocol described by Wang and Storm [24] and Fantoni and Simoneau [49]: a mass of 0.1g of sample was dissolved in 4 ml of THF; PVC was then precipitated by addition of methanol (volume of THF/methanol: 1/2.5 v/v). The obtained solution was filtered. The obtained filtrate (separated from PVC) was dried at 80°C and the residue dissolved in 5 ml of chloroform. DEHP used in the identification and quantification of the plasticizer which was present in the composition of the formulation of the blood pockets was a substance of low volatility produced by the General Company of Plasticizers (SGP) from Tunisia.

GC analysis was performed on gas chromatograph apparatus with automatic injector connected to mass spectrometer (Perkin Elmer EP-5MS) with capillary tube of 0.25 mm internal diameter, 30 m length and a film of 0.25 μm (polar phase:5% diphenyl, dimethyl 95% polysiloxane) used under the following conditions: 90°C held for 3 min, heated up to 280°C and held for13 min. Carrier gas was helium with a pressure of 31 kPa. The injection temperature was 300°C with splitless mode, ionization potential of 70 eV and interval scan between 50 uma and 450 uma. The volume of injection was 1.0 μl.

A calibration curve for DEHP was produced by mixing DEHP standard in chloroform at concentrations that covered the concentration range found in the polymer extracts; the resulting line was linear with correlation coefficient of 0.999.

3. Determination of Phthalates in Saliva

3.1. Materials

A suspension PVC (Petvinil S39/71) produced by SO.G.I.S.SPA(Italy) was used. Zn and Ca stearates from Aldrich (France), DEHP from SGP (Tunisia) and stearic acid (stearine JOS) from SO. G. I.S.SPA (Italy) were used as received. ESO was especially prepared as described previously [50]. The level of oxirane oxygen was 6.4 %.

3.2. Preparation of PVC Films

Formulations containing 1 wt % of Zn stearate and and 1 wt % of Ca stearate, 5 wt % of ESO, 1 wt % of stearic acid and 15 wt % of plasticizer (semi rigid formulation) and 45 wt % of plasticizer (plasticized formulation) were prepared. PVC and additives were mixed in a two- roll mill at 140°C for 20 min and melt compressed at 170 °C for 5 min under a pressure of 300 kN/m^2 in order to obtain the desired thickness (2 ± 0.1 mm).

3.3. Migration Testing

The migration tests with synthetic saliva were performed at 37°C. The samples used are specimens of dimensions (5 × 2) cm immersed in the synthetic saliva which was prepared according to [14].

The surface of PVC samples subjected to migration tests is 10 cm^2, migration tests were carried out over a period of 1, 3 and 6 hours. Each specimen was immersed in 100 ml of synthetic saliva with and without agitation. The tests are designed without agitation to mimic the action of sucking the objects and with agitation to simulate the action of chewing. For that purpose, 20 beads of diameters 4 and 30 mm were introduced into the cell migration.

Cell migration was a bottle of 150 ml capacity, hermetically sealed, immersed in a water bath. After each test, the sample was cleaned and then weighed.

The rate of the mass variation was determined as a function of time following equation (1).

3.4. FTIR Analysis

The PVC circular samples were dissolved in THF. After evaporation of the solvent, a polymeric film was recovered and analyzed with a JASCO FTIR-430 spectrophotometer.

3.5. Scanning Electron Microscopy Analysis

The scanning electron microscopy is a powerful technique for observing surface topography. The PVC samples were analyzed by a scanning electron microscope JEOL JSM-6340F type.

4. Determination of Phthalates in Soil

4.1. Materials

A suspension PVC (Shintech SE 1200 grade 1 produced by INC- USA) was used. Zn and Ca stearates complex from Prolabo, Commercial plasticizer (DEHP from SGP, Tunisia) and stearic acid (stearine JOS from SO. G. I.S.SPA, Italy) were used as received. ESO was especially prepared as described previously [50].

4.2. Preparation of PVC Films

Formulations containing 2 wt % of Zn and Ca stearates complex, 5 wt % of ESO, 1 wt % of stearic acid and 40 wt % of plasticizer were prepared. PVC and additives were mixed in a two- roll mill at 140 °C for 20 min and melt compressed at 170 °C for 5 min under a pressure of 300 kN/m^2 in order to obtain the desired thickness (2 ± 0.1 mm).

4.3. Soil Burial Degradation Test

Square samples of 10 cm were buried in the ground at a depth of 1 m for 24 months (February 2007- February 2009); the site was located near a landfill. A sample was taken off every 03 months. It was completely washed with water and dried for 24 hours at ambient temperature (20-25°C) before being weighed and characterized (FTIR spectroscopy, density, glass transition temperature, tensile properties and shore D hardness).

The biodegradation process was investigated by scanning electron microscopy. The rate of variation of the mass was determined as a function of time following the equation:

$$\tau(\%) = \frac{m_t - m_0}{m_0} \times 100 \tag{2}$$

where: m_0 , is the initial mass sample at the time t_0 and m_t, is the mass of the sample at the time t (after soil burial).

4.4. FTIR Analysis

The PVC samples were dissolved in THF, after evaporation of the solvent and drying at 40°C during 24 h, a polymeric film was recovered and analyzed with a NICOLET (550) spectrophotometer at a resolution of 4 cm^{-1}.

4.5. Physical Properties

The density was measured according to the ISO/ R1183. Differential scanning calorimetry was performed on a DSC (Q10) thermal analyser, using the following temperature program: (1) initial temperature = 120 °C ; (2) isotherm for 5 min; (3) cooling to -30 °C at 20 °C/min; (4) isotherm for 5min; (5) heating from -30 to 120 °C at 10 °C/min. The results of Tg correspond to step 5.

4.6. Mechanical Properties

Measurements of tensile properties were undertaken using an INSTRON model 1185 (France) testing machine according to the ISO/ R27 at a crosshead speed of 100 mm/s. Shore D hardness was measured according to ISO 868.

4.7. Morphology Analysis

The biodegradation process was investigated by using a light microscope model Optica DM-15 and 20.Read phoneticallyDictionary - View detailed dictionary.

5. Biodegradation of DEHP in Soil

This part describes the validation and application of a simple respirometer system designed to assess mineralisation of the plasticizer DEHP under defined conditions. In laboratory experiments, a natural soil which contains sand, silt, clay and organic matter is generally used. The soil used in this study was collected near a landfill.

For biodegradation in soil, there are validity criteria such as [51, 52]:

- The release of CO_2 by the witness soil must be less than 20% of the average value of the CO_2 produced by the other samples;
- C / N must be included between 10/1 and 40/1;
- Temperature: it is important to be in the usual conditions for the development of the microorganisms in the soil. In our case, the temperature was set at 28°C;
- Concentration of substrate in the soil: the amount of substrate used should be sufficient for the release of CO_2. In practice, a mass greater than 0.5 mg of sample per gram of soil is used;
- Aeration of the medium is necessary since we want to be in aerobic condition. It is done with air devoid of CO_2. The ultimate biodegradation is to run the production of CO_2, that is to say, the mineralization of organic matter.

The release of CO_2 during mineralization of an organic substrate is used to estimate the overall metabolic activity of micro-population (bacteria, fungi) of the soil. The activity of microorganisms is accompanied by oxygen consumption and a concomitant production of CO_2.

The rate of biodegradation is calculated according to the following equation:

$$\% \text{ biodégradation of substrate} = \frac{CO_{2(soil + substrate)} - CO_{2(soil)}}{CO_{2th}} \times 100 \quad (3)$$

CO_2 (soil + substrate) is the cumulative amount of CO_2 released by the soil-substrate (mg);

CO_2 (soil) is the average amount of CO_2 released by the control soil in mg;

CO_2 th: is the theoretical content of CO_2 that may be produced by the test material (substrate) in mg. It is calculated according to the following equation [33]:

$$CO_{2\,th} = M_{TOT} \times C_{TOT} \times \frac{44}{12} \quad (4)$$

M_{TOT}: is the mass of total solids of the material test introduced into the composting vessels at the beginning of the essay in grams;

C_{TOT}: is the relative content of total organic carbon of the total solids contained in the test material, in grams per gram of soil;

44 and 12 are the molecular and atomic weights of carbon dioxide and carbon, respectively.

RESULTS AND DISCUSSION

1. Determination of Phthalates in Liquid Fatty and Aqueous Simulants

1.1. Preliminary Study of the Rate of Mass Variation

The rates of mass variation (τ) as a function of time give information's about the phenomenon which occurred between the samples and the food simulants used. An increase means that the food simulant penetrated the sample while a decrease means that some additives migrated in the food simulant. Hence τ gives information's about the overall migration that occurred. Figures 1 and 2 illustrate the variation of τ as a function of time for the four food simulants used at 45 °C with and without agitation, respectively. It can be noted that the highest rates of mass variations were obtained in sunflower oil indicating that migration of additives occurred. For aqueous simulants a very weak increase of τ was observed indicating penetration the food simulant in the PVC discs. On the other hand, the highest values of τ were obtained during the migration tests with agitation and for the formulations containing the highest amounts of plasticizer.

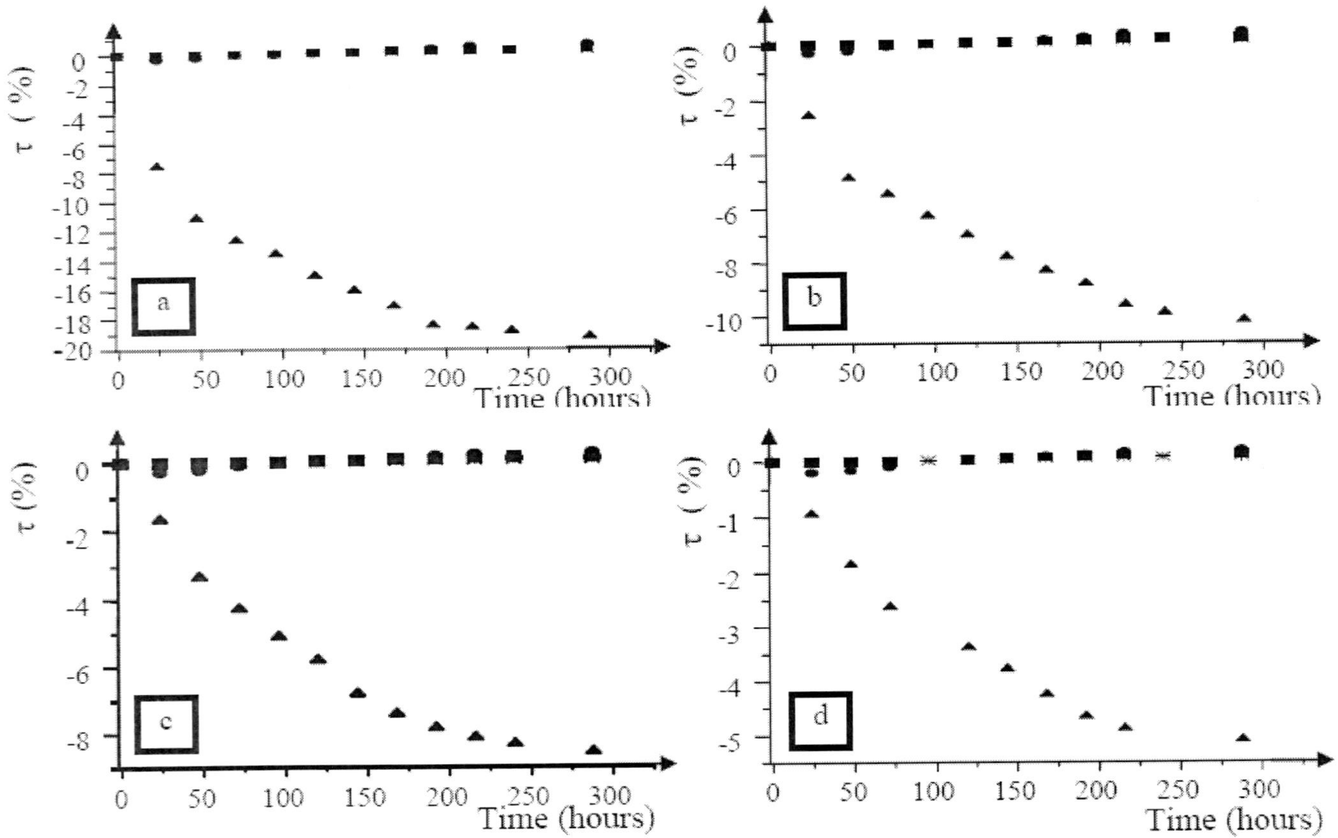

Figure 1. Effect of the nature of the food simulant on the rate of mass variation τ at 45°C with agitation: a) 60 % DEHP ; b) 50%DEHP ; c) 40 % DEHP ; d) 30 % DEHP: ▲ sunflower ● ethanol ▬ acetic acid ✳ water .

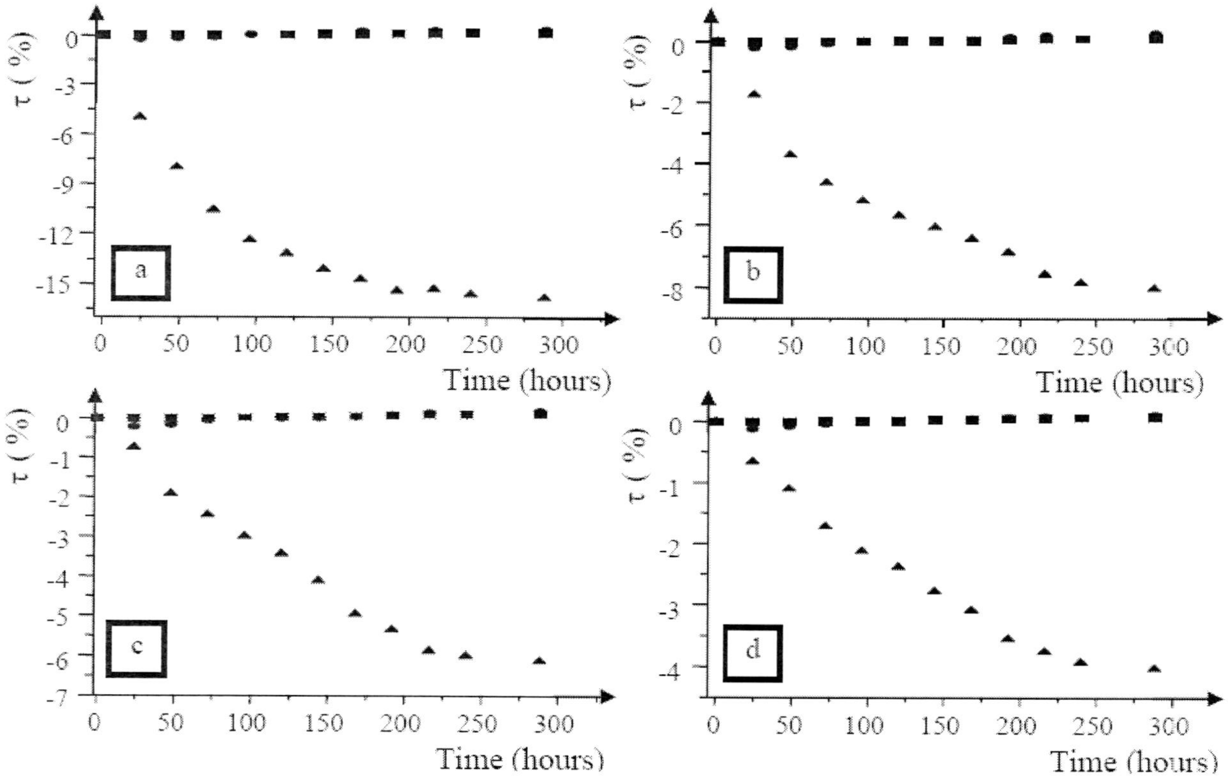

Figure 2. Effect of the nature of the food simulant on the rate of mass variation τ at 45°C without agitation: a) 60 % DEHP ; b) 50 % DEHP ; c) 40 % DEHP ; d) 30 % DEHP: ▲ sunflower ● ethanol ■ acetic acid * water .

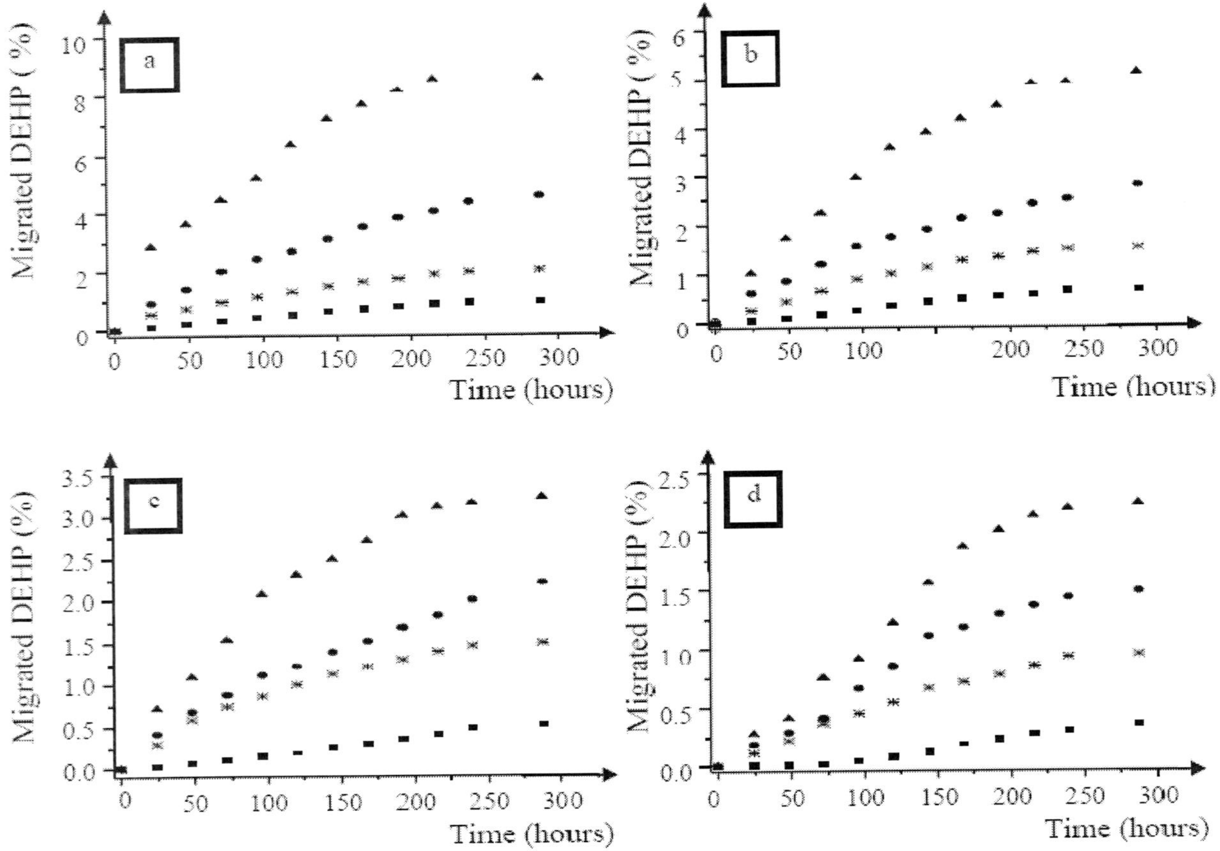

Figure 3. Effect of the initial concentration of DEHP on its specific migration in sunflower oil: a) 45 °C with agitation; b) 45 °C without agitation; c) 25 °C with agitation; d) 25 °C without ▲ 60% DEHP, ● 50% DEHP, * 40% DEHP, ■ 30% DEHP.

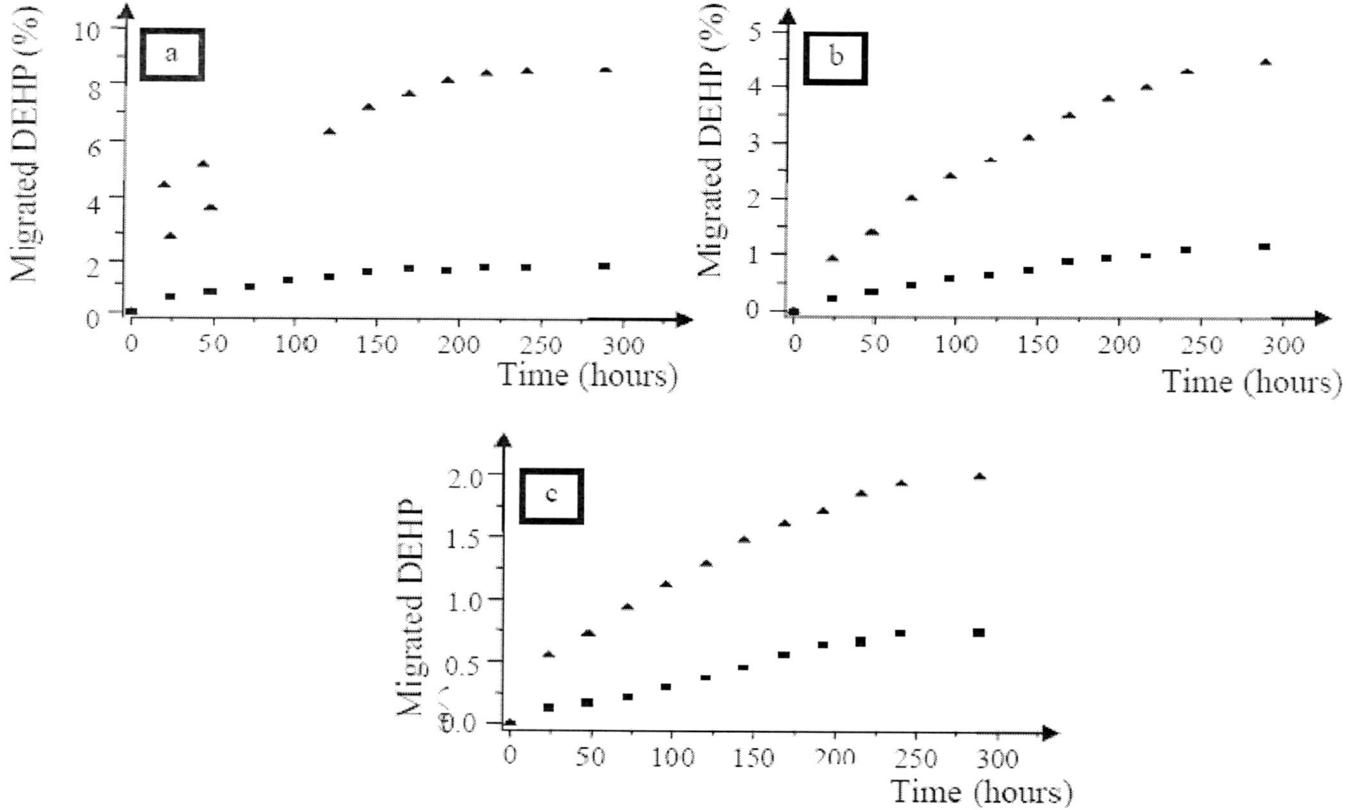

Figure 4. Effect of the nature of food simulant on the migration of DEHP at 45 °C with agitation: a) 60 % DEHP ; b) 50 % DEHP ; c) 40 % DEHP: ▲ sunflower oil, ■ ethanol.

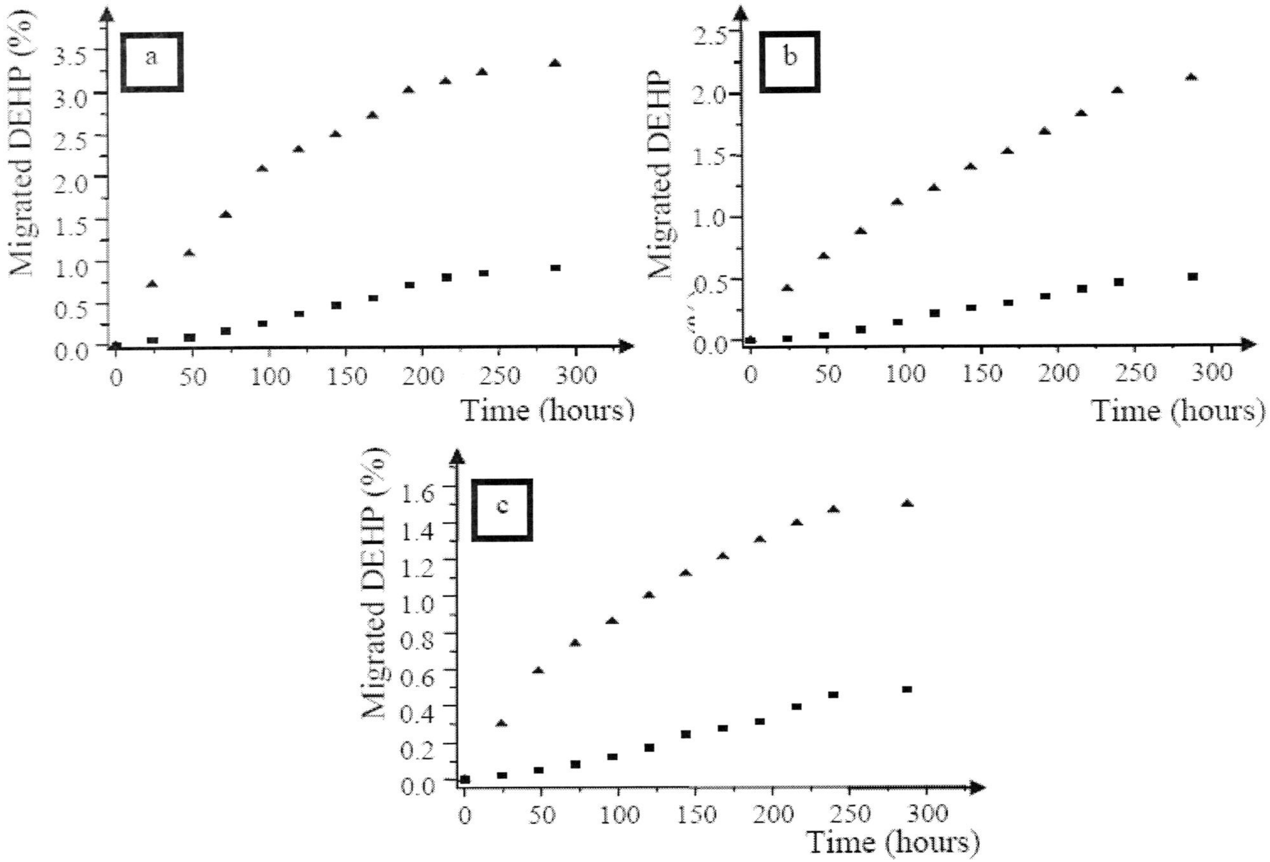

Figure 5. Effect of the nature of food simulant on the migration of DEHP at 25 °C with agitation: a) 60 % DEHP ; b) 50 % DEHP ; c) 40 % DEHP: ▲ sunflower oil, ■ ethanol.

This is expected since the first factor allows to renew the gradient of concentration at the surface of the PVC samples which favours additives migration and since the solubility of DEHP is better in sunflower oil than in aqueous simulants. Furthermore, the migration of DEHP can involve the migration of the other additives which are present in the formulation like stabilizer and lubricant.

1.2. GC Determination of DEHP Migration

The effect of the initial concentration of DEHP on its specific migration in sunflower oil is illustrated by figure 3. It is observed that the amount of migrating DEHP is related to its initial concentration in the PVC discs, to the temperature of migration testing and to the presence or absence of agitation. It is obvious that the mobility of the plasticizer molecules increased with increasing temperature and that the migration is favoured by agitation.

Figures 4 and 5 illustrate the effect of the nature of food simulant on the migration of DEHP at 45°C and 25°C with agitation, respectively. In all the cases, it is observed that the amounts of migrating DEHP are more pronounced in sunflower oil than in aqueous ethanol. This is due to the better solubility of DEHP in sunflower oil.

Furthermore, for a same initial concentration of plasticizer, the amounts of migrating DEHP are higher at 45°C in the two food simulants. This is due to the fact that the mobility of the plasticizer molecules increased with increasing temperature.

1.3. FTIR Investigation

The direct analysis of the spectra of food simulants hasn't permitted to show the migration of additives because the overlapping of characteristic bands and the very low amounts of migrated additives.

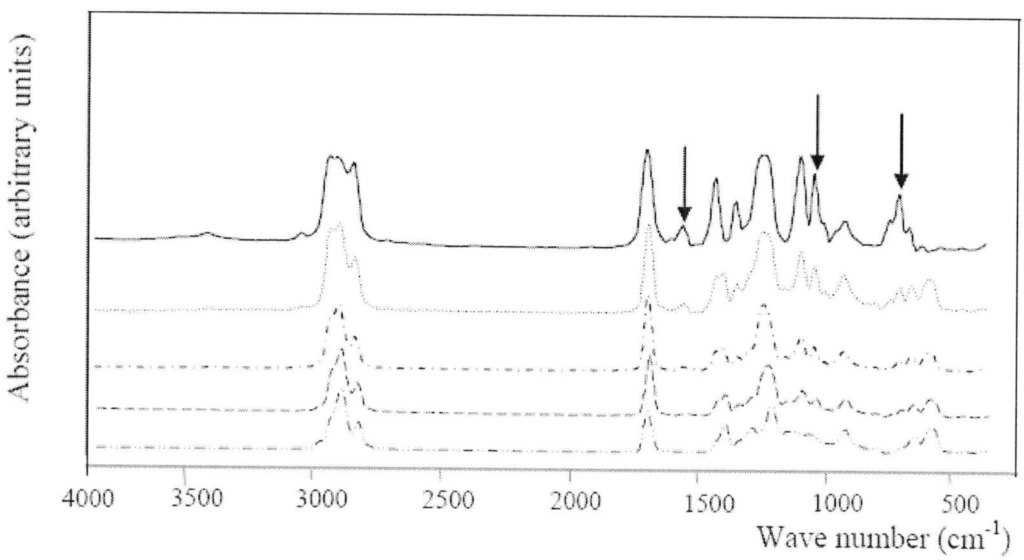

Figure 6. FTIR spectra of DEHP and PVC films obtained from the discs which were submitted to the migrating test with sunflower oil at 45°C with agitation. The initial content of DEHP in the PVC discs was 60 %: ——— DEHP, ... PVC 0 day, -.-.-. PVC1 day, ——- PVC 5 days, —··—·· PVC 9 days.

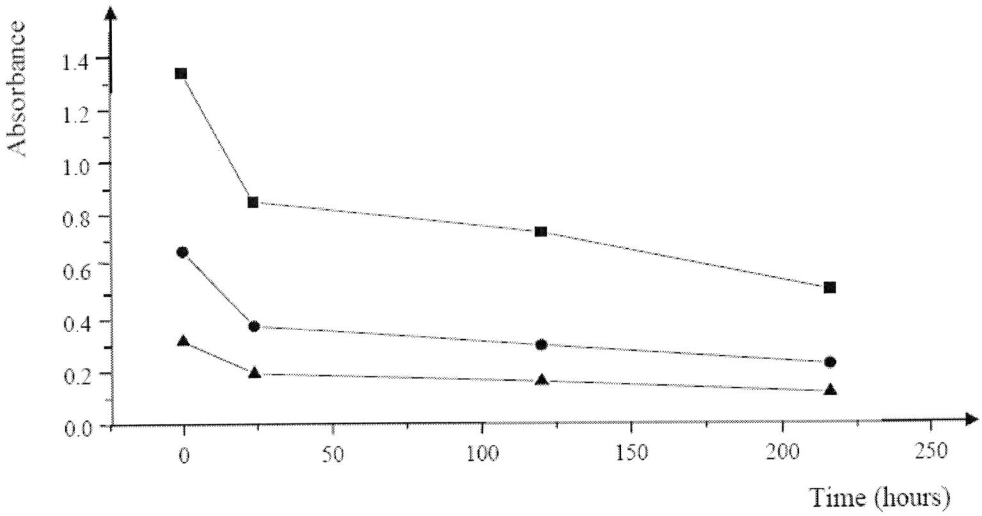

Figure 7. Variation of the absorbances of characteristic bands of DEHP as a function of time: —▲— 1585 cm^{-1} , —■— 1128 cm^{-1} , —●— 742 cm^{-1}.

On the other hand, the investigation of the PVC films spectra evidenced the DEHP migration as shown by figure 6.

A progressive decrease until disappearance of the plasticizer bands located at 1585 cm^{-1}, 1128 cm^{-1} and 741 cm^{-1} was observed and Figure 7 illustrates the variation of their respective absorbances as a function of time.

1.4. DSC Analysis

It is known that the presence of a plasticizer decreases the glass transition temperature (Tg) of PVC. For that purpose, the Tg's of virgin PVC and of four formulations containing 0, 10, 20 and 30 % of DEHP were determined. The results are given in table 1. It can be noted that the values of Tg decreased with increasing the amount of the plasticizer.

Table 1. Glass transition temperatures (Tg) for various formulations of PVC

Sample	Tg (°C)
Virgin PVC	81.00
0 % DEHP	69.40
10 % DEHP	53.50
20 % DEHP	33.30
30 % DEHP	15.40

An eventual migration of DEHP will result in an increase of Tg. Hence, the Tg's of PVC discs, whose initial content of DEHP was 30 % and which were submitted to migrating test in sunflower oil at 45 °C with agitation, were determined. The results are given in table 2. The Tg varied from an initial value of 15.40 °C to 48.50 °C after 10 days of contact with sunflower oil. This last value is close to the one obtained for a PVC containing 10 % DEHP and evidenced clearly the migration of the plasticizer.

Table 2. Tg's values after different times of contact with sunflower oil

Sample	Tg (°C)
0 day	15.40
4 days	26.20
10 days	48.50

2. Determination of Phthalates in Blood

From the chromatograms of the extract from the control blood bag and DEHP standard analyzed by the CG/SM under the same operating conditions (figure 8), it can be noted that the peak having the m/z equal to 149 with tr = 29.53 min, has practically the same time of retention as DEHP standard (tr = 29.46 min), which means that the plasticizer used in the manufacture of these bags is the DEHP. These results are confirmed in figure 9 which presents the mass spectra of the DEHP from the control blood bag and the standard of the di-2 ethyl hexyl phtalate which is also a ramified dioctyl phthalate (DOP). Through these spectra, it can be confirmed that the plasticizer which was used in the composition of the blood bag is the DEHP.

The results of the specific peak of phthalates in the extracts from the control blood bag and from the bag which was in contact with the red globules during 45 days at +4°C, with the platelets during 5 days at +18°C and with plasma for 7 months at – 30°C are summarized in table 3. In first, the comparison of the four chromatograms surface, indicates that surfaces of the specific peak of phthalates (DEHP) in the extract of the control bag is more significant than in the extract of the bag put in contact of plasma, followed-up by the surface of the same peak in contact of the red globules and finally by the surface of the same peak in the case of the bag put in contact of the platelets.

Table 3. Surfaces of the specific peak of DEHP in the extract of the control blood bag and in the bags which were submitted to the tests of migration

Surface of the specific peak of DEHP (µv.s)			
Control bag	Bag put in contact of red globules	Bag put in contact of plates	Bag put in contact of plasma
2278469	2131250	1753149	2262327

Table 4. Residual concentrations of DEHP in the extracts of the control bag and those of the bags which were submitted to the tests of migration

Residual concentrations of DEHP in the extracts (ppm)			
Control bag	Bag put in contact of red globules	Bag put in contact of plates	Bag put in contact of plasma
3.73	3.23	2.70	3.67

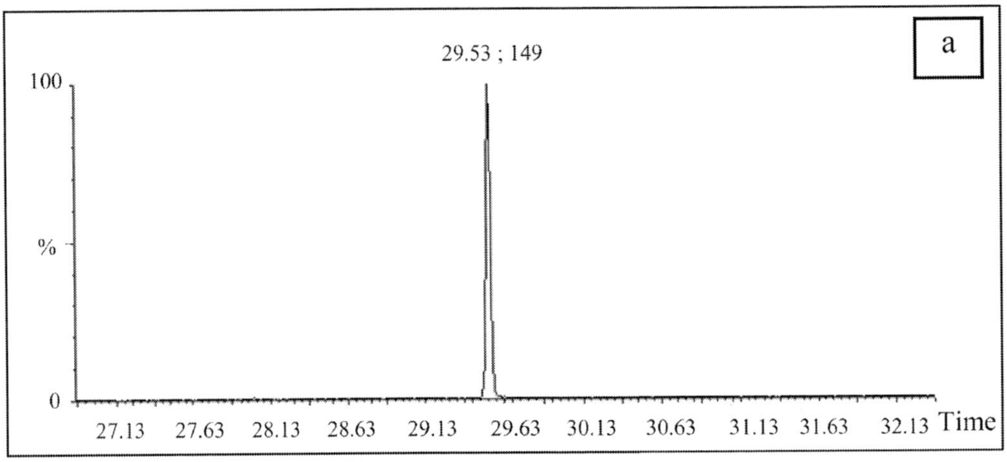

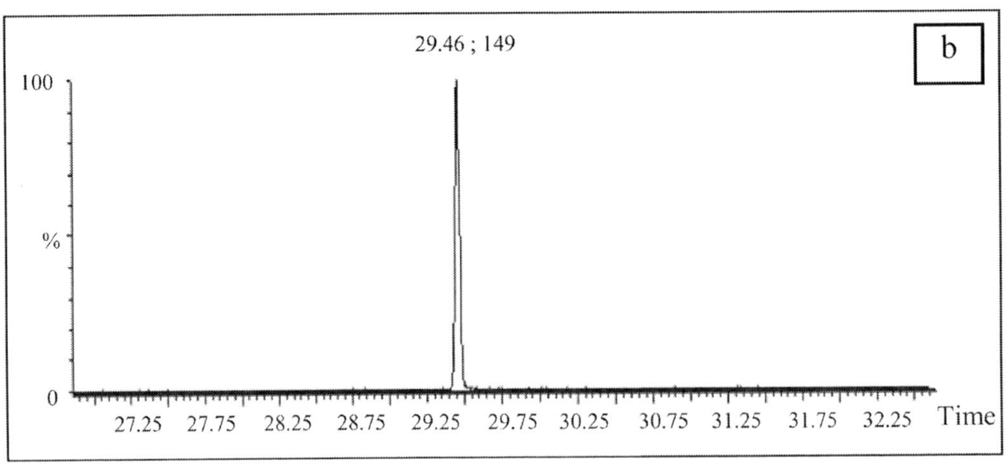

Figure 8. Chromatograms: -a- of the extract from the control blood bag. -b- of the standard DEHP.

Tables 3 and 4 show that the reduction in the surfaces of the specific peaks of the DEHP corresponds to the reduction in the residual concentrations of this plasticizer in all the bags having undergone the tests of migration and it could be due only to the migration of the DEHP in the three blood derivatives. In addition, the results of the two tables highlighted the influence of the temperature on the phenomenon of migration.

3. Determination of Phthalates in Saliva

3.1. Rate of Mass Variation

The mass variation data for plasticized PVC with and without agitation are given in Figure 10. The decrease of the rate of mass variation indicates that migration of some additives occurred in the fatty simulant. On the other hand, the highest values of (τ) were obtained during the migration tests with agitation and for the formulations containing the highest amount of plasticizer.

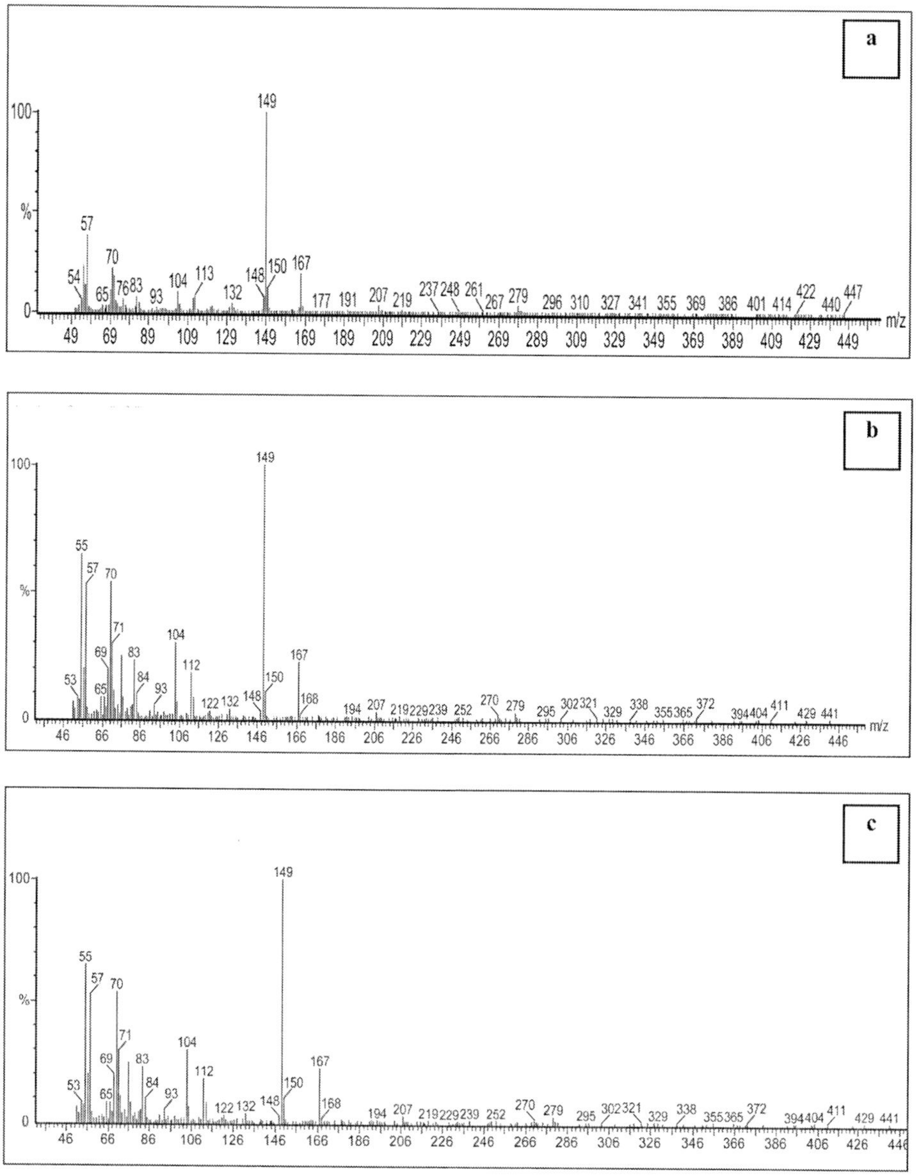

Figure 9. Spectra of mass: - a- of the extract of the control blood bag. - b- of the standard DEHP. - c- of DEHP (DOP) according to the literature [24].

The rate of the mass variation allowed to estimate the overall migration in saliva simulant which was 1.27 mg/dm^2 and 1.96 mg/dm^2 for semi rigid rigid PVC with and without agitation, respectively, and 1.93 mg/dm^2 and 2.16 mg/dm^2 for plasticized PVC with and without agitation, respectively. All these values are lower than the allowable overall migration: 10 mg/dm^2.

3.2. FTIR Investig7ation

The direct analysis of the spectra of saliva simulant did not show the migration of additives because the overlapping of characteristic bands and the very low amounts of migration additives.

However, the investigation of the PVC films spectra allowed to confirm that some migration of DEHP and other additives (ESO, Zn and Ca stearates) occurred. The comparison of the spectra of PVC alone and of additives, with the spectrum of PVC with additives (Figure 11) allowed the identification of some characteristic bands which are related to the additives present in the formulation as shown in Table 5.

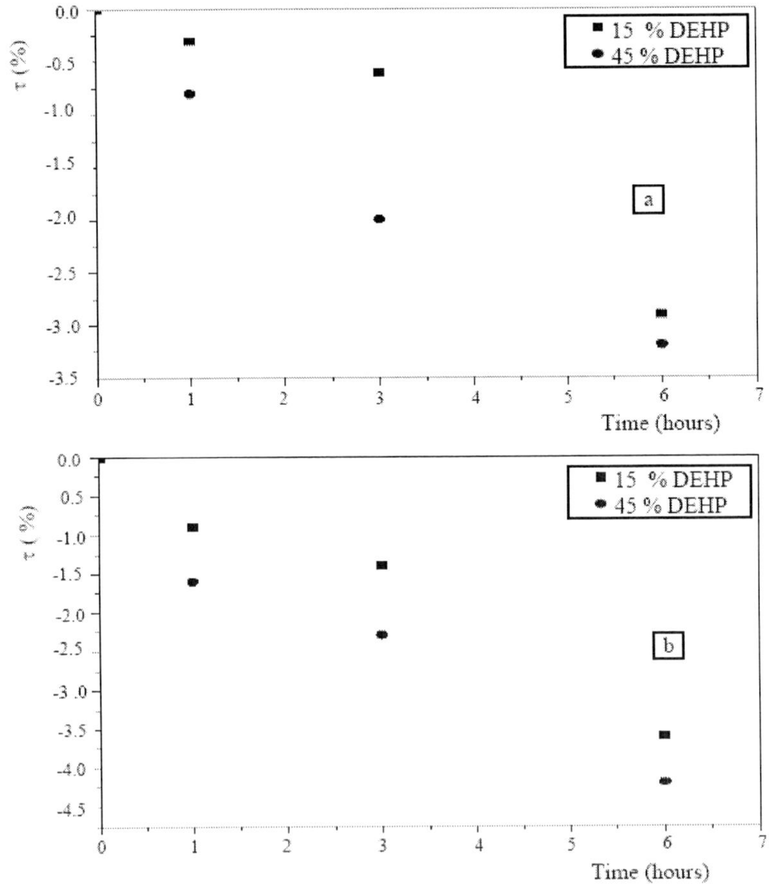

Figure 10. Effect of the plasticizer initial concentration on the rate of mass variation at 37°C: -a- Without agitation, -b- With agitation.

A semi-quantitative estimation of the migration of DEHP was done. For that purpose, the following absorbance ratio was calculated:

A1729/A1432: DEHP and ESO migration

Table 5. Characteristic bands of the used additives present in PVC films [53]

N°	Wave number (cm^{-1})	Assignement	Additive
1	1729	C=O (ester)	DEHP, ESO
2	1578	CO_2^- (carboxylic acid salt)	Ca stearate
3	1538	CO_2^- (carboxylic acid salt)	Zn stearate
4	1461	CH_2 (methyl, methylene)	ESO Zn stearate

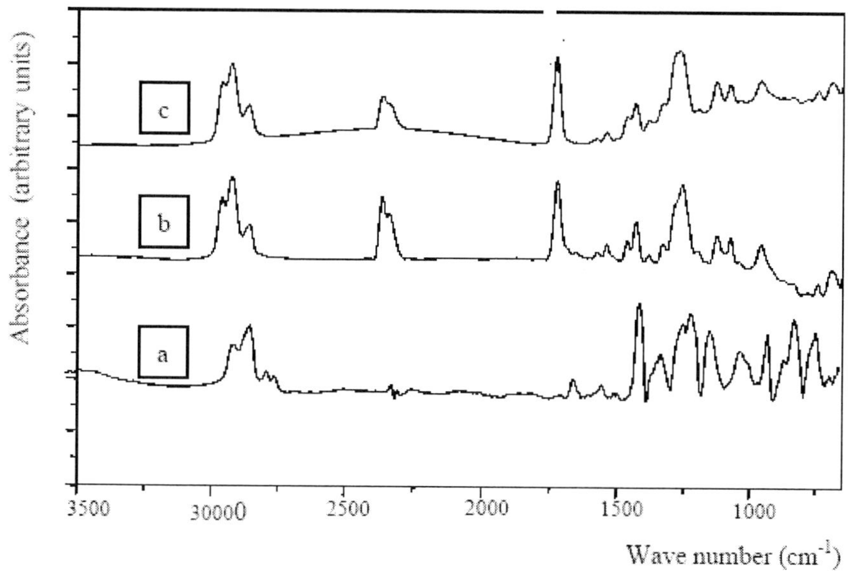

Figure 11. FTIR spectra of PVC alone (a) and PVC containing 15% (b) and 45 % (c) of DOP.

The band at 1427 cm $^{-1}$ is due to the vibration of CH_2 of PVC [54] and was taken as a reference band. The variations of the carbonyl ratios of absorbance as a function of time of contact with saliva simulant, are given in Figs. 12 and 13. An initial increase of all the absorbance ratios is first observed indicating the penetration of the saliva simulant in the PVC discs. It is followed by a decrease of all the curves with contact time. This can be directly related to a phenomenon of migration of DEHP and additives (ESO, Zn and Ca stearates) in the saliva simulant used. On the other hand, Figures 12 and 13 show that the absorbance's ratios obtained for migration testing with agitation are relatively lower than those obtained without agitation.

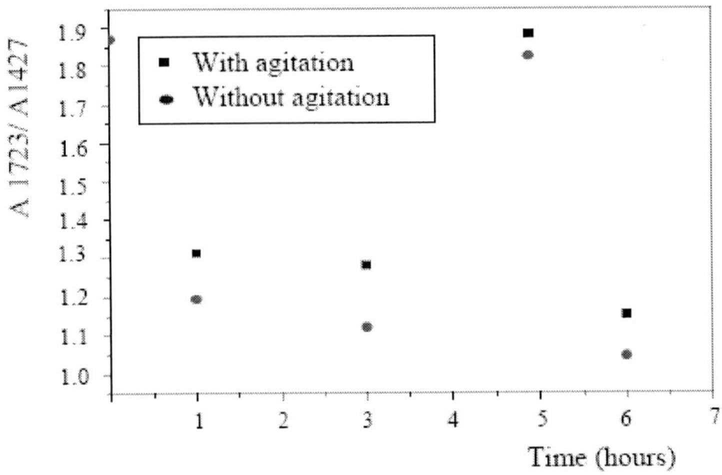

Figure 12. Variation of absorbance ratio as a function of time of contact of the formulation containing 45% of DEHP at 37 °C.

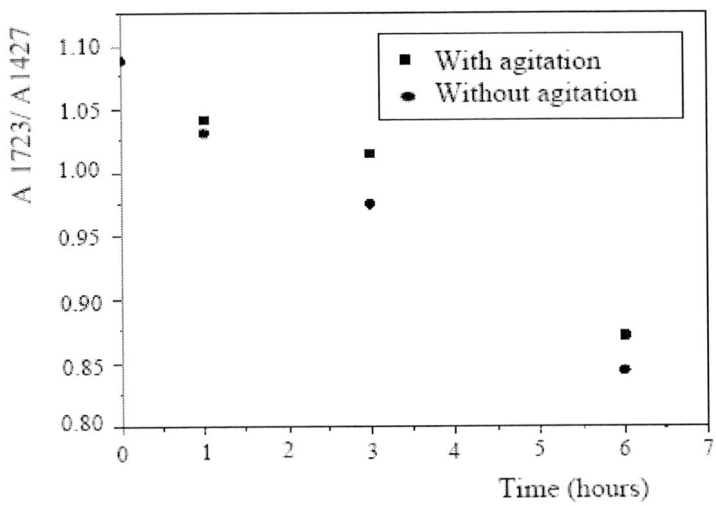

Figure 13. Variation of absorbance ratio as a function of time of contact of the formulation containing 15% of DEHP at 37 °C.

3.3. Scanning Electron Microscopy

Analysis by scanning electron microscopy of the samples taken after 3 and 6 hours of contact with saliva with and without agitation (Figures 14 and 15) showed the emergence of depressions (holes); these areas are more notable in the case of the formulation plasticized with 45% DEHP. These holes are especially more important when the contact time increased. The appearance of preferred areas (horizontal or vertical), which was not observed for the formulation plasticized with 45% DEHP, was noted in the case of the formulation containing 15 % DEHP.

All these observations indicate that there was an initial absorption of the liquid medium by the samples which allowed the release of a certain amount of additives leaving the appearance of empty areas. The migration is more important in the case of the formulation containing 45% DEHP.

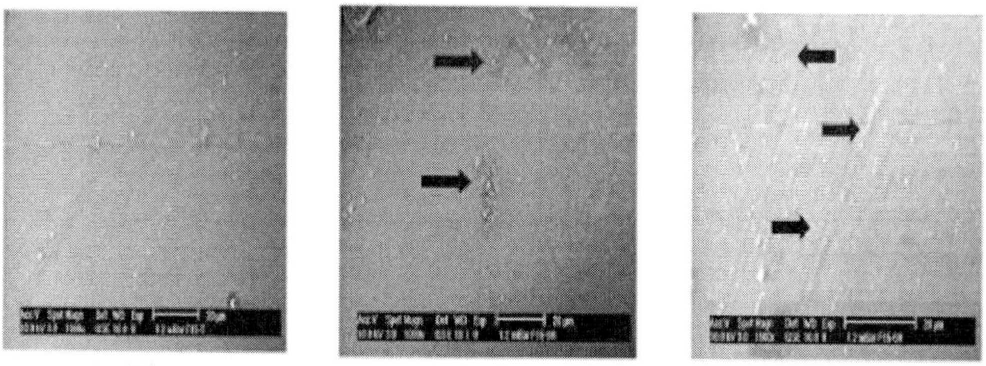

a: 0h b: 3h c: 6h

Figure 14. Scanning electron microscopy analysis of plasticized PVC after contact with saliva with agitation.

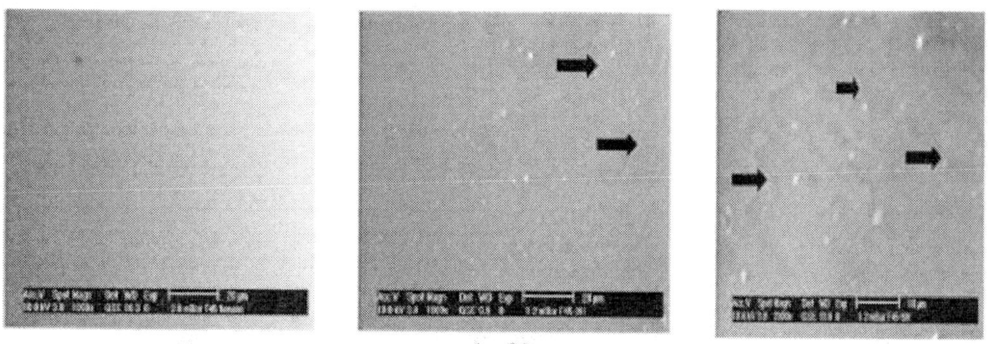

a: 0h b: 3h c: 6h

Figure 15. Scanning electron microscopy analysis of rigid PVC after contact with saliva without agitation.

4. Determination of Phthalates in Soil

4.1. Variation of Mass

The rates of mass variation (τ) as a function of time give information about the phenomenon which occurred between the samples and the soil. Globally, an increase means that the water penetrated the samples, whereas a decrease means that some additives migrated into the soil. However, even when there's mass loss, water could possibly have diffused into the specimens. The additives such as DEHP and ESO will absorb water over time. Table 6 displays that, globally, the variation of mass is negative which indicates that more additives were extracted from the specimens than the amounts of water absorbed. However, fluctuations are observed which confirms the diffusion of water in the specimens. This feature is related to the quantities of rainfalls measured in the area (about 850 mm/year).

Table 6. Mass variation as a function of time of soil burial

Time (months)	Mass variation (%)
0	0
3	-0.0046
6	-0.0077
9	-0.0094
12	-0.0077
15	0.0214
18	-0.0481
21	-0.0196
24	-0.0156

4.2. Variation of Physical and Mechanical Properties

Table 7 displays the variations of density and glass transition temperature (Tg) with time of soil burial. It can be noted that the density remained practically constant but the glass transition temperature increased with the time of soil burial. This result indicates the loss of the plasticizer [1].

Table 7. Variations of density and glass transition temperature as a function of time of soil burial

Time (months)	Density	Glass transition temperature (°C)
0	1.23	9.94
6	1.23	9.95
12	1.23	9.97
15	1.21	9.99
18	1.24	10.10
21	1.22	10.40
24	1.23	11.01

Table 8. Variations of elongation, stress at break, and Shore D hardness with time of soil burial

Time (months)	Elongation at break (%)	Stress at break (Mpa)	Shore D hardness
0	358.33 ±1.33	20.33± 0.00	40 ± 0.0
3	325.00 ±0.25	20.83 ± 0.16	40 ± 0.3
6	316.66 ± 2.33	20.00± 0.00	40 ± 0.4
9	275.00 ± 0.28	26.13 ± 0.00	49 ± 0.2
12	325.00 ± 1.38	22.79 ± 0.80	49 ± 0.2
15	329.00 ± 0.50	19.00 ± 0.29	45 ±0.3
18	291.00 ± 1.40	21.48 ± 0.88	54± 0.4
21	284.00 ± 1.22	22.38 ± 1.47	54± 0.2
24	284.00 ± 1.45	22.53± 0.18	54 ± 0.5

Table 8 shows that elongation at break decreased with time of soil burial, whereas stress at break and shore D hardness increased. These variations can be explained by the loss of the plasticizer which reduced the flexibility of the samples and also by the degradation of the plasticized PVC composition. The degradation could include that of the plasticizer which in turn could lead to the decreased compatibility over time. On the other hand, the void spaces (pores) resulting from the loss of the plasticizer are filled with water in soil which explain the increase of the elongation and the decrease of the stress at break and shore D hardness in the period corresponding to the twelfth and the fifteenth months.

4.3. FTIR Characterization

To check the presence of degradation structures in the PVC matrix, the samples were analyzed by FTIR spectroscopy. Figure 16 shows the FTIR spectra of the plasticized PVC samples after various times of soil burial. The appearance of two new bands located at 2368 cm^{-1} and 2341 cm^{-1} can be observed on the PVC spectra. These bands are related to structures of the type X =Y= Z (where X, Y, Z can represent atoms of C, N, O and S) [53, 55]. Their presence indicates that some biodegradation of the plasticizer occurred, and they are probably due to the metabolites of the plasticizer degradation by microorganisms [55, 56] and enzymatic hydrolysis [57].

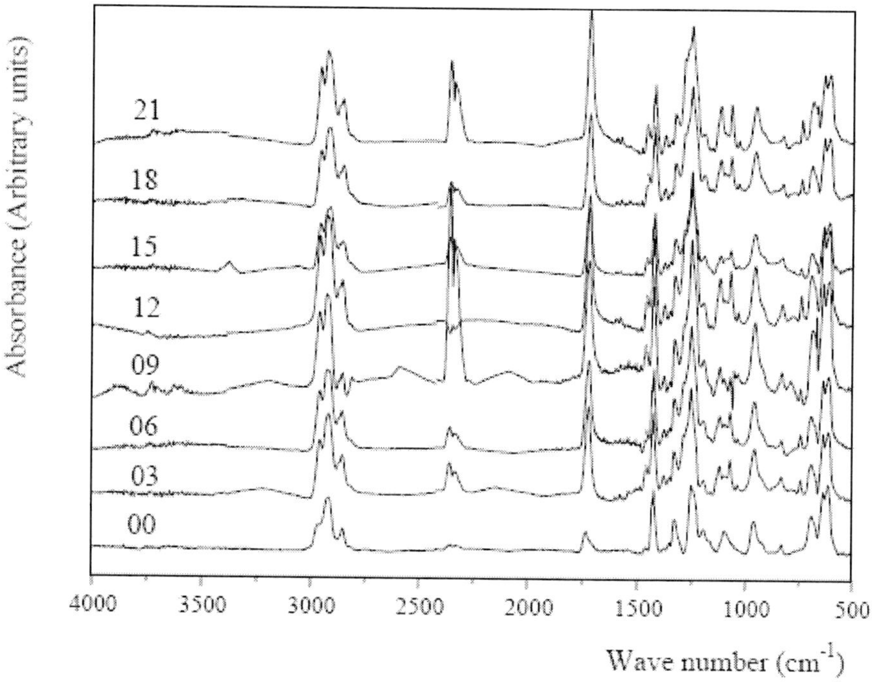

Figure 16. Infrared spectra of the plasticized PVC samples after various times of soil burial (months).

To investigate the eventual modifications which occurred, the spectra of PVC alone and of additives present in the various formulations were compared with the spectrum of PVC with additives. This allowed the identification of some characteristic bands which are related to the additives present in the formulation as shown in Table 6 [25, 53 and 58]. A semi quantitative estimation of the modifications which occurred was done. For that purpose, the following absorbance's ratios were calculated:

A1726/A1436: DEHP, ESO and their metabolites;
A1466/A1436: DEHP, ESO, Zn and
Ca stearates complex, and their metabolites;

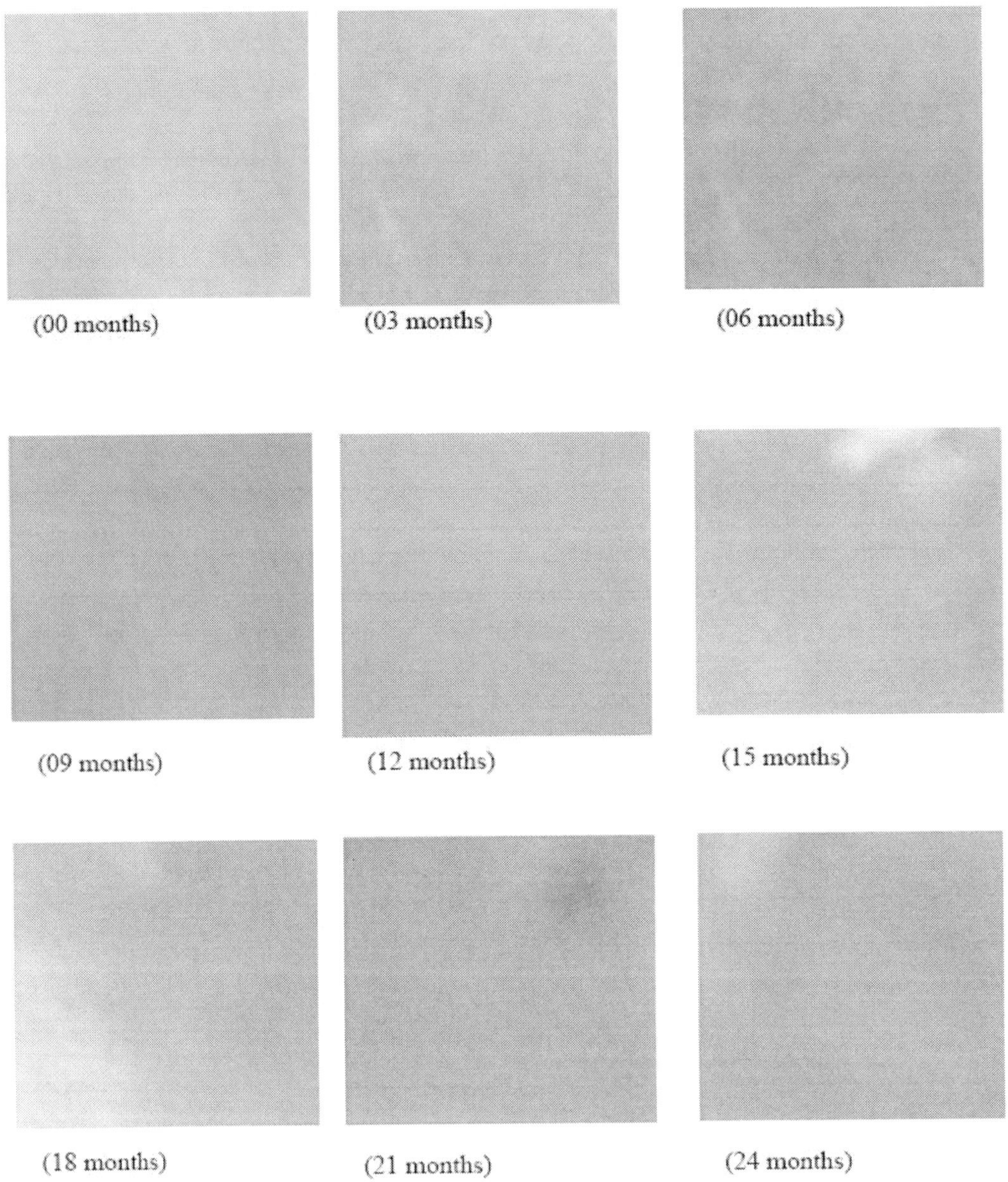

Figure 17. Photographs of the plasticized samples after various times of soil burial.

The band at 1436 cm^{-1} is due to the vibration of CH_2 of PVC [54] and was taken as a reference band. The variations of these four ratios of absorbance's as a function of time are given in Table 9. Globally, the results show that the absorbance's ratios varied with the time of soil burial for all the formulations considered. This means that the concentrations of the corresponding additives (DEHP, ESO, Zn and Ca stearates complex) varied in the PVC samples. This feature may be due to the migration of these additives in the soil or to the metabolites due to the microorganisms attack.

Table 9. Variations of absorbances ratios as a function of time of soil burial

Time (months)	00	03	06	09	12	15	18	21	24
1726/1436	1.130	1.194	1.175	1.045	1.477	1.209	1.408	2.229	0.152
1466/1436	0.178	0.289	0.137	0.263	0.259	0181	0.299	0.272	0.098

4.4. Visual Observation

After 21 months of soil burial (Figure 17), the samples showed the appearance of isolated spots which confirm the microorganisms attack.

5. Biodegradation of DOP in Soil

5.1. Daily Production of CO_2

The curve of the daily production of CO_2 obtained after incubation is shown in Figure 18. The results show that, in soils containing DEHP, the respirometric activity pass by two phases. The first one in which there is a release of CO_2 corresponds to the degradation of the most labile substances; this phase lasts until the twelfth day.

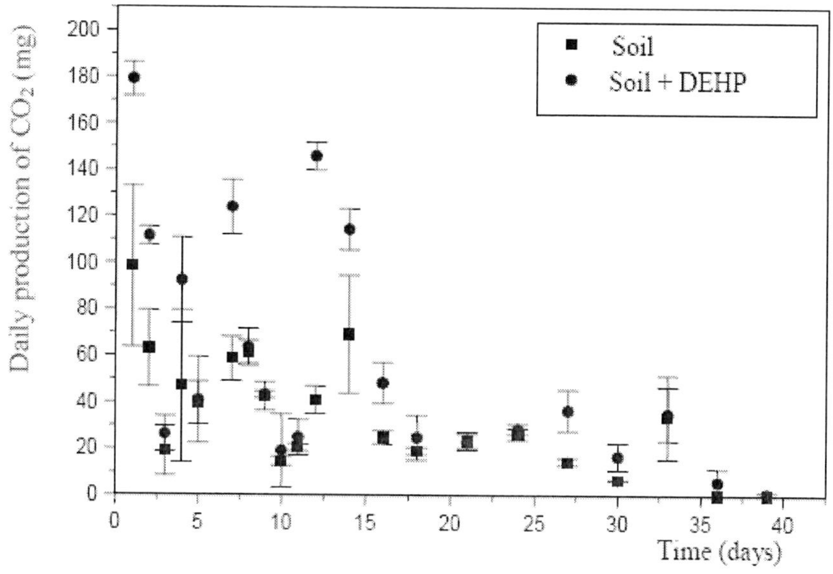

Figure 18. Daily production of CO2 as function of time.

The largest amounts of CO_2 produced were recorded during the first two days and the twelfth day. From the twelfth day, the decay phase of the biological activity begins, due to the decrease of easily metabolized compounds.

Mineralization continues, but slowly and it is characterized by a low release of CO_2 from the biodegradation of the most resistant substance.

5.2. Cumulative Production of CO_2

The cumulative production of CO_2 after incubation is represented in Figure 19. The results showed that the cumulative amount of CO_2 increased with the time of incubation.

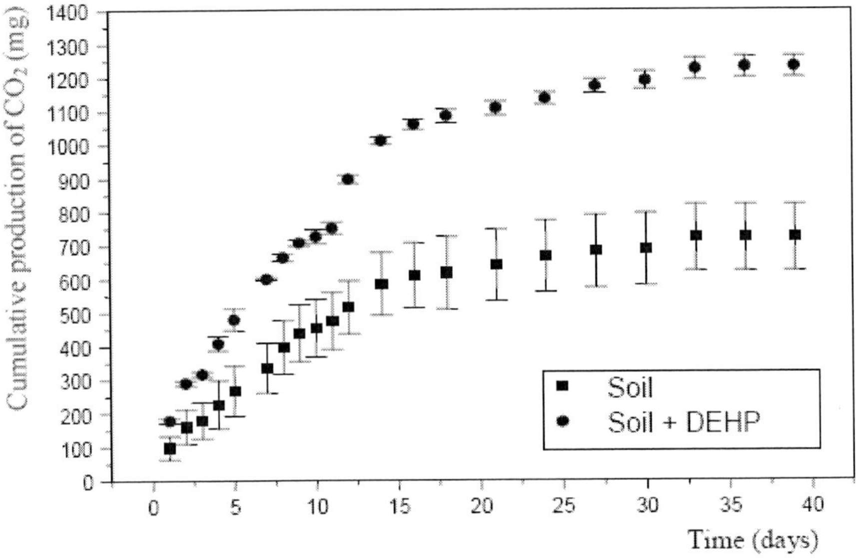

Figure 19. Cumulative production of CO2 as function of time.

5.3. Biodegradation Rate

The rate of biodegradation is calculated using the equations 3 and 4. Figure 20 shows the evolution of the biodegradation rate as function of time.

The following various phases were observed:

- Degradation phase: from the first day until the seventh day;
- Stationary phase: from the seventh day until the ninth day
- Degradation phase: from the ninth day until the twenty-seventh day;
- Stationary phase: from the twenty-seventh day until the thirty- ninth day.

The results showed that after the thirty-ninth day of incubation, the rate of biodegradation of DEHP is equal to 37.71%. This result is related to the structure of the substrate. Indeed, the DEHP has an aromatic ring; therefore it is more resistant compared to other plasticizers. Our results agree with those found by Jackson et *al.,* [35] in an experiment involving the attack of the DOP and bis 2-ethylhexyl adipate (BEHA) by *Rhodococcus rhodochrou.* Indeed, half the concentration of DOP was found in the solution at the end of the experiment while the BEHA was completely assimilated by the bacteria.

Several studies have been conducted on the biodegradation of phthalate esters by bacteria, fungi and algae [36-38]. Phthalate esters are metabolized by *Pseudomonas testosteroni* and *Pseudomonas capacia*, gram negative to form the 4, 5 dihydroxyphtalate [39,

40]. *Pseudomonas fluoresences* FS1 have a great ability to degrade phthalate esters in monoesters. It was showed that the biodegradation depends on the affinity between microorganisms and the plasticizer [41]. On the other hand, it was found that the process of biodegradation of phthalates in alkaline medium is important[42].

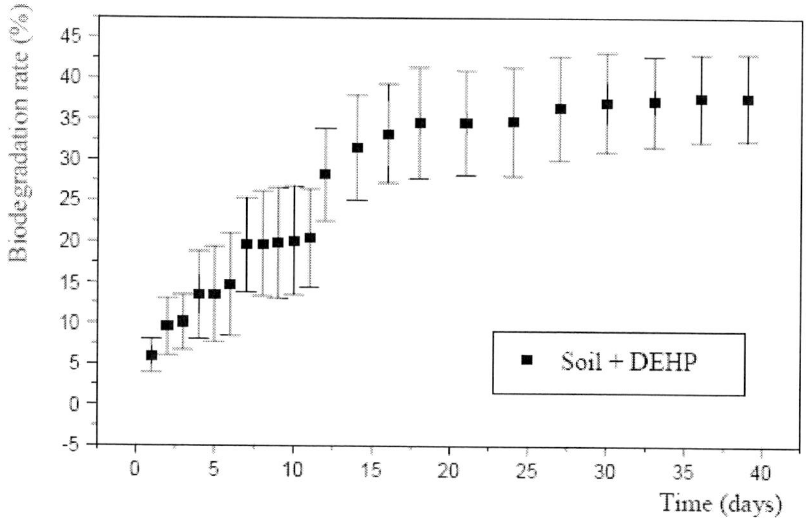

Figure 20. Biodegradation rate as function of time.

Conclusion

Based on the results of these different studies, the issues to be noted are the following:

- Concerning the behavior of DEHP in contact with liquid fatty and aqueous simulants, all the various analytical methods used (mass variation, GC, FTIR spectroscopy, DSC) showed that the migration of the plasticizer DEHP occurred in the fatty simulant (sunflower oil) while the aqueous simulants (15% v/v aqueous ethanol, 3 % aqueous acetic acid and distilled water) penetrated in the plasticized PVC samples. The phenomena are influenced by the temperature, the agitation and the initial concentration of the plasticizer.
- Concerning the behavior of DEHP in contact with blood, the GC/MS analysis of the control bag and of the bags which were submitted to the tests of migration allowed to confirm that the DEHP is the plasticizer used in the formulation of the blood bag and showed that its migration occurred in red globules, platelets and plasma. The phenomenon is influenced by the temperature of contact.
- Interactions between PVC and saliva simulant were evidenced: penetration of the saliva and migration of DEHP occurred. The rate of the mass variation allowed to estimate the overall migration in saliva simulant which was 1.27 mg/dm^2 and 1.96 mg/dm^2 for semi rigid rigid PVC with and without agitation, respectively, and 1.93 mg/dm^2 and 2.16 mg/dm^2 for plasticised PVC with and without agitation,

respectively. All these values are lower than the allowable overall migration: 10 mg/dm^2.

- Interactions between plasticized PVC and soil occurred. Migration and biodegradation of DEHP and additives lead to changes in mechanical and physical properties. These phenomena were confirmed by Fourier transform infrared spectroscopy (FTIR).
- A rate of biodegradation of DEHP equal to 37.71% was found.

All these results evidenced the migration of DEHP in the various media. The adverse health effects observed for some of these compounds, which are now ubiquitous environmental contaminants, justify the debate in the regulatory world about what to do about phthalates. So, their control in samples coming from different sources is of great social and economic interest and it would be desirable to have available new and simpler analytical techniques.

Finally, if DEHP toxicity in humans is confirmed, even only partially, it would be advisable to replace current PVC plasticizers with safer materials.

References

[1] Mersiowsky. *Prog Polym Sci.* 2002, 27, 2227-2277.

[2] J. Murphy, *Additives for plastics handbook*; Second ed., Elsevier Advanced Technology, Oxford, 2003.

[3] E. Kondyli, P.G. Demertzis and M.G. Kontominas. *Food Chemisry*. 1992, 45, 163- 168.

[4] R. Chamberlain and A.C. Harrison. *Plastics Rubbers Textiles*. 1972, 3 n°2, 331-341.

[5] D. Messadi, J. M. Vergnaud and M. Hivert. *J. Appl. Polym. Sci.* 1981, 26, 667-677.

[6] J.L. Taverdet and J.M. Vergnaud. *Eur Polym J.* 1986, 22 n°12, 959-962.

[7] Y. Khatir, J.L. Taverdet and J.M.Vergnaud. *J. Polym. Eng.* 1988, 8n°1-2, 111- 129.

[8] Mercer, L. Castle, J. Comyn and J. Gilbert. *Fd Additives and Contaminants*. 1990, 7 n°4, 497-507.

[9] Cooper and P.A. Tice. *Fd Additives nd Contaminants*. 1995, 12, n°2, 235- 244.

[10] J. H. Petersen, E.T. Naamansn and P. A. Nielsen. *Fd Additives and Contaminants*. 1995, 12 n°2, 245-253.

[11] L. Oi- Wah and W. Siu- Kay. *J. Chromatogr A*. 1996, 737, 338- 342.

[12] L. Oi- Wah and W. Siu- Kay. *Packaging Technology and Science*. 1996 9, 19- 27.

[13] J. Hojslev, L. Lillemark and L. Lund. *Fd Addtives and Contaminants*. 1997, 14 n°4, 345- 353.

[14] Steiner, L. Scharf, F. Fiala and J.Washuttl. *Fd Additives and Contaminants*. 1998, 15, n° 7, 812-817.

[15] J. Hojslev and T. Breindhal. *Fd Additives Contaminants*. 1998, 15 n°5, 600- 608.

[16] S.E. Djilani, T. Bouchami, F. Krid, N. Boudiaf and D. Mesadi. 2000, 36, 1981- 1987.

[17] L. Monney, M. Jamois-Tasserie, C; Dubois, P. Lallet, F. Villa and C. Renaud. *Polym Degr. and Stab.* 2001, 72, 459- 468.

[18] R. Zelko, A. Orban, J. Nagy, G. Csoka and I. Racz. *J. Thermal. Anal. Calor*. 2002, 68, 531- 537.

[19] T. Kovacic and Z. Mrklic.*Thermochim Acta*. 2002, 381, 49- 460.

[20] S. Yeong-Tarng, H. Tzu-Ting, L. Tzu-I and Y. Hang-Shung. *J. Appl. Polym. Sci.* 2003, 90, 4032- 4037.

[21] J. H. Kim, S.H. Kim, C.H. Lee, J.W. Nah and A. Hahn. *Bul. Korean Chem. Soc.* 2003, 24 n°3, 345- 349.

[22] J.C. Huang and R. Deanin. *J. Appl. Polm. Sci.* 2004, 91,146-156.

[23] A. Marcilla and J.C. Garcia- Quesada. *J. Anal. Appl. Pyrolysis*. 2004, 71, 457- 463.

[24] Q.Wang and B.K.Storm. *Polymer. Testing.* 2005, 24, 290- 300.

[25] M. O. Boussoum, D. Atek and N. Belhaneche-Bensemra. Polym. Degrad and Stab. 2006, 91, 579-5 84.

[26] Fankhauser- Noti, G. Grob. *Trends in Food Science and Technology*. 2006, 17, 105-112.

[27] M.T. Khorasani and H. Mirzadeh. *Radiation Physics and Chemistry*. 2007, 76, 1011-1016.

[28] N. Ladjane and N. Belhaneche-Bensemra, *J. Appl. Polym. Sci.*2009, 111, 525- 531.

[29] D. Atek, N. Belhaneche-Bensemra and M. Turki. *Int. J. Polym Mat.* 2010, 59, 342-352.

[30] M. O. Boussoum and N. Belhaneche-Bensemra. *J. Appl. Polym. Sci.* 2012, 124 n° 2, 1241– 1248.

[31] M. O. Boussoum, Y. R. Hocine and N. Belhaneche- Bensemra, *Int. J. Envir. Studies*. 2012, 11, 913- 923.

[32] K. Figge. Fd. Cosmet. *Toxicol.* 1972, 10, 815- 823.

[33] J. Gilbert, M. W. Sharmani and L. Rossi. *Fd Additives and Contaminants*. 1994, 11 n°4, 97- 518.

[34] R.Kavlock, K.Boekelheide. Reproductive Toxicology. 2002, 529-653.

[35] O.Earls, I. P. Axford and J.H; Braybrook. *J. Chromatogr.* A. 2003, 983, 237- 246.

[36] S. O. Oldman-Ghazi, F. Hatcher and M. M. Whalen. *Chemio-Biological Interaction*. 2003, 146, 1-18.

[37] Gomez-Hens and M.P. Aguilar-Caballos. *Trends in Anal Chem*. 2003, 22 n°11, 847-857.

[38] G. Latini, C. De Feliceand A. Verroti. *Reproductive Toxicology*. 2004, 19, 27- 33.

[39] C.S. Brazel and M. Rahman. *Prog. Polym. Sci.* 2004, 29, 1223- 1248.

[40] R. Ito, F. Seshimo, N. Miura, M. Kawaguchi, K. Saito and H. Nakazawa; *J. Pharmaceuical and Biomedical Analysis*. 2006, 41, 455- 460.

[41] C. F.Wilkinson and J.C. Lamb. *Regul Toxical Pharm* 1990, 30, 140- 151.

[42] M. Allsop, D.Santillo and P. Johnston. Technical Note 14/00. *Greenpeace Research Laboratories*, Exeter, 2000.

[43] E. A. Tehrany and S. Desobry. *Fd Chem*. 2007, 101, 1714- 1718.

[44] R. J. Jaeger. *New. Engl. J. Med.* 1972, 58-65.

[45] R.Kavlock and K.Boekelheide. *Reproductive Toxicology*. 2002, 529- 653.

[46] R. Ito. *Int. J. of Pharmaceutics*. 2005, 104- 112.

[47] N. Haider and S. Karlsson. *Polym Degrad and Stab.* 1999, 64, 321-328.

[48] S. Jonsson, J.O. Ejlertsson, A. Ledin, I. Mersiowsky and B. H. Svensson. *Wat. Res.* 2003, 37, 609-617.

[49] L. Fantoni, C. Simoneau, *Fd Addiives and Contaminants*. 2003, 20 n°11, 1087- 1096.

[50] M. T. Benaniba, N. Belhaneche- Bensemra and G. Gelbard, *Polym. Degrad. and Stab.* 2001, 74, 501-505.

[51] Norme Internationale, ISO/CD 17556. 1999. Détermination of ultime aerobic biodegradability in soil by measuring the oxygen demand in a resperometer or amount of carbon dioxid released.

[52] Norme Internationale ISO 14855. 1999. Evaluation de la biodégradabilité aérobie ultime et de la désintégration des matériaux plastiques dans des conditions contrôlées de compostage: Méthode par analyse du dioxyde de carbone libéré.

[53] G.Socrates. Infrared characteristic group frequencies. New York: John Wiley and Sons, 1980.

[54] S. Krimm, V. L. Fol., J. J. Shipman and A. R. Berens. *J. Polym. Sci.* 1963. 1, A1, 2621-2640.

[55] N. Lardjane, N. Belhaneche- Bensemra. *J. Appl. Polym. Sci.* 2009, 111, 525- 505.

[56] R.Gattin, C. Poulet, A. Copinet, Y. Couturier. *Biotech Let.* 2001, 22, 1471- 1476.

[57] Z. Saadi, *Thèse de doctorat, Laboratoire de Chimie Organique et Macromoléculaire, CNRS, France* (2008).

[58] D. Atek. N. Belhaneche-Bensemra, *Eur Polym J.* 2005, 41,. 707- 714.

[59] M. jackson, D. Labeda and L. Becker. *J. Ind. Microbiol.* 1996, 16, 301- 304.

[60] K. Sivamurthy, B. Gpujar. *J Fermen Bioeng.* 1989, 68, 375–377.

[61] C. Juneson, OP. Ward and A.Singh. *Process Biochem.* 2001, 37, 305–313.

[62] C. Juneson, OP. Ward, A. Singh. *World J. Microbiol. Biotechnol.* 2002, 18, 7- 10.

[63] F. Zeng, K. Cui, J. Fu, G. Sheng, HF. Yang. *Water Air Soil Pollut*, 2002, 140, 297- 305.

[64] F. Zeng, K. Cui, X. Li, J Fu and G. Sheng. *Process Biochem.* 2004, 39, 1125- 1129.

[65] J. S. Webb, C.H. Van der MEI, M. Nixon, I. M. Eastwood, M. Greenhalgh, S. J. Read, G. D. Robson and P. S. Handley. *Appl. Environ. Microbiol.* 1999, 65, 3575- 3581.

[66] S. Murad, F. Hasan, A. Ali Shah, A. Hameed and S. Ahmed. *Pak. J. Bot.* 2007, 39, 1833- 1841

In: Phthalates
Editors: Gerardo L. Moretti and Drago Romano

ISBN: 978-1-62081-994-4

Chapter 5

REMOVAL OF PHTHALATES FROM ENVIRONMENT

Sha Li[1], Shan Wu[1], Hua-Bin Li[1], Qing Hao[2] and Xiang-Rong Xu[2,*]

[1]School of Public Health, Sun Yat-Sen University, China
[2]Key Laboratory of Marine Bio-resources Sustainable Utilization, South China Sea Institute of Oceanology, Chinese Academy of Sciences, China

ABSTRACT

Phthalates are synthesized in massive amounts, and widely used in building materials, home furnishings, transportation, clothing, paints, lacquers, cosmetics, food and medical products packaging. Release of phthalates into the environment during the manufacturing, use and disposal has caused serious concerns, since some of them are suspected to be mutagenic, hepatotoxic and carcinogenic. Especially, phthalates have been considered as endocrine-disrupting chemicals, and could interfere with reproductive system and behavior of humans and wild life through disturbance of the endocrine system even at a very low concentration level. The removal of phthalates from environment has attracted increasing attention in recent years. In this chapter, removal of phthalates from environment by various physical, chemical and biological techniques are reviewed, and special attention is paid to adsorption, advanced oxidation process and biodegradation of phthalates. The review provides an account on our research work combined with a reference of the literature published in the past ten years.

1. INTRODUCTION

Phthalate esters or phthalates are the dialkyl or alkyl aryl esters of 1, 2-benzendicarboxylic acid (phthalic acid). The basic chemical structure of phthalates is benzene dicarboxylic acid with two side chains, which can be alkyl, benzyl, phenyl, cycloalkyl, or

[*] Corresponding author. Tel.: +86-20-89100753, fax: +86-20-89100753, *Email address:* xuxr2000@yahoo.com.

alkoxy groups (Liang et al., 2008). The chemical structures of several phthalates are shown in Figure 1.

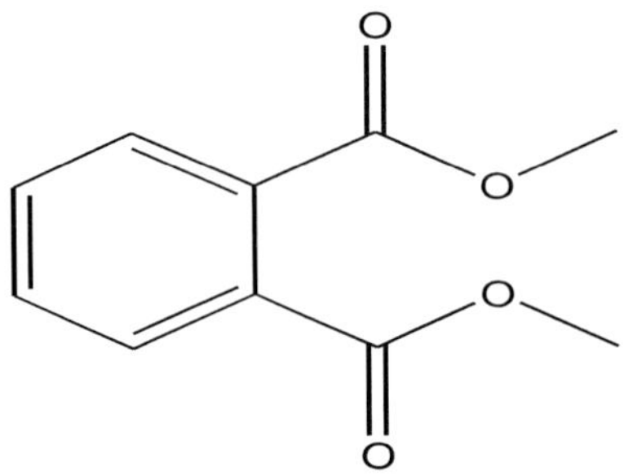

DMP

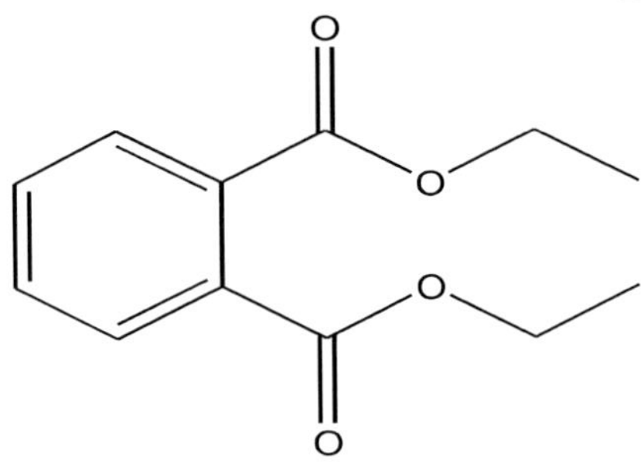

DEP

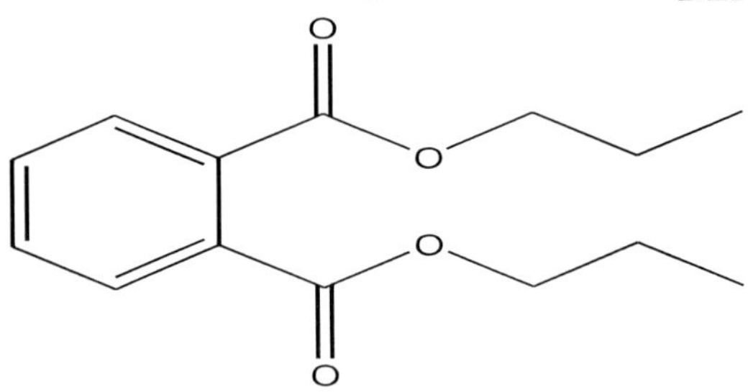

DPP

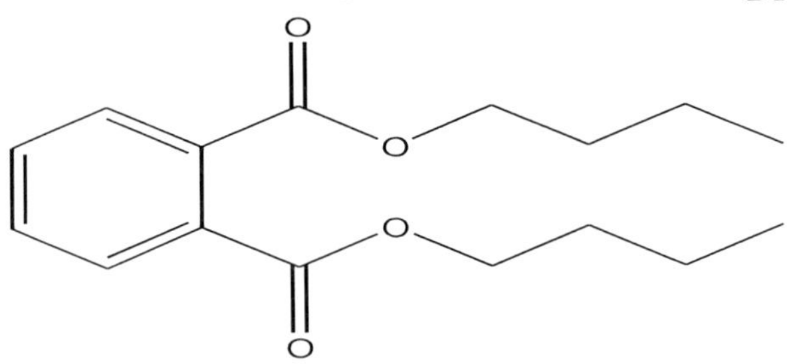

DBP

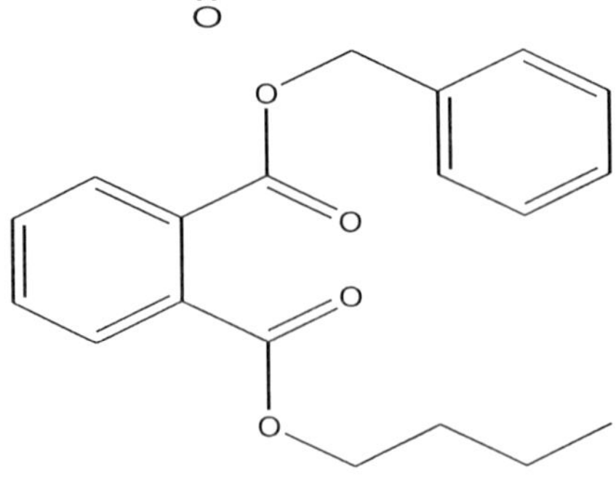

BBP

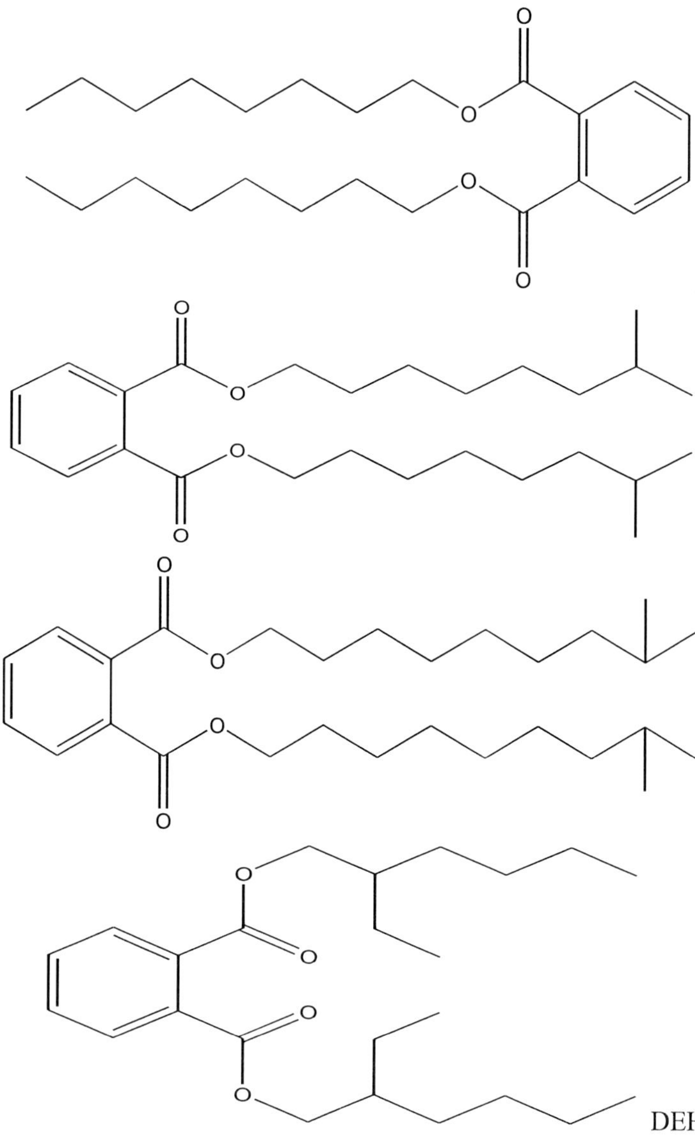

DOP

DINP

DIDP

DEHP

Figure 1. The chemical structures of several phthalates. DMP: dimethyl phthalate; DEP: diethyl phthalate; DPP: dipropyl phthalate; DBP: dibutylphthalate; BBP: benzyl butyl phthalate; DOP: Di-n-octyl phthalate; DINP: diisononyl phthalate; DIDP: diisodecyl phthalate; DEHP: di(2-ethylhexyl) phthalate.

Most of them are colorless liquids with a faint odor and insoluble in water. Phthalates are generally added to plastics to make them last longer and/or make other plastics more flexible. They are found in many products, such as child rearing products, household furnishings, building materials, footwear, rainwear, blood bags, dialysis tubing, lacquers, coatings of capsules, wood finishes and vinyl flooring. Phthalates are also used as solvents in paints, inks, adhesives, sealants, air fresheners, and scented products. When used as plasticizers, phthalates are not chemically bonded to the plastics polymer and can therefore eventually migrate from the plastics into the environment. Some phthalates have been shown to be hepatotoxic, teratogenic, and carcinogenic (Matsumoto et al., 2008). Especially, phthalates have been considered as endocrine-disrupting chemicals.

They could interfere with reproductive system and behavior of humans and wild life through disturbance of the endocrine system even at a very low concentration level (Li et al., 2011). Consequently, release of phthalates into the environment during the manufacturing, use and disposal is a severe problem, and exposure of phthalates to animals and human beings cause serious concerns. Human exposure to phthalates may occur via several routes, including oral, dermal, inhalation, and intravenous (Blount et al., 2000). Usually, people are exposed to phthalates by using products that contain the chemicals, breathing phthalate contaminated air, or living in close proximity to a manufacturing plant that uses phthalates. Women have a higher rate of exposure to phthalates because these compounds are found in cosmetic products, deodorants and fragrances that are used by women daily. Many studies that have been conducted in the last few years indicated that phthalate metabolites could be detected in virtually all humans tested (Fromme et al., 2007). Generally speaking, health effects are dependent on the amount an individual is exposed to, the type of phthalate and the age or timing of the exposure. In this chapter, the occurrences of phthalates in environment as well as the removal and degradation of phthalates from environment by various physical, chemical and biological techniques are reviewed, and special attention is paid to adsorption, advanced oxidation process and biodegradation of phthalates.

2. Occurrences of Phthalates in Environment

Phthalates have been liberated and detected in various environments in recent years. Afshari et al. (2004) have studied on the emission of phthalates from PVC and other materials. The results showed that polyolefine covered with wax resulted in an air concentration of 22 $\mu g/m^3$ of dibutylphthalate (DBP), which is two orders of magnitude larger than any other materials, but did not emit di(2-ethylhexyl) phthalate (DEHP). The other materials resulted in max concentration of approximately 1 $\mu g/m^3$ of DEHP and low emissions of DBP. The concentration of DEHP in each chamber slowly increased to a rather stable level after 150 days. DBP concentrations in the chambers with PVC skirting, PVC flooring, polyolefin and floor wax reached their quasi-static equilibrium after 60 days.

A two-room model is developed to estimate the emission rate of DEHP from vinyl flooring and the evolving gas-phase and adsorbed surface concentrations in a realistic indoor environment. The predicted indoor air DEHP concentration at steady state is 0.15 $\mu g/m^3$. What's more, the room developed in the project reaches steady state within about one year, while the adjacent room reaches steady state about three months later. It was found that ventilation rate has a strong influence on DEHP emission rate while total suspended particle concentration has a substantial impact on gas-phase concentration (Xu et al., 2009). Al-Saleh et al. (2004) have studied on phthalates residues in plastic bottled waters, and DMP, DEP, DBP, DEHP and benzyl butyl phthalate (BBP) were measured by headspace solid-phase micro-extraction followed by gas chromatography-mass spectrometer detection. Most of these phthalates were detected in the selected bottled water samples that might be either leached from the plastic packaging materials or contamination during bottling processes.

The levels of BBP were the highest at 4 °C storage (average 4.592 ± 3.081 μg/L with a range of 1.194 – 21.128 μg/L), and approximately 76% of the bottled waters had BBP above the limit of quantification of 0.994 μg/L. In addition, the presence of phthalates (DMP, DEP, DBP, BBP, DEHP) in bottled waters under poor storage conditions was investigated by liquid-liquid extraction with gas chromatography-mass spectrometry after exposure outdoors under realistic conditions for 15 – 30 days. It is suggested that the estimated exposure to phthalates via consumption of drinking water was very low (Diana et al., 2011). Despite the levels of phthalates in tested bottled waters were low, one should not neglect these chemicals that may cause endocrine disruption through several mechanisms, especially to potentially vulnerable populations such as infants and pregnant women.

The urban water system is believed to be an important sink for phthalates. The presence of eight phthalates was analyzed in urban storm water and sediment from three catchment areas in Sweden. High concentrations of DEHP, diisodecyl phthalate (DIDP) and diisononyl phthalate (DINP) (≤ 48, 66 and 200 μg/g dw, respectively) were found in the sediment. Aqueous concentrations of the pollutants varied considerably: branched NP was detected in concentrations up to 1.2 μg/L, whereas DEHP, DIDP, and DINP were the most frequently detected phthalates in concentrations up to 5.0, 17 and 85 μg/L, respectively. Calculation of catchment area emission factors from an urban highway environment revealed that as much as 2.1 kg of total phthalates may be emitted per hectare and year. The results indicated that all monitored phthalates are present in Swedish urban water systems and the long-chained phthalates are believed to occur at higher concentrations than other phthalates because of their higher environmental persistence and increasing use in Sweden (Bjorklund et al., 2009). In another study, DMP, DEP, DBP, BBP, DEHP and DOP were analyzed in raw and treated wastewater as well as in surface runoff samples from traffic roads. It was found that DEHP was the most abundant compound. The DEHP inflow concentrations ranged 3.4 – 34 μg/L and effluent concentrations 0.083 – 6.6 μg/L. Additionally, environmental quality standard (EQS) for DEHP in surface waters was exceeded only in a few effluent samples, whereas nearly all road runoff samples were higher than the EQS (Clara et al., 2010).

Moreover, a study was carried out to investigate the hazardous organic pollutants coming from decomposed municipal solid waste (MSW). Among these compounds detected, phenolics and phthalates are highly hazardous in nature and occurred in relatively higher concentrations. Concretely, phthalates were found in concentrations more than 200 mg/kg in MSW (Swati et al., 2008).

Among the phthalic acid esters targeted by the United States Environmental Protection Agency as priority pollutants, DEHP is the major pollutant identified at a high concentration level in lagooning sludge, at about 28.67 mg/kg, and in activated sludge, at about 6.26 mg/kg. Other phthalic acid esters, such as DBP and DMP, show very low concentrations (Amir et al., 2005). The levels of dissolved and particle-associated phthalates were measured in sea-surface micro-layer and sub-surface water samples collected in the Leghorn marine environment in September and October 1999. The results showed that the concentrations greatly varied among the sampling sites, and the highest level of phthalates was detected to be 177 μg/L (Cincinelli et al., 2001).

To sum up, for its wide-ranging applications in industry, phthalates have become highly prevalent in the environment and can now be found in food, water, and air. How to remove these pollutants effectively from our surroundings has become the hot research areas.

3. Removal and Degradation of Phthalates

Currently, different methods have been suggested for the removal of phthalate from the environment, such as adsorption, advanced oxidation process and biodegradation.

3.1. Adsorption

Adsorption, the binding of molecules or particles to a surface, has been used in some pollutants treatment and now it is increasingly used on large scale as an economical and efficient separation technique for removal of phthalates from wastewater. Huang et al. (2008) have studied the adsorption capacity of SBA-15 silica molecular sieves and its modified materials to phthalate esters (dimethyl and diethyl phthalate) by dynamic adsorption experiments with high performance liquid chromatography. Octyl (C-8) and octadecyl (C-18)-modified meso-porous were prepared in their study. It was found that the modified materials could increase the adsorption of phthalate esters compared to SBA-15 particles, and the adsorption capacity increased with the increased length of alkyl chain on SBA-15. The maximum adsorption capacity for diethyl phthalate was 3.9 (C-8-SBA-15) and 4.3 (C-18-SBA-15) times higher than that of SBA-15 particles, respectively.

In addition, effects of copper on the sorption of phthalate esters to yellow river sediment were investigated. The experimental results showed that the sorption isotherms of phthalates could be reasonably described by the Freundlich equation. Higher sorption equilibrium constant was obtained for di-n-butyl phthalate due to its greater hydrophobicity. The existence of copper would enhance the sorption of DBP. Moreover, strong sorption of copper to sediment was found and attributed to abundant carbonates in the Yellow River sediment.

After carbonates were removed, notable effects of copper on the phthalate sorption were observed due to the decrease of copper sorption and the increase of aqueous copper concentration. With 153 mg/L copper added, the partition coefficient decreases by 52% for DEP and increases by 79% for DBP. Primary factors that may influence interaction between the sorption of copper and DEP and DBP were also investigated, such as complexation between copper and phthalate, and phthalate hydrophobicity. The complexation between phthalate esters and copper was substantiated by polarogram and fluorescence spectrum, and the calculated molar ratio of complexation (copper: phthalate) was found to be 2:1 (Xu et al., 2007).

In the literature, other techniques have been also tested. For example, some studies have found that nanofiltration (NF) is a promising membrane technique with a large number of applications in the field of process water, drinking water or wastewater treatment. NF membranes can separate small organic micro-pollutants and multivalent ions from liquid streams. The rejection property of NF membranes for phthalates was examined, and it was found that higher desalting membranes can effectively reject these hazardous organic pollutants. Because the efficiency of coagulation and adsorption on a granulated activated carbon process was not good enough to remove the organic pollutants completely, NF and coagulation-NF integrated systems were proposed as a second stage of the water treatment. The application of NF in the second step of water treatment made it possible to remove the pollutants to a high degree: (1) > 63%, NF and (2) > 81.5% coagulation-NF (Bodzek et al., 2006).

3.2. Advanced Oxidation Processes

Advanced oxidation processes, which generate hydroxyl free radicals in water, have been also studied in various pollutants treatment widely. The commonly used oxidation processes in both theory and practice for the removal and degradation of organic pollutant compounds include UV-H_2O_2 process, Fenton process, ozonation, photo-degradation, electro-coagulation, and sono-chemical degradation.

3.2.1. UV- H_2O_2 Process

The removal efficiency of DMP from water using the UV-H_2O_2 photo-oxidation process was studied. Pure water samples, spiked with 20 ppm DMP were treated by the combination of UV photolysis as well as hydrogen peroxide oxidation.

In the experiments, the concentration of hydrogen peroxide was varied from 34 to 136 ppm and a low-pressure mercury UV lamp of 100 mW power output was used to provide the necessary radiation. The effects of initial concentration of H_2O_2, UV exposure time, pH, and temperature were investigated. The results showed that about 60% of DMP was removed directly by activation caused by UV light radiation after an exposure time of 1 h. Furthermore, the removal efficiency increased when the DMP-spiked water was dosed with H_2O_2 prior to irradiating with UV light. More than 98% of DMP was removed after 45 min when the UV-irradiated solution was dosed with 136 ppm of H_2O_2. The results also showed that lowering the pH and increasing the temperature enhanced the removal of DMP by UV-H_2O_2 process (Tawabini et al., 2004).

3.2.2. Fenton Degradation

The Fenton reaction involving H_2O_2 and Fe (II) salts in the acidic aqueous solution was widely used to treat polluted water. During past several years, some studies were carried out to improve Fenton degradation. The application of photo-Fenton treatment was investigated in order to obtain the maximum organic matter removal and solve the by-products issue as much as possible. When a solar chamber as light source was used (light intensity = 500 W/m^2), the best operational conditions were the following: $[H_2O_2] = 3703$ mg/L, $[Fe^{2+}] = 92.1$ mg/L, initial pH 2.7 – 3.0, reaction time = 100 min. The results showed that main initial organic substances were eliminated after the reaction and no significant by-products were identified (Sarasa et al., 2006).

Moreover, Swiderska-Dabrowska et al. (2008) have studied the use of iron-rich sludge for oxidation of phthalates in Fenton process. DBP was removed from its aqueous emulsions (200 mg/L) by oxidation with H_2O_2 at pH 3 for 2 h in presence of $FeSO_4$ or Fe-rich sludge from water treatment plants as catalysts with following coagulation with NaOH (up to pH 8) and sedimentation for 1 h. The conversion of DBP to Bu and Et acetates, BuOH, PrCOOH, PrEt and Me acetate was evidenced by chromatogram.

A substantial decrease of organic matter content in the aqueous phase (90%) was achieved when the Fe-rich sludge was used, but the adsorption of DBP on sludge particles was not excluded. During the Fenton reaction, pollutants are decomposed directly by hydroxyl free radicals. But in some cases, toxic by-products could be produced. DMP, DEP, and DPP by-products formed during the Fenton reaction were studied. Fenton degradation of selected phthalates yielded numerous transformation products, such as hydroxylated phthalates. The hydroxylation reaction occurred at the aromatic ring of phthalates and yielded mono- and dihydroxylated phthalates. For monohydroxylated phthalate, 3-hydroxy- and 4-hydroxy-dialkylphthalates are the main transformation products. In addition to hydroxylated derivatives, aliphatic chain degraded mono- and dihydroxylated phthalates were also detected (Tay et al., 2011).

3.2.3. Ozonation

The rate of phthalates degradation increased exponentially with increasing temperature in the range studied (20 to 50°C) and in proportion with applied ozone dosages. It was found to increase with increasing pH of the solution from 2.6 to 11.0, and decreased exponentially with increasing initial concentrations of the phthalates. Degradation of phthalates via ozonation was found to develop dominantly through direct reaction with hydroxyl radicals (Tay et al., 2010). However, some studies found that ozone-based advanced oxidation processes might be not effective in degrading phthalates for its relative stable to ozone oxidation (Ning et al., 2007).

Therefore, researchers attempted to add some activator to promote the reaction. de Oliveira et al. (2011) have studied DEP removal by a water treatment process based on the coupling of ozone (O_3) and activated carbon (AC). The influence of AC properties on the process efficiency and on the coupling mechanism was analyzed. Results showed that degradation efficiency depends both on textural properties (microporous and external surfaces favor this treatment) and chemical functions (both acid and basic functions favor radical hydroxyl generation). Moreover, it was demonstrated that AC acts more as a radical initiator and promoter and a reaction support than as an adsorbent material. The influence of pH on the reaction efficiency and mechanism was also proved: in acidic conditions ($pH < 5$) radical reactions are due to O_3/AC interactions, and they are due to indirect ozonation in the bulk liquid for higher pH.

3.2.4. Photodegradation

Photocatalytic oxidation (PCO) relies on hydroxyl free radical, which is a strong oxidizing agent, for the oxidative degradation of pollutants. In previous studies, DEHP was effectively adsorbed from aqueous solution by biosorption onto chitinous materials. However, biosorption can only pre-concentrated DEHP from the aqueous phase for further treatment, yet cannot degrade it. Therefore, the degradation (and detoxification) of DEHP adsorbed onto chitinous material by photocatalytic oxidation is attempted. Adsorption was carried out in the physicochemical conditions optimized in the previous study, with 500 mg/L chitin A and 40 mg/L DEHP at initial pH 2, 22 ± 2 °C and 150 rpm agitation for 5 min. After optimization of PCO, a 61% removal efficiency of 10 mg/L of DEHP was achieved within 45 min under 0.65 mW/cm^2 of UV-A with 100 mg/L TiO_2, and 10 mM of H_2O_2 at initial pH 12. The results showed that UV-A and TiO_2 are essential for the degradation of DEHP by PCO. The degradation intermediates/products were identified by GC-MS analysis. GC-MS results showed that the di(2-ethylhexyl) side chain was first degraded, producing phthalates with shorter side chains. Further reaction produced phathalic anhydride and aliphatic compounds such as alkanol and ester. The toxicities of parental and degradation intermediates in the solution phase and on chitinous materials were followed by the Microtox (R) test. Results indicated that the toxicity can be removed after 4 h treatment by PCO. Thus the decontamination of DEHP by integrating biosorption and PCO is feasible (Chan et al., 2007).

3.2.5. Electro Coagulation

Treatment of phthalates by electro-coagulation employing stainless steel electrodes was investigated using DMP as a model compound. DMP was completely destructed within 30 min up to the high initial concentration of 100 mg/L while total mineralization was also obtained within a couple of hours. The applied current density of 22.5 mA/cm^2 and electrolyte (NaCl) concentrations varying between 1000 and 1500 mg/L as chloride resulted in the highest treatment performance. Desorption experiments were carried out, and the reaction rates were obtained for DMP. Activated sludge inhibition experiments revealed that toxicity could be significantly reduced by electro-coagulation application (Kabdasli et al., 2009).

3.2.6. Sono-Chemical Degradation

The sono-chemical degradation of aqueous solutions containing low concentrations of six phthalate esters at an ultrasonic frequency of 80 kHz has been investigated. Experiments were carried out at an overall initial phthalate concentration of 240 μg/L, values of electric power of 75 and 150 W, temperatures of 21 and 50 °C and in the presence of NaCl to study the effect of various operating conditions on degradation. Ultrasonic treatment was found capable of removing the four higher molecular mass phthalates (DBP, BBP, DEHP and DOP) within 30-60 min of irradiation. The rest (DMP and DEP) were more recalcitrant and nearly complete removal could be achieved only after prolonged irradiation time. The relative reactivity of phthalates was explained in terms of their hydrophobicity. In addition, solid-phase micro-extraction coupled with GC-MS proved to be a powerful analytical tool to monitor the sono-chemical degradation of phthalate esters at low μg/L concentration levels, minimizing the risk of secondary contamination during sample preparation, a major parameter to consider during phthalates analysis (Psillakis et al., 2004).

3.3. Biodegradation

Numerous studies indicated that phthalates can be degraded by bacteria and fungi under aerobic, anoxic, and anaerobic conditions (Liang et al., 2008). The major species have been isolated and identified for biodegradation of phthalates, such as *Sphingomonas sp., Pseudomonas sp.,* and *Actinomycetes sp.* (Kapanen et al., 2007). In this section, major species of phthalate degrading bacteria, biodegradation of phthalates under various conditions, purification and properties of degrading enzymes, and pathways of the degradation are summarized and discussed.

3.3.1. Major Species of Phthalate-Degrading Bacteria

Thirty-two strains of phthalic acid ester (PAEs)-degrading bacteria were isolated from thirteen geographically diverse sites by enrichment using mixtures of PAEs as the sole source of carbon and energy. Sequence analyses of the 16S rRNA gene indicated that these isolates were from six genera (*Arthrobacter, Gordonia, Rhodococcus, Acinetobacter, Pseudomonas,* and *Delftia*).

Furthermore, HPLC analysis showed that the degradation rate of dimethyl phthalate varied from 48.32 to 100% (Liang et al., 2010). Thus, among PAEs-degrading bacteria, there is a high level of genetic diversity and great potential to clean up phthalates-contaminated environments. In another study, *Bacillus subtilis* strain 3C3 was isolated as an organic-solvent-tolerant bacterium and capable of utilizing diethyl phthalate as a sole carbon source. The biodegradability was significantly enhanced with the supplementation of yeast extract as a co-metabolic substrate (Navacharoen et al., 2011). The results suggested that *B. subtilis* strain 3C3 has potential application as a bioaugmented bacterial culture for bioremediation of phthalates. In addition, Liang et al. (2008) reported that a Gram negative bacterium JDC-16 was isolated from river sludge, which can grow well on the substrate of phthalic acid esters as the sole source of carbon and energy. Based on the morphology, physiological and biochemical properties and analysis of 16S rRNA gene sequence, it was preliminarily identified belonging to the genus *Acinetobacter*. The optimal pH and temperature for DEP degradation by *Acinetobacter sp* JDC-16 is 8.0 and 35 °C, respectively.

Moreover, *Rhodococcus sp.* JDC-11, capable of utilizing di-n-butyl phthalate (DBP) as the sole source of carbon and energy, was isolated from sewage sludge and confirmed mainly based on 16S rRNA gene sequence analysis by Jin et al. (2010). The optimum pH, temperature, and agitation rate for DBP degradation by *Rhodococcus sp.* JDC-11 were 8.0, 30 °C, and 175 rpm, respectively. In addition, high concentrations of glucose increased its degradation. Meanwhile, a substrate utilization test showed that JDC-11 was also able to utilize other phthalates. Through a sequence analysis, they suggested that the phthalate dioxygenase gene of JDC-11 was highly homologous to the large subunit of the phthalate dioxygenase from *Rhodococcus coprophilus* strain G9.

A bacterium *Micrococcus luteus* was isolated from soil contaminated with DEHP at a poly(vinyl chloride) manufacturing site. The microorganism degraded DEHP through the formation of a mono-2-ethylhexyl phthalate (MEHP), which was subsequently metabolized to phthalic acid (PA), as detected by GC-MS analysis. *Micrococcus luteus* was able to degrade almost 90% of the initial DEHP within 12 d. The results suggested that a new strain of *Micrococcus luteus* with a strong ability to degrade DEHP into nontoxic metabolites might contribute to decontamination of environmental phthalates and consequently risk reduction of human exposure to DEHP (Baek et al., 2009).

In another study, degradation of BBP by a co-culture of *Gordonia sp.* strain MTCC 4818 and *Arthrobacter sp.* strain WY was investigated (Chatterjee et al., 2008). In this study, the limitations of the individual species in metabolizing BBP were overcome, leading to the development of a consortium capable of complete utilization of this ester. In the degradation of BBP by the co-culture, the presence of multiple esterases was demonstrated in both species by activity staining of non-denaturing gels, indicating their roles in the degradation process. In addition, 10 white rot fungi capable of degrading three phthalates (DEP, DMP and BBP) was isolated (Hwang et al., 2008). When the fungal mycelia were added together with 100 mg/L of phthalate into yeast extract-malt extract-glucose (YMG) medium, *Pleurotus ostreatus, Irpex lacteus, Polyporus brumalis, Merulius tremellosus, Trametes versicolor,* and *T. versicolor MrP1* and *MrP13* (transformant of the Mn-repressed peroxidase gene of *T. versicolor*) could remove almost all of the 3 kinds of phthalates within 12 days of incubation.

In both culture conditions, *P. ostreatus* showed the highest degradation rates for the 3 phthalates tested. BBP was degraded with the highest rates among the 3 phthalates by all fungal strains. The estrogenic activity of 100 mg/L DMP decreased during biodegradation by *P. Ostreatus*. Furthermore, a white-rot basidiomycete *Phlebia tremellosa*, which showed good growth in media containing various hormone-mimicking compounds, degraded benzylbutylphthalate and diethylphthalate by up to 30% and 80%, respectively. When this fungus was grown under suitable conditions for degrading phthalic esters, the laccase activity and its transcript level were both highly increased (Yeo et al., 2008).

A bacterial isolate, *Pseudomonas* sp. P1, was found to degrade phthalic acid in agar plate assay as an evidence by the formation of clear zone around the colony. The strain was tested for the growth and tolerance limit on different concentrations (10 – 3000 ppm) of phthalic acid in mineral salt medium with and without glucose. In oil mineral salt agar plates, containing phthalic acid as a sole source of carbon, rich growth along with the hydrolyzing zone was observed up to the maximum concentration of 2800 ppm without glucose and up to the maximum concentration of 2900 ppm of phthalic acid with glucose.

Transformational studies were carried out in mineral salt liquid medium containing varying concentrations (100 – 500 ppm) of phthalic acid. The highest degradation percentage of phthalic acid was found at 37 ^{o}C and pH 8. In the mineral salt medium without glucose, phthalic acid was degraded to 72% at 500 ppm after 48 h of incubation (Murad et al., 2007). Another study showed that *Pseudomonas* sp. strain PP4 could utilize phthalate isomers as carbon source (Phale et al., 2005). Additionally, two bacterial strains *Arthrobacter* sp. and *Sphingomonas paucimobilis* were obtained from soil by enrichment cultures using DMP or monomethylphthalate (MMP) as sole carbon source, respectively (Vega et al., 2003).

3.3.2. Biodegradation of Phthalates under Various Conditions

In natural environments, complete degradation of phthalates is always carried out syntrophically by several members of microorganisms (Gu et al., 2005). Phthalates degradation under various environments, such as river, soil and sediment, has been extensively studied. Hashizume et al. (2002) reported that almost 100% of diethyl phthalate (DEP), di-isobutyl phthalate and DBP, and approximately 70% of DEHP in Tempaku River could be degraded. According to the results, all eight isolates from Tempaku River water (R1-R7, D1) did not degrade DMP, but showed biodegrading ability for the other PAEs. The DBP-degrading ability was particularly high for the isolates R1-R3 and D1 of *Acinetobacter lwoffii*.

In another study, soil was treated with DEP or DEHP at 0.1 to 100 mg/g to assess the impact of phthalates on soil microorganisms. Diethyl phthalate was biodegraded rapidly in soil, and was not persistent in the environment. The DEHP, although biodegradable in aqueous solution, was recalcitrant in soil, because of poor bioavailability (only 10% degraded by 70 d at 20 ^{0}C) and was predicted to account for the majority of phthalate contamination in the environment.

Addition of DEP or DEHP to soil at a concentration similar to that detected in non-industrial environments (0.1 mg/g) had no impact on the structural diversity (bacterial numbers, fatty acid methyl ester analysis) or functional diversity of the microbial community (Cartwright et al., 2000). In addition, microbial degradation rates were measured for eight phthalate esters (DEP, DPP, DBP, DPhP, BBP, DHP, DCP, and DEHP) found in 14 surface water and six sediment samples taken from rivers in Taiwan. Under aerobic conditions, average degradation half-lives for DEP, DPP, DBP, DPhP, BBP, DHP, DCP and DEHP were 2.5, 2.8, 2.9, 2.6, 3.1, 9.7, 11.1 and 14.8 days, respectively. Under anaerobic conditions, respective average half-lives were 33.6, 25.7, 14.4, 14.6, 19.3, 24.1, 26.4 and 34.7 days. Aerobic degradation rates were up to 10 times faster than anaerobic degradation rates (Yuan et al., 2002).

The microbial degradation by acclimated activated sludge of three phthalates DMP, DBP and di-n-octyl phthalate (DOP) was investigated (Wang et al., 1996). The results demonstrated that DMP and DBP with the shorter alkyl-side chain phthalic acid esters could be rapidly degraded nearly 90% within 3 d for DMP and 8 d for DBP. However, the degradation rate of DOP was slow. The degradation of PAEs can be described by the first-order reaction model. In another study, a planar bioreactor, equipped with a polypropylene membrane on which a lipase was immobilized, was employed in a bioremediation process involving water polluted by DMP (Mita et al., 2010). In addition, Oliver et al. (2007) reported about the laboratory trickle filter microcosms (continuous and re-circulating flow). It was found that adsorption was critical for the removal of both DEP (77.5%) and DEHP (55.7%) in continuous flow microcosms.

The proportion of phthalates removed by biodegradation in the continuous flow microcosms was estimated. Re-circulating flow microcosms improved the removal of DEHP compared to continuous flow microcosms. Microcosm biofilm used for an enrichment culture on phthalate media isolated a varied group of microbes including Grain negative and Gram positive bacteria, yeasts and fungi. Moreover, Marttinen et al. (2004) reported that the sequencing batch reactor could efficiently remove 95% of DEHP, together with NH_4-N, from water originating from sewage sludge. On the other hand, the type of inoculum used may play a very important role for this quick degradation, since the reactor was inoculated with an acclimated mixed culture composed of *Brevibacterium iodinum*, *Rhodococcus luteus*, and *Bacillus brevis*, which can utilize DEHP as a sole source (Juneson et al., 2001). Anaerobic reactors were successful in the degradation of DMP (Liang et al., 2007) and phthalate isomers, such as PA (Tur and Huang, 1997) and TA in wastewater (Kleerebezem et al., 1997; Wu et al., 2001; Kleerebezem et al., 2005). Liang et al. (2007) reported that more than 99% of DMP (600 mg/L) and 93% of COD were effectively removed in a UASB reactor at an 8-h hydraulic retention time.

3.3.3. Purification and Properties of Degrading Enzymes

An esterase that specifically hydrolyzes medium-chain (C_3-C_5) monoalkyl phthalates was purified from phthalate-grown *Micrococcus* sp. YGJ1. The purified enzymes showed similar properties with respect to molecular mass (60 kDa), subunit molecular mass (27 kDa), N-terminal amino acid sequence, optimal pH (about 7.5), temperature-dependence, substrate specificity, and inhibitor susceptibility. Diethylpyrocarbonate, p-chloromercuribenzoate, Hg^{2+}, and Cu^{2+} strongly inhibited the enzymes, while phenylmethylsulfonyl fluoride produced weak inhibition, and various metal chelating reagents were ineffective (Maruyama et al., 2005). Another esterase was obtained from *Ochrobactrum anthoropi* 6-2b. It was purified 19.1 fold to electrophoretic homogeneity with a yield of 26%.

The purified esterase had a specific activity of 32.2 μmol/min/mg protein based on the hydrolysis of DET at pH 7.0 and 30 °C. The enzyme showed a maximum activity at pH 8.0

and was stable at pH 6.0 – 8.0 with optimum stability at pH 6.0. Its optimal reaction temperature was 50 °C, and it was stable below 40°C. It was able to hydrolyze a number of p-phthalates and o-phthalates (Xu et al., 2006).

3.3.4. Pathways of the Degradation

Generally, the phthalates biodegradation pathways consist of two processes: primary biodegradation from PDEs to PMEs and then to PA and ultimate biodegradation from PA to CO_2 and/or CH_4 (Staples et al., 1997). Primary degradation consists of different type of pathways, including de-esterification or dealkylation, β-oxidation and trans-esterification (Liang et al., 2008). PA is degraded by two dioxygenase-catalyzed pathways, forming the common intermediate protocatechuate (Eaton et al., 1982; Nomura et al. 1992).

Moreover, the recent researches give us some other clues about the biodegradation pathways. In the process of biodegradation of diester plasticizers mediated by *Rhodococcus Rhodochrous*, it is demonstrated that the first and rate-limiting step in their co-metabolism by *Rhodococcus rhodochrous* is the hydrolysis of one ester bond. This mechanism is different from that proposed earlier for direct biodegradation of di-2-ethylhexyl phthalate (Sauvageau et al., 2009). In the process of utilizing phthalate isomers by *Pseudomonas* sp. strain PP4 and C5, the metabolic studies suggested that phthalate isomer degrading pathways converge at 3, 4-dihydroxybenzoic acid. Phthalate dioxygenases responsible for the degradation of the respective phthalate isomers were induced specifically, which suggested that probably there are three different phthalate dioxygenases.

Glucose-grown cells lost the phthalate degradation property indicating probable involvement of plasmid, which is expressed and maintained selectively in the presence of phthalate isomers (Phale et al., 2005). In addition, Hashizume et al. (2002) reported that the alkyl chains in the diesters are also decomposed in addition to monoester formation from DBP or DEHP at the first step. For the degradation of plasticizers by the yeast *Rhodotorula rubra* J-96-1, the metabolite observed when D(EG)DB was being degraded was tentatively identified as diethylene glycol monobenzoate by analogy to the D(PG)DB system. In contrast, no metabolites were observable and toxicity did not increase in the media during the degradation of B(EH)A, DOP, or DOTP by *R. rubra* (Gartshore et al., 2003).

CONCLUSION

Phthalates possess hepatotoxic, teratogenic, and carcinogenic characteristics, and are detected in various environments. They could be degraded and removed from the environment by adsorption, advanced oxidation process and biodegradation. Octyl (C-8) and octadecyl (C-18)-modified meso-porous could be used for adsorption of phthalates. The commonly used advanced oxidation processes for degradation of phthalates include UV-H_2O_2 process, Fenton process, ozonation, photo-degradation, electro-coagulation, and sono-chemical technique. Furthermore, the biodegradation is a promising method for the treatment of phthalates, and the major species include *Sphingomonas* sp., *Pseudomonas* sp., and *Actinomycetes* sp. In the future, more microorganisms that could effectively degrade phthalates should be isolated and identified. In addition, adsorption, advanced oxidation process and biodegradation could be integrated a system for treatment of wastewater containing high concentration of phthalates.

REFERENCES

Afshari, A., Gunnarsen, L., Clausen, P. A. and Hansen, V. (2004). Emission of phthalates from PVC and other materials. *Indoor Air*, 14, 120-128.

Al-Saleh, I., Shinwari, N. and Alsabbaheen, A. (2011). Phthalates residues in plastic bottled waters. *J. Toxicol. Sci.*, 36, 469-478.

Amir, S., Hafidi, M., Merlina, G., Hamdi, H., Jouraiphy, A., El Gharous, M. and Revel, J. C. (2005). Fate of phthalic acid esters during composting of both lagooning and activated sludges. *Proc. Biochem.,* 40, 2183-2190.

Baek, J. H., Gu, M. B., Sang, B. I., Kwack, S. J., Kim, K. B. and Lee, B. M. (2009). Risk Reduction of Adverse Effects Due to Di-(2-Ethylhexyl) Phthalate (DEHP) by Utilizing Microbial Degradation. *J. Toxicol. Environ. Health-Part A-Current Issues*, 72, 1388-1394.

Bjorklund, K., Cousins, A. P., Stromvall, A. M. and Malmqvist, P. A. (2009). Phthalates and nonylphenols in urban runoff: Occurrence, distribution and area emission factors. *Sci. Total Environ.,* 407, 4665-4672.

Blount, B. D., Silva., M. J., Caudill, S. P., Needham., L. L., Pirkle, J. L., Sampson., E. J., Lucier, G. W., Jackson, R. J. and Brock, J. W. (2000). Levels of seven urinary phthalate metabolites in a human reference population. *Environ. Health Persp.*, 108, 10-15.

Bodzek, M. and Dudziak, M. (2006). Elimination of steroidal sex hormones by conventional water treatment and membrane processes. *Desalination*, 198, 24-32.

Cartwright, C. D., Thompson, I. P., Burns, R. G., (2000). Degradation and impact of phthalate plasticizers on soil microbial communities. *Environ. Toxicol. Chem.*, 19, 1253-1261.

Chatterjee, S. and Dutta, T. K. (2008). Metabolic cooperation of *Gordonia* sp. strain MTCC 4818 and *Arthrobacter* sp. strain WY in the utilization of butyl benzyl phthalate: effect of a novel co-culture in the degradation of a mixture of phthalates. *Microbiology-SGM*, 154, 3338-3346.

Clara, M., Windhofer, G., Hartl, W., Braun, K., Simon, M., Gans, O., Scheffknecht, C. and Chovanec, A. (2010). Occurrence of phthalates in surface runoff, untreated and treated wastewater and fate during wastewater treatment. *Chemosphere.,* 78, 1078-1084.

Cincinelli, A., Stortini, A. M., Perugini, M., Checchini, L. and Lepri, L. (2001). Organic pollutants in sea-surface microlayer and aerosol in the coastal environment of Leghorn - (Tyrrhenian Sea). *Mar. Chem.,* 76, 77-98.

de Oliveira, T. F., Chedeville, O., Fauduet, H. and Cagnon, B. (2011). Use of ozone/activated carbon coupling to remove diethyl phthalate from water: Influence of activated carbon textural and chemical properties. *Desalination*, 276, 359-365.

Diana, A. and Dimitra, V. (2011). Alkylphenols and phthalates in bottled waters. *J. Hazard. Mater.,* 185, 281-286.

Eaton, R. W. and Ribbons, D. W. (1982). Metabolism of dibutylphthalate and phthalate by *Micrococcus* sp. strain 12b. *J. Bacteriol.,* 151, 48-57.

Fromme, H., Bolte, G., Koch, H. M., Angerer, J., Boehmer, S., Drexler, Mayer, H. and Liebl, R. B. (2007). Occurrence and daily variation of phthalate metabolites in the urine of an adult population. *Int. J. Hyg. Environ. Health,* 210, 21-33.

Gu. J. D., Li, J. and Wang, Y. (2005). Biochemical pathway and degradation ofphthalate ester isomers by bacteria. *Water Sci. Technol.*, *52*, 241-248.

Gartshore, J., Cooper, D, G. and Nicell, J. A. (2003). Biodegradation of plasticizers by Rhodotorula rubra. *Environ. Toxicol. Chem.*, 22, 1241 1251.

Hashizume, K., Nanya, J., Toda, C., Yasui, T., Nagano, H. and Kojima, N. (2002). Phthalate esters detected in various water samples and biodegradation of the phthalates by microbes isolated from river water. *Biol. Pharmac. Bull.,* 25, 209-214.

Huang, H. Y., Yang, C. L., Zhang, H. X. and Liu, M. C. (2008). Preparation and characterization of octyl and octadecyl-modified mesoporous SBA-15 silica molecular sieves for adsorption of dimethyl phthalate and diethyl phthalate. *Micropor. Mesopor. Mater.*, 111, 254-259.

Jin, D. C., Liang, R. X., Dai, Q. Y., Zhang, R. Y. and Wu, X. L. (2010). Biodegradation of di-n-butyl phthalate by *Rhodococcus* sp. JDC-11 and molecular detection of 3, 4-phthalate dioxygenase gene. *J. Microbiol. Biotechnol.*, 20, 1440-1445.

Juneson, C., Ward, O. P. and Singh, A. (2001). Biodegradation of bis-(2-ethylhexyl) phthalate in a soil slurry-sequencing batch reactor. *Proc. Biochem.*, 37, 305-313.

Kabdasli, I., Keles, A., Olmez-Hanci, T., Tunay, O. and Arslan-Alaton, I. (2009). Treatment of phthalic acid esters by electrocoagulation with stainless steel electrodes using dimethyl phthalate as a model compound. *J. Hazard. Mater.*, 171, 932-940.

Kapanen, A., Stephen, J. R., Bruggemann, J., Kiviranta, A., White, D. C. and Itavaara, M. (2007). Diethylphthalatein compost: Ecotoxicological effects and response of the microbial community. *Chemosphere*, 67, 2201-2209.

Kleerebezem. R., Mortier, J., Pol, L. W. H. and Lettinga, G. (1997). Anaerobic pre-treatment of petrochemical effluents: Terephthalic acid wastewater. *Water Sci. Technol.*, 36, 237-248.

Kleerebezem, R., Beckers. J., Pol, L. W. H. and Lettinga, G. (2005). High rate treatment of terephthalic acid production wastewater in a two stage anaerobic bioreactor. *Biotechnol. Bioeng.*, 91, 169-179.

Liang, D. W., Zhang, T., Fang, H. H. P. and He, J. Z. (2008). Phthalates biodegradation in the environment. *Appl. Microbiol. Biotechnol.*, 80, 183-198.

Liang, R. X., Wu, X. L., Dai, Q. Y., Jin, D. C. and Wang, Y. Y. (2010). Genetic diversity of phthalic acid esters-degrading bacteria isolated from different geographical regions of China. *Int. J. Gen. Molec. Microbiol.*, 97, 79-89.

Liang, D. W., Zhang, T. and Fang, H. H. P. (2007). Anaerobic degradation of dimethyl phthalate in wastewater in a UASB reactor. *Water Res.*, 41, 2879-2884.

Mita, L., Sica, V., Guida, M., Nicolucci, C., Grimaldi, T., Caputo, L., Bianco, M., Rossi, S., Bencivenga, U., Eldin, M. S. M., Tufano, M. A., Mita, D. G., Diano, N. (2010). Employment of immobilised lipase from Candida rugosa for the bioremediation of waters polluted by dimethylphthalate, as a model of endocrine disruptors. *J. Mole. Catal. B-Enzy.,* 62, 133-141.

Maruyama, K., Akita, K., Naitou, C., Yoshida, M., Kitamura, T., (2005). Purification and characterization of an esterase hydrolyzing monoalkyl phthalates from *Micrococcus* sp. YGJ1. *J. Biochem.,* 137, 27-32.

Marttinen S. K., Ruissalo, M. and Rintala, J. A. (2004). Removal of bis-(2-ethylhexyl) phthalate from reject water in a nitrogen-removing sequencing batch reactor. *J. Environ. Manag.*, 73, 103-109.

Matsumoto, M., Hirata-Koizumi M. and Ema, M. (2008). Potential adverse effects of phthalic acid esters on human health: a review of recent studies on reproduction. *Regul. Toxicol Pharm.,* 50, 37-49.

Murad, S., Hasan, F., Shah, A. A., Hameed, A. and Ahmed, S. (2007). Isolation of phthalic acid degrading *Pseudomonas* sp. P1 from soil. *Pakistan J. Botany*, 39, 1833-1841.

Ning, B., Graham, N., Zhang, Y. P., Nakonechny, M. and El-Din, M. G. (2007). Degradation of endocrine disrupting chemicals by ozone/AOPs. *Ozone-Sci. Eng.,* 29, 153-176.

Nomura, Y., Nakagawa, M., Ogawa, N., Harashima, S. and Oshima, Y. (1992). Genes in pht plasmid encoding the initial degradation pathway of phthalate in *Pseudomonas putida. J. Ferm. Bioeng*., 74, 333-344.

Oliver, R., May, E. and Williams, J. (2007). Microcosm investigations of phthalate behaviour in sewage treatment biofilms. *Sci. Total Environ.,* 372, 605-614.

Phale, P. S. (2005). Biodegradation of carbaryl and phthalate isomers by soil microorganism. *Environmental Health Risk III Book Series*: *WIT Transactions on Biomed. Health*., 9, 427-436.

Psillakis, E., Mantzavinos, D. and Kalogerakis, N. (2004). Monitoring the sonochemical degradation of phthalate esters in water using solid-phase microextraction. *Chemosphere*, 54, 849-857.

Sarasa, J., Llabres, T., Ormad, P., Mosteo, R. and Ovelleiro, J. L. (2006). Characterization and photo-Fenton treatment of used tires leachate. *J. Hazard. Mater.,* 136, 874-881.

Sauvageau, D., Cooper, D. G., Nicell, J. A. (2009). Relative rates and mechanisms of biodegradation of diester plasticizers mediated by *Rhodococcus rhodochrous*. *Canadian J. Chem. Eng*., 87, 499-506.

Staples, C. A., Peterson, D. R., Parkerton, T. F. and Adams, W. J. (1997). The environmental fate of phthalate esters: a literature review. *Chemosphere*, 35, 667-749.

Swati, M., Rema, T. and Joseph, K. (2008). Hazardous organic compounds in urban municipal solid waste from a developing country. *J. Hazard. Mater.,* 160, 213-219.

Swiderska-Dabrowska, R., Piaskowski, K. and Schmidt, R. (2008). Use of iron-rich sludges for oxidation of phthalates in Fenton process. *Przemysl Chemiczny*, 87, 587-589.

Tawabini, B. S. and Al-Suwaiyan, M. S. (2004). Removal of dimethyl phthalate from water by UV-H_2O_2 process. *J. Environ. Engin. Sci*., 3, 289-294.

Tay, K. S., Rahman, N. A. and Bin Abas, M. R. (2010). Ozonation of a mixture of dialkylphthalates in aqueous solution. *J. Chem. Technol. Biotechnol.,* 85, 726-729.

Tay, K. S., Rahman, N. A. and Bin Abas, M. R. (2011). Fenton degradation of dialkylphthalates: products and mechanism. *Environ. Chem. Lett.*, 9, 539-546.

Vega, D. and Bastide, J. (2003). Dimethylphthalate hydrolysis by specific microbial esterase. *Chemosphere*, 51, 663-668.

Wang, J. L., Ping, L. and Yi, Q. (1996). Biodegradation of phthalic acid esters by acclimated activated sludge. *Environ. Int*., 22, 737-741.

Wu, J. H., Liu, W. T., Tseng, I. C. and Cheng, S. S. (2001). Characterization of microbial consortia in a terephthalate-degrading anaerobic granular sludge system. *Microbiol*., 147, 373-382.

Xu, N., Ni, J. R. and Sun, W. L. (2007). Effects of copper on the sorption of phthalate esters to yellow river sediment. *Water Air Soil Pollut.,* 184, 207-216.

Xu, S. C., Komatsu, C., Takahashi, I. and Suye, S. (2006). Purification and properties of diethyl p-phthalate esterase from *Ochrobactrum anthoropi* 6-2b. *Sen-1 Gakkanshi*, 62, 226-231.

Xu, Y., Hubal, E. A. C., Clausen, P. A. and Little, J. C. (2009). Predicting residential exposure to phthalate plasticizer emitted from vinyl flooring: A mechanistic analysis. *Environ. Sci. Technol.,* 43, 2374-2380.

Yeo, S., Kim, M. K. and Choi, H. T. (2008). Increased expression of laccase by the addition of phthalates in *Phlebia tremellosa*. *FEMS Microbiol. Lett*., 278, 72-77.

Yuan, S. Y., Liu, C., Liao, C. S. and Chang, B. V. (2002). Occurrence and microbial degradation of phthalate esters in Taiwan river sediments. *Chemosphere,* 49, 1295-1299.

Reviewed by Prof. Ji-Dong Gu in the School of Biological Sciences, The University of Hong Kong.

In: Phthalates
Editors: Gerardo L. Moretti and Drago Romano

ISBN: 978-1-62081-994-4

Chapter 6

GAS PHASE FRAGMENTATION REACTIONS OF PHTHALATES IONS

Natarajan Ravi[1], Victor Ibeanusi[2] and Yassin Jeilani[3,*]
[1]Department of Physics,
[2]Environmental Science and Studies Program,
[3]Department of Chemistry, Spelman College, Spelman Lane, SW, Atlanta, GA, US

ABSTRACT

The widespread industrial use of phthalates as plasticizers led to major concern in exposure to phthalates; therefore, many analytical approaches based on mass spectrometry (MS) have been developed to analyze phthalates in various matrices. The MS of phthalates have been reported under various ionization modes including electron ionization (EI), positive chemical ionization (PCI), and negative chemical ionization (NCI). Understanding the gas phase chemical reactions under the various ionization conditions provides useful structural information. Under EI, PCI, and NCI modes, fragmentation mechanisms of selected phthalates were studied by collision induced dissociations (CID). Relatively low energy CID spectra with collision energy range of 5-10 eV provided useful dissociations for phthalates. The CID of the molecular ion or the protonated molecular ion peak of phthalates depending on the ionization mode showed multipathway fragmentations. Under EI and PCI conditions, two major pathways were identified as loss of alkoxy group and McLafferty rearrangement pathways. Ion structures and reactions along these pathways were studied by density functional theoretical (DFT) methods and showed important features of the ion structures. In many cases, both cyclic and open ion structures had to be considered because the benzene ring framework of phthalates allows the design of energetically stable cyclic ions. The DFT studies led to a better understanding of the mechanisms in terms of energetics and transition states involved along the fragmentation pathways. Under the EI mode, new strategies using computational modeling of cation-molecule complexes were developed for the well-known McLafferty rearrangement of phthalates leading to protonated phthalic anhydride at m/z 149. The DFT results allowed the identification of two types of complexes as: hyperconjugated carbocation-molecule and radical carbocation-molecule complexes. The

[*] Corresponding author, email: yjeilani@spelman.edu.

EI ionization made possible to study McLafferty rearrangement of an odd-electron M+•; on the other hand, the PCI spectra showed McLafferty rearrangement product from an even electron $[M+H]^+$. The DFT results confirmed that McLafferty rearrangement is energetically feasible from either $M^{+\bullet}$ or $[M+H]^+$. Under methane PCI conditions, CID spectra of $[M+H]^+$, $[M+C_2H_5]^+$ and $[M+C_3H_5]^+$ of phthalates were used to design fragmentation mechanisms and to provide useful structural information. The PCI spectra showed the protonated phthalic anhydride as well as adducts of phthalic anhydride with $C_2H_5^+$ and $C_3H_5^+$. These adducts are very useful for structural identification purposes.

Structure of Phthalates

Phthalates are commonly referred to the di-ester derivatives of phthalic acid (*ortho*-dicarboxylic acid). The monoester derivatives of phthalic acid, phthalate monoesters, are also part of the phthalate family. The monoesters are the biological metabolites of phthalates.[1] Other groups related to phthalates are terephthalates and isophthalates. Terephthalates and isophthalates are based on terephthalic acid (*para*-dicarboxylic acid) and isophthalic acid (*meta*-dicarboxylic acid), respectively. In terms of mass spectrometry (MS), phthalates and terephthalates show similar electron ionization (EI) mass spectra with a base peak at *m/z* 149; however peak intensities in terephthalates spectra are different to some extent from the corresponding phthalates. Because phthalates are more often targeted in MS analysis than terephthalates, it became common to associate *m/z* 149 with phthalates when it should be related to both phthalates and terephthalates. It is important to note that the *m/z* 149 is identified as protonated phthalic anhydride for phthalates and it has a different structure when it is generated from terephthalates. Mass spectra of isophthalates are quite different than those of phthalates.

The *m/z* 149 is not the base peak in the mass spectra of isophthalates. Studies of gas phase fragmentation reactions of phthalates significantly contribute to the understanding of the pattern of fragmentations observed in both terephthalates and isophthalates; the pattern of fragmentations in phthalates, terephthalates, and isophthalates have been reported.[2] In principle, MS approaches under negative chemical ionization (NCI) have been shown to distinguish phthalates from terephthaltes and isophthalates;[3] also, MS approaches have been reported to distinguish dimethyl isophthalate from dimethyl terephthalate.[4]

Recently, major advances have been made toward understanding the pathways of these fragmentations. The focus of this chapter is the gas phase fragmentation reactions of phthalates under MS conditions. Figure 1 shows structure of common phthalates that have been targeted in various environmental matrices. Phthalates commonly targeted in various environmental matrices by MS methods include: dimethylphthalated (DMP), diethylphthalated (DEP), dibutylphthalate (DBP), benzylbutylphthalate (BBP), di-(2-ethylhexyl)-phthalate (DEHP) and di-(*n*-octyl)-phthalate (DnOP). All phthalates show the diagnostic *m/z* 149 with the exception of dimethylphthalate (DMP); instead, DMP shows a characteristic *m/z* 163 peak.

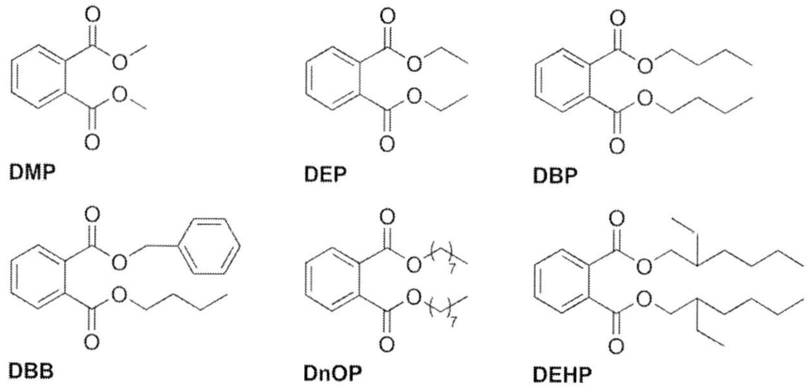

Figure 1. Structure of common phthalates.

Figure 2 shows the year-by-year increase in publications reporting use of mass spectrometry (MS) for the analysis of phthalates. As expected, gas chromatography (GC) continues to be the major technique for phthalates analysis although high pressure liquid chromatography (HPLC) approaches are becoming increasingly popular; therefore, understanding unimolecular decomposition reactions for both $M^{+\bullet}$ and $[M+H]^+$ is important, depending on the ionization methods used. Different MS ionization methods reported for phthalate analysis including: Electron ionization (EI),[5-7] positive chemical ionization (PCI), [8] negative chemical ionization,[3] and electrospray ionization (ESI).[9] Under positive ionization modes, a characteristic *m/z* 149 ion is common in phthalates mass spectra. This observation suggests a common fragmentation path to *m/z* 149. For DMP, *m/z* 163 is observed instead of *m/z* 149. For negative chemical ionization (NCI) MS, *m/z* 148 is observed as the base peak in the mass spectra of phthalates, and it has been identified as negatively charged phthalic anhydride.[3]

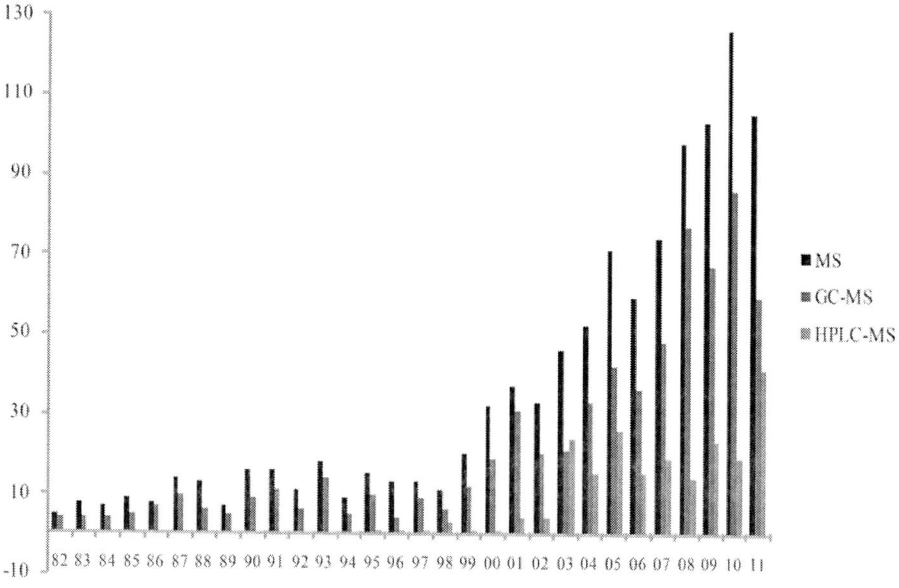

Figure 2. Number of phthalate publications per year by mass spectrometry, gas chromatography/mass spectrometry, or high pressure liquid chromatography/mass spectrometry.

FRAGMENTATION REACTIONS UNDER ELECTRON IONIZATION MS

Tandem Mass Spectrometry

Figure 3 shows collision induced dissociation (CID) spectra of selected phthalates. These spectra were obtained by a triple quadrupole mass spectrometer (QqQ).[6] QqQ methods provide the means to study CID at low collision energies. Under electron ionization mass spectrometry (EI-MS), phthalates are ionized to radical cation ($M^{\bullet+}$), however the full scan spectra of phthalates do not show a peak corresponding to $M^{\bullet+}$. QqQ-MSMS of $M^{\bullet+}$ is possible and provides useful strategies to study the corresponding fragmentation patterns.

Fragmentation reactions of phthalates are dominated by reactions at the acyl group rather than the rigid aromatic ring. For ethyl and higher alkyl chain esters, bond cleavages involving the alcohol moiety are dominant. The CID spectra of $M^{\bullet+}$ showed both [M - CH_2=CH-CHR'] $^{\bullet+}$ and [M - $^{\bullet}$OR]$^+$ with structures assigned as 4 and 9, respectively (Figure 3: a-c). Scheme 1 shows the two pathways leading to *m/z* 149 through 4 and 9 corresponding to the McLafferty+1 rearrangement and the loss of alkoxy radical pathways, respectively. The R_1 group (Scheme 1) is equal to R_2 for DBP, DnOP, and DEHP; for BBP, R_1 is a benzyl group and R_2 is a butyl group.

The pathways were studied by both MSMS and theoretical calculations at a B3LYP[10,11] level of theory using 6-311G(d,p) basis sets.[12] To determine energetically stable structures, both cyclic and acyclic ions were considered in these calculations. Spectra were obtained either by product or precursor ions scanning. DBP, DEHP, and DnOP fragment by the two major pathways; BBP showed a characteristic fragmentation pathway. Each of these pathways deserves a closer look to both spectra and the corresponding theoretical calculations.

Structure of 1 (*m/z* 149)

Protonated phthalic anhydride is an important ion used for the identification of phthalates. Back in 1959, McLafferty proposed the formation of *m/z* 149 from $M^{+\bullet}$ through McLafferty rearrangement step.[13] Since then, the structure of *m/z* 149 has been reported either with a closed anhydride ring[5,7,14,15] or without anhydride ring.[6] For the closed ring anhydride, protonation at either carbonyl oxygen[15] or middle oxygen[5,7,14] has been reported. With recent advances in computational methods, it was possible to determine the most stable structure of *m/z* 149.

Phthalic anhydride has a five-membered anhydride ring that can be protonated at either carbonyl oxygen (1a) or middle (1b) oxygen of the anhydride ring.[6] In this study, the calculations were reported at B3LYP/6-311G(d,p) level of theory. Figure 4 shows the optimized structures of 1a and 1b. The bond length between C2-O2 (Figure 4) was 1.72 Å and 2.44 Å for 1a, and 1b, respectively; these distances suggested an open anhydride ring for both 1a and 1b. For comparisons, Figure 4-d shows the optimized closed structure of phthalic anhydride with a C2-O2 bond length of 1.40 Å; and, this C2-O2 bond length is significantly shorter than that observed in either 1a or 1b. The 1a with the initial protonation at the carbonyl-O led to the more stable structure than 1b by 6.5 kcal/mol. Note that the difference between 1a and 1b is the orientation of COOH group, and rotation around a single bond (C3-C4) is allowed (Figure 4); free energy calculations were based on the low energy structure 1a.

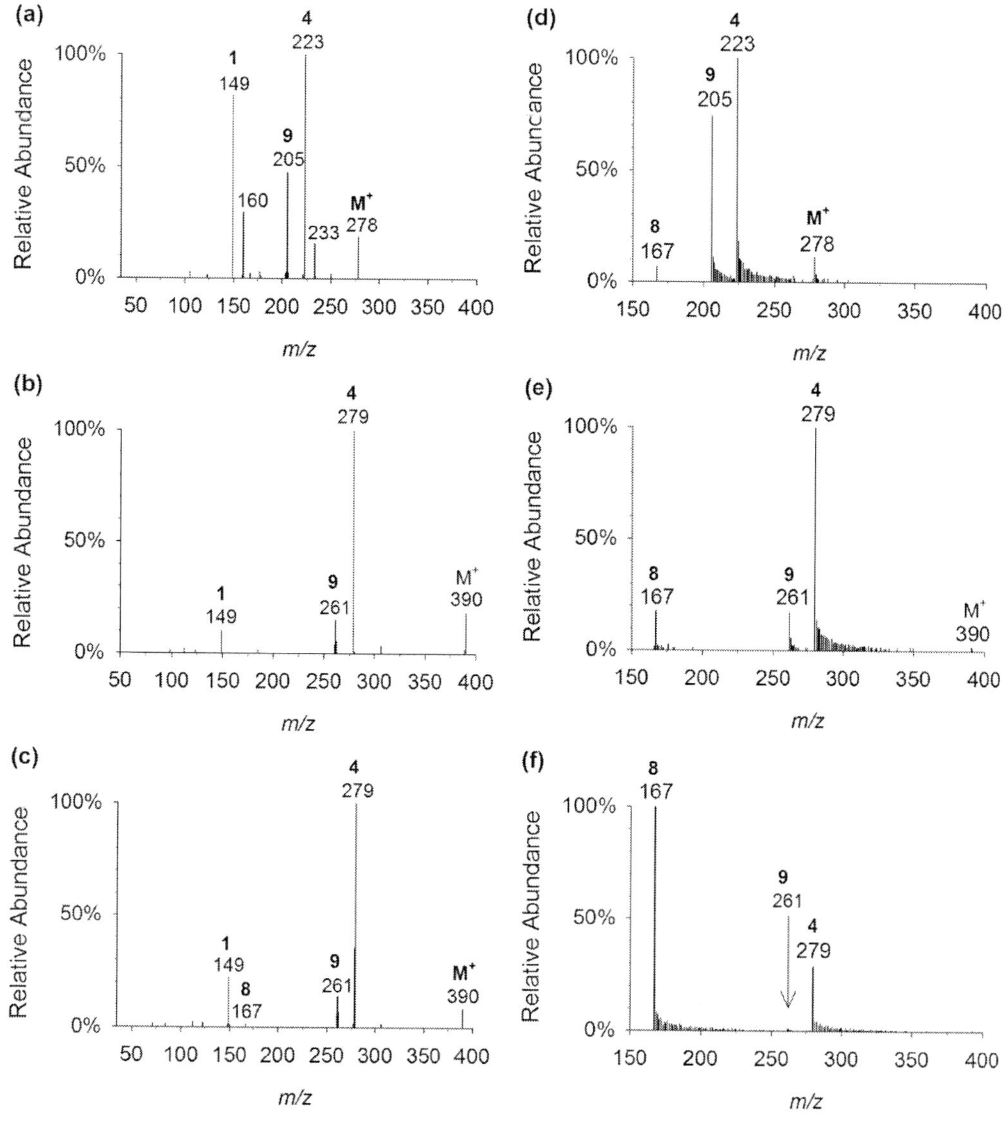

Figure 3. MSMS spectra: (a) DBP: CID spectrum $M^{\bullet+}$ at 5 eV, (b) DnOP: CID spectrum $M^{\bullet+}$ at 5 eV, (c) DEHP: CID spectrum $M^{\bullet+}$ at 5 eV, (d) DBP: 1 (m/z 149) precursor ions spectrum at 10 eV, (e) DnOP: 1 (*m/z* 149) precursor ions spectrum at 10 eV, and (f) DEHP: 1 (*m/z* 149) precursor ions spectrum at 10 eV.

The highest occupied molecular orbitals (HOMOs) of 1 showed significant orbital overlap between the carbonyl O of COOH group with the C of the *ortho* CO group (Figure 5) when protonation takes place at the carbonyl oxgyen. The coefficients of the HOMO-2 showed that the overlap is contributed predominantly by the p-orbital of the O. These orbital interactions are manifested in the bond angles. The C1C2O1 (Figure 4) bond angle was 145°, that is neither linear (180°) nor perfectly trigonal planar (120°); the C1C2O1 bond angle for phthalic anhydride was 131°. Finally, these theoretical studies suggested that the structure of the protonated phthalic anhydride 1 is an open structure.

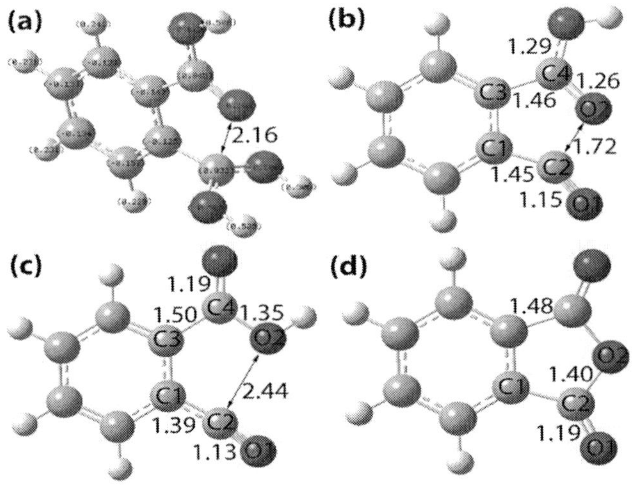

Figure 4. Optimized ion structures: (a) phthalic anhydride protonated at carbonyl oxygen (1a), *m/z* 149, (b) phthalic anhydride protonated at middle oxgyen (1b), and (c) phthalic anhydride.

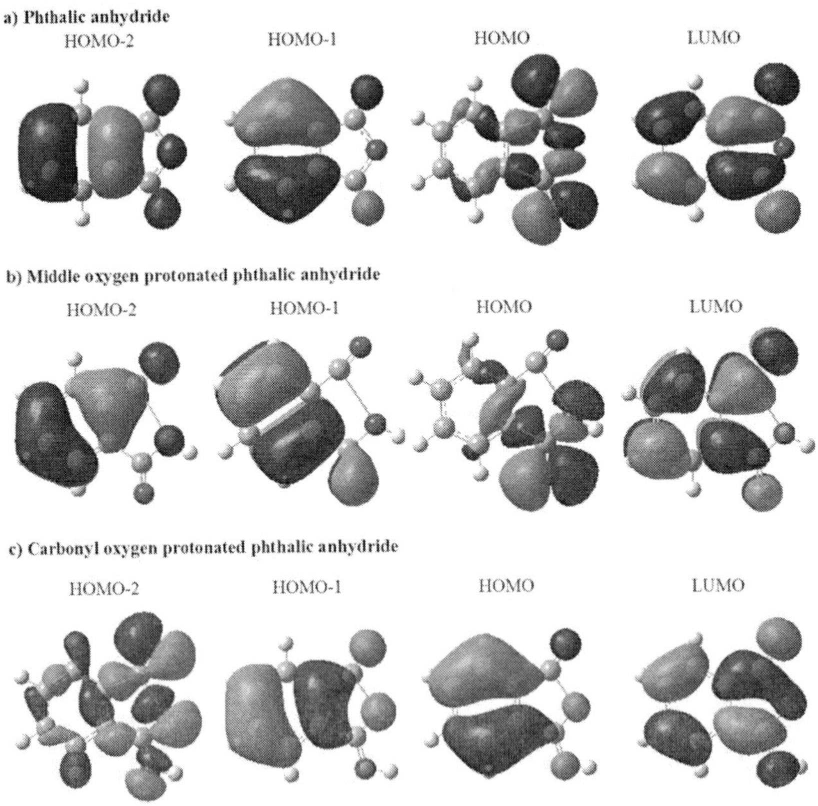

Figure 5. HOMO and LUMO orbitals from B3LYP/6-311G(d,p) optimized structures of (a) phthalic anhydride, (b) middle-oxygen protonated phthalic anhydride, and (c) carbonyl-oxygen protonated phthalic anhydride.

Collision Induced Dissociation of 1 (M/Z 149)

Several low-mass ions have been associated with the dissociation of 1 at *m/z* 149 such as *m/z* 121.[7] The product ion, *m/z* 121, was recently confirmed by CID studies.[6] At low collision energies (from 5 to 20 eV), CID spectra of **1** from *di*-(*n*-octyl)phthalate (DnOP) did not show significant fragmentation. At higher collision energies (from 30 to 50 eV), the CID spectra showed peaks at *m/z* 121, 93, and 65 due to loss of CO and 2CO, respectively; these speaks were shifted by four mass units when deuterated-DnOP was used with all four aromatic hydrogens replaced with deuterium. The *m/z* 149 ion is quite stable and its fragmentation requires relatively high energy (Figure 6); in particular the low mass peak at *m/z* 65 is only abundant at high collision energy. Using magnetic sector instrument, Yinon reported CID of 1 to give *m/z*: 148, 121, 104, and 93 due to loss of H, CO, COOH, and 2CO, respectively.[5]

As expected, the collision energy has significant effect on the CID spectra. Structurally informative product ion spectrum of $M^{+\bullet}$ from phthalates requires the use of low collision energies. At high collision energy, the *m/z* 149 dominates the spectrum; at even higher energy, *m/z* 149 starts to fragment to low mass fragments (Figure 6). For phthalates, the best collision energy for CID of $M^{+\bullet}$ that provide good structural information was 5 eV.

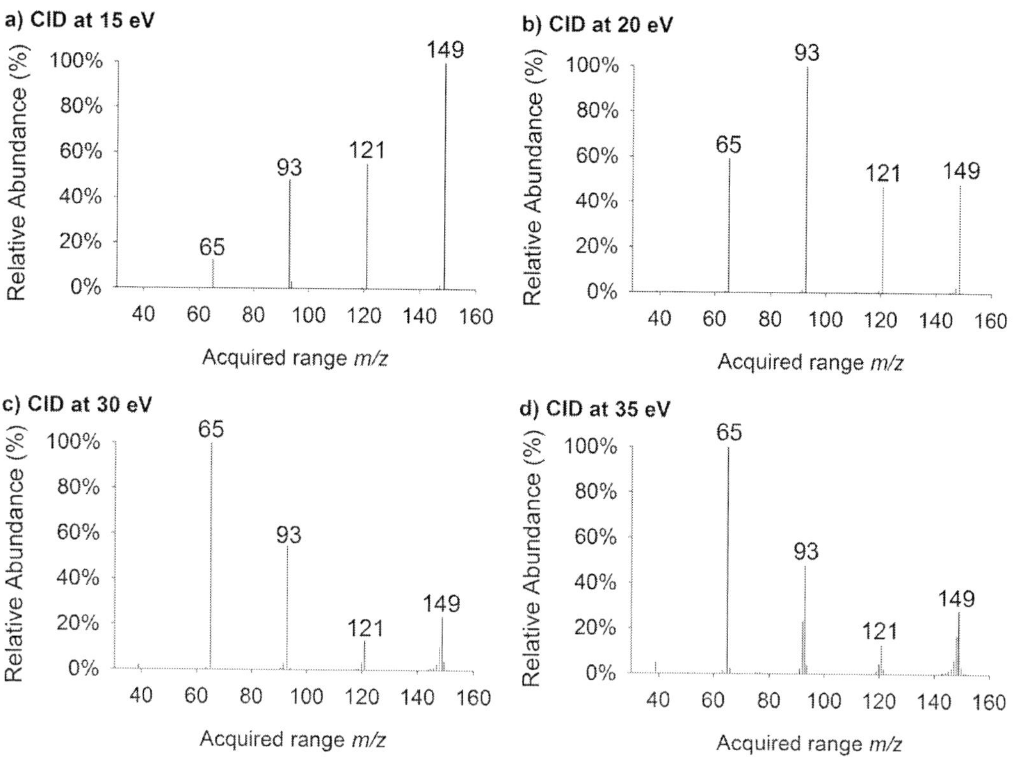

Figure 6. CID spectra of protonated phthalic anhydride *m/z* 149 (**1**).

Mclafferty+1 Rearrangement Pathway

The odd-electron of $M^{+\bullet}$ resides on a carbonyl oxygen, and it can induce the formation of a new bond to a hydrogen atom attached to γ-carbon of the carbonyl. This transfer is favored because it proceeds through a six-membered transition state (TS). The rearrangement is well known as McLafferty rearrangement, and it is observed for many carbonyl compounds possessing γ-hydrogens.[16,17]

When McLafferty rearrangement is accompanied by transfer of additional hydrogen, it is known as McLafferty+1 rearrangement. Fred McLafferty described the "McLafferty+1" rearrangement as a stepwise transfer of two hydrogens from the alkoxy alkyl chain of an ester to the oxygens of the ester group (or similar functional groups).[17] The transfer of the second hydrogen was facilitated by a radical cation-neutral molecule complexes. Such complexes have been postulated as intermediates under mass spectrometric conditions.[18] Fragmentations through both types of rearrangements led to thermodynamically stable products.[19] The product of the McLafferty rearrangement was a stable distonic ion.

In this mechanism, the first step is the traditional McLafferty rearrangement to give a distonic ion (Scheme 1). The second step is the transfer of a second hydrogen from the alkyl group to the ester group oxygen leading to the elimination of an alkene radical from the $M^{\bullet+}$. The transfer of the second hydrogen proceeds through an ion molecule complex (Scheme 1). Ion molecule complexes as intermediates are well documented in the gas phase fragmentation mechanisms.[20,21] In this mechanism, the charge remains at the oxygen of the ester group while a stabilized allylic radical is eliminated (Scheme 1). Common phthalates that undergo McLafferty+1 rearrangement are *di*-(*n*-butylphthalate (DBP), benzylbutylphthalate (BBP), *di*-(2-ethyl-*n*-hexyl)-phthalate (DEHP) and *di*-(*n*-octyl)-phthalate (DnOP).

Transition states of the McLafferty rearrangement were reported.[6] The distonic ion 2 is formed through a six-membered ring transition state (TS) involving the transfer of a β-hydrogen of the alkyl group to the ester carbonyl oxygen. TS structures for DBP, DnOP, and DEHP were found using QST2 methods as implemented in the Gaussian software package. The B3LYP intrinsic reaction coordinates (IRC or the steepest descent from the transition state leading to the product in the forward direction and to the reactant in the reverse direction) was calculated for all transition state structures. The IRC results confirmed that the transition states are those for a McLafferty reaction.

Figure 7 shows TS structures and bond distances of the six-membered ring transition state. Table 1 shows activation energies for McLafferty rearrangement TS. Low activation energy of 0.4 kcal/mol was determined for DEHP (Table 1).[6] The activation energies were reported using B3LYP/6-311G(d,p) level of theory; the effects of basis sets were tested using 4-31G(d,p), 6-311G.

The structure of the transition state for DBP and DnOP had the migrating H at 1.3 Å from the carbon it departs, and it is at 1.4 Å from the oxygen it joins (Figure 7-a,b). For DEHP, the H was closer to the carbon it departs at a distance of 1.2 Å, and it was 1.5 Å from the oxygen it joins (Figure 7-c). Table 2 shows positive values of the overlap populations of the TS structures between the migrating H and the C it departs and O it joins are positive that indicated bonding characteristics between these atoms. The bonding between the migrating H and O it joins was found to be stronger for DBP and DnOP than for DEHP. The migrating H in the transition state maintained a stronger bond with C than with O. Natural population analysis (NPA) of the transition states structures provided more insight about structure of ions. The NPA charge analysis (Table 2) showed less negative charge on the C of DEHP than of DBP and DnOP. This is consistent with the fact that the C in DEHP is a tertiary C, and it is a secondary C in DBP and DnOP. The NPA charge on the migrating H was quite similar for all transition states. Also, the spin density was found distributed around the part of the molecule where the rearrangement takes place; also, the spin density was delocalized over the aromatic ring and the ester group involved in the rearrangement step.[6] The transfer of the second H from the McLafferty product 2 was proposed to take place through an ion molecule complex. Computational models of these complexes showed important structural features. As the two fragments within the complex start to separate the charge is partitioned according to Stevenson's rule; the positive charge to remains on the O (Scheme 1).

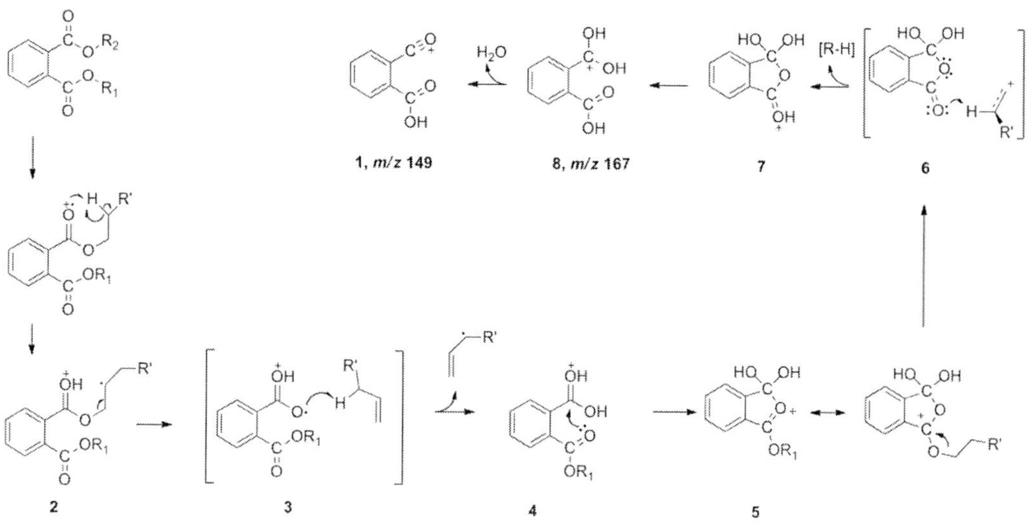

Scheme 1. McLafferty rearrangement pathways under EI-MS.

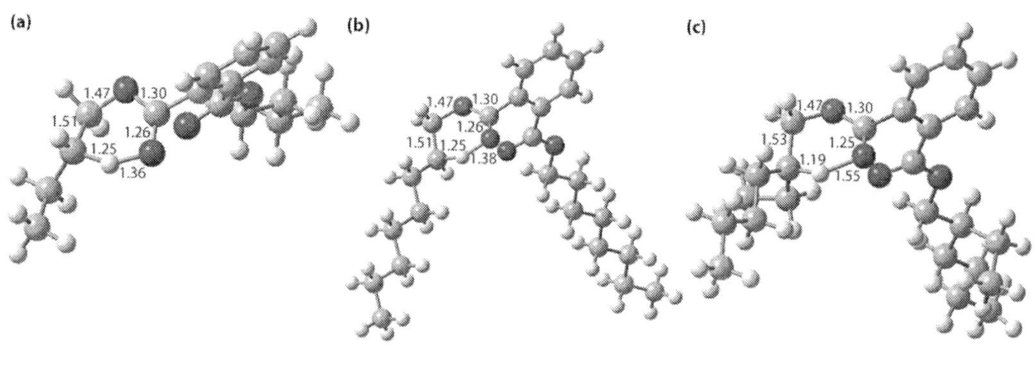

Figure 7. Optimized transition state structures: (a) DBP, (b) DnOP, and (c) DEHP.

Optimized models of the radical cation-neutral molecule complexes of DBP, DnOP, and DEHP (Figure 8) showed a preferred interaction between the alkene and the charged fragment through the Cs and Hs of the alkene's double bond. The closest distance between a H from the alkene and the O radical of the second partner was in the range of 2.37-2.40 Å (Figure 8).

Table 1. Free energy calculations for EI Fragmentation Mechanisms

	ΔG^a			$\Delta S_{total}{}^b$
Compound	4-31G(d,p)	6-311G	6-311G(d,p)	6-311G(d,p)
$M^{\cdot+}\rightarrow$TS				
DBP	5.2	4.9	5.5	-5.4
DnOP	5.9	6.1	6.2	-6.0
DEHP	0.0	-1.3	0.4	-4.8
TS→2				
DBP	-23.3	-25.2	-23.2	6.0
DnOP	-23.2	-25.9	-23.3	5.1
DEHP	-25.9	-28.9	-26.3	9.6
2→3				
DBP	20.8	17.3	15.9	6.3
DnOP	19.5	16.4	15.4	6.3
DEHP	16.5	14.7	13.1	1.4
3→4				
DBP	-29.8	-29.0	-25.9 (2.6)[e]	28.9
DnOP	-28.3	-22.0	-24.4 (3.7)[e]	32.1
DEHP	-16.8	-13.4	-15.9 (10.4)[e]	31.6
4→6				
DBP	59.2	53.1	50.1	7.2
DnOP	57.6	47.2	48.9	6.7
DEHP	36.7	28.7	31.9	14.2
6→7				
DBP	-23.5	-21.2	-24.7	34.5
DnOP	-22.0	-19.5	-23.6	35
DEHP	-11.1	-6.5	-12.6	29.5
8→1	12.0	6.9	0.4	40.1
$M^{\cdot+}\rightarrow$9				
DBP	11.9	7.6	10.1	36.9
BBP[c]	18.2	13.4	16.0	39.5
BBP[d]	17.2	13.4	14.6	42.2
DnOP	13.1	9.2	11.3	37.1
DEHP	8.4	3.8	6.9	40.9
9→10				
DBP	24.4	21.0	21.0	8.9
BBP	24.4	21.0	21.0	8.9
DnOP	21.9	18.3	18.6	10.8
DEHP	6.5	1.0	3.4	13.0
10→1				
DBP	-11.6	-12.2	-13.6	32.1
BBP	-11.6	-12.2	-13.6	32.1
DnOP	-9.3	-9.6	-11.2	32.2
DEHP	2.8	4.9	0.9	29.9
$M^{\cdot+}\rightarrow$11	-22.9	-21.8	-22.4	-3.3
11→12	16.7	16.4	13.7	38.9

[a] In units of kcal/mol at 298K.
[b] In units of cal/mol K at 298K.
[c] loss of butoxy radical.
[d] loss of benzylox radical.
[e] In parenthesis corresponds to free energy for the loss of a vinyl radical.

Molecular orbital (MO) analysis of 3 indicated that part of the LUMO for DBP and DnOP, had an overlap region between the two partners of 3; this suggested that the LUMO is not antibonding between the O and the C of the alkene group. Such bonding characteristic of a LUMO has been previously described.[22] The formation of the DBP and DnOP complexes required energy of about 16 kcal/mol (Table 1). The excess energy of the complex relative to the distonic ion 2 is sufficient to allow the reorientation of the two partners that make up the complex.[21,23] As expected, the formation of all complexes was entropically favored. The calculated increase of the entropy was contributed mainly by an increase in the vibrational entropy contribution of the complex.

Table 2. Natural population analysis of bonds around the migrating H

Description	DBP	DnOP	DEHP
Overlap population between migrating H and C it departs	0.269	0.268	0.308
Overlap population between migrating H and O it joins	0.115	0.115	0.075
Bond order between migrating H and C it departs	0.573	0.577	0.670
Bond order between migrating H and O it joins	0.371	0.360	0.243
NPA charge of the migrating H in the TS	0.339	0.340	0.331
NPA charge of C in the TS as H migrates	-0.269	-0.271	-0.148
NPA charge of O in the TS as H joins it	-0.442	-0.444	-0.414

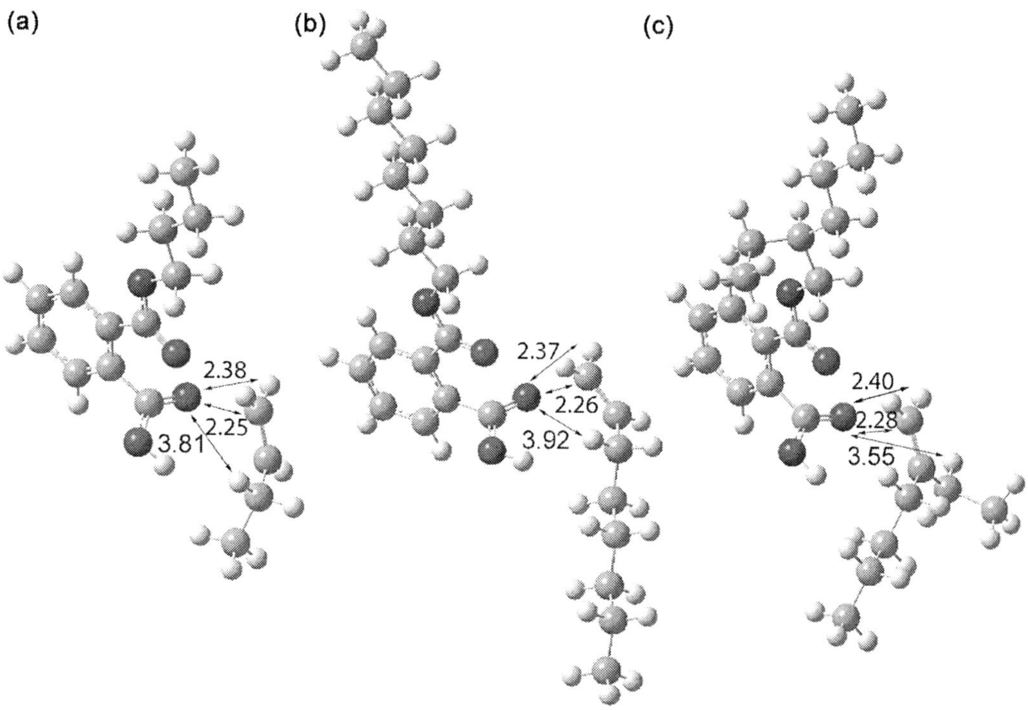

Figure 8. Optimized structures of radical cation molecule complex 3 for: (a) DBP, (b) DnOP, and (c) DEHP.

The product of the McLafferty+1 rearrangement, 4, is a closed shell ion with two resonance structures. Two structures of 4 were optimized: (a) type 1: with hydroxyl group H-bonded to the *ortho*-carbonyl oxygen, and (b) type 2: with the plane of the HOCOH group almost perpendicular to the plane of the ester group.The type 1 structure of 4 with H-bond was found to be energetically favored over type 2 structure;[6] all free energy calculations were based on type 1 structure.

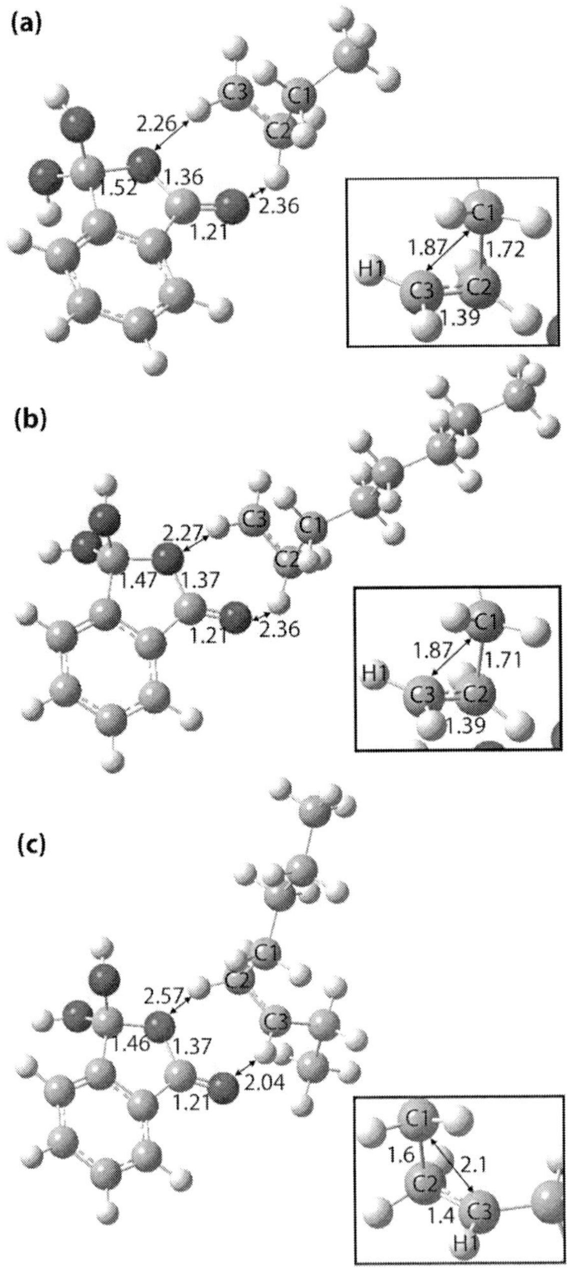

Figure 9. Optimized structure of hyperconjugated radical cation molecule complex 6 for: (a) DBP, (b) DnOP, and (c) DEHP.

Next step is the cleavage of R1 group from the McLafferty rearrangement product ion 4. In this step, R1 in 4 is lost with a net transfer of one H from R1 to an O (Scheme 1). The H transfer has been previously reported for closed shell cations; however, the mode of transfer has not been addressed adequately.[24,25] The transfer of the H has been reported by two mechanisms: McLafferty type rearrangement[24] or a transfer through ion molecule complexes.[25] Therefore, both types of H transfers, the McLafferty rearrangement and the cation molecule complexes, were considered: (a) transfer of the H from a β position of the alkyl chain through a cyclic transition state, and (b) modeling cation molecule complexes that project the two partners of the complex to a H transfer process. Transfer through stable carbocation-neutral molecule complexes was the preferred mechanism to account for the H transfer.[6] The initial step is an intramolecular carbonyl-O attack at the protonated carbonyl-C of 4 leading to 5 (Scheme 1). Intuition suggests that the intramolecular attack proceeds through the carbonyl O attack instead of the alkoxy O; similar mechanisms have been previously considered.[8,26]

Cleavage of the intermediate 5 led to complex 6 as shown in Scheme 1. The two partners in complex 6 are a primary carbocation species and a gem-diol analog of phthalic anhydride. In principle, the primary carbocation can undergo a rearrangement from primary to secondary carbocation facilitating the formation of the hyperconjugated cation; this type of rearrangement is common in carbocations chemistry;[27] computational modeling of 6 with a hyperconjugated carbocation center led to stable ion molecule structures involving hyperconjugated carbocation. Hyperconjugation has been attributed to the stability of carbenium ions (R_3C^+).[28] In case of phthalates, hyperconjugative stability is attributed to the stability the carbocation-neutral molecule complex. Formation of 6 required 50, 49, and 32 kcal/mol for DBP, DnOP, and DEHP, respectively (Table 1). The excess energy was proposed to be sufficient for the partners of the complex to reorient in space which is consistent with previous studies.[21,22]

Figure 9 shows the most important bond lengths of the hyperconjugated carbocation molecule complexes with a carbenium ion center at C3. The C1-C2 bond is significantly elongated relative to a typical C-C bond of 1.53 Å; such elongation of a β C-C (relative to the carbenium ion center) has been previously suggested.[28] Also, the calculated C2-C3 bond lengths are significantly shorter than a normal C-C single bond length (Figure 9). The C2-C3 bond length of 1.39 Å was slightly longer than a typical C=C bond length.

The structure of 6 for DEHP was quite different from DBP and DnOP because C1 of DEHP, as shown is normally attached to C3 (Figure 9-c); however, optimization led to a stable structure showing the migration of an alkyl group from C1 to C2. This observation demonstrated substituent effects of the hyperconjugated carbocation leading to an alkyl group migration to form a more stable carbocation; group migration in DEHP led to a secondary carbocation while such migration in DBP and DnOP does not enhance the stability of the carbocation.

Loss of Alkoxy Group Pathway

Loss of an alkoxy group from $M^{\bullet+}$ can be initiated by the attack at a carbonyl center by an oxygen from *ortho*-COO group. Intuition suggests that the carbonyl-O of *ortho*-COO group ought to be more nucleophilic than the alkoxy-O of an ester group; computational modeling showed evidence to support the nucleophilic attack by the oxygen. Computational studies have shown carbonyl-O of $M^{\bullet+}$ to have more negative charge than the corresponding alkoxy-O;[6] therefore, the mechanism for loss of the alkoxy group is best described to proceed via a carbonyl-O attack on the *ortho*-carbonyl-C (Scheme 2) followed by loss of a neutral alkoxy radical. In an earlier study, loss of an alkoxy radical was proposed as the first step, followed by an attack by the *ortho*-alkoxy-O on the CO C.[5] One of the –OCO was found in a trigonal plane that is perpendicular to the *ortho*-carbonyl-O with a dihedral angle of about 100°.[6] Such geometry further supports the carbonyl-O attack mechanism.

Loss of •OR from $M^{\bullet+}$ leads to 9. All CID spectra showed 9 as a product ion (Figure 3a-c). Similarly, all precursor ion scanning spectra (Figure 3-d-f) showed a peak of 9 as a precursor of 1 (*m/z* 149). Note the 9 peak for DEHP in the spectrum (Figure 3-f) was significantly smaller than that observed in DBP and DnOP spectra. Structure of 9 can be modeled either as a cyclic or an acyclic structure. The cyclic structure of 9 was found to be the energetically favored structure, and it was used for free energy calculations.[6]

An ion molecule complex was proposed in the fragmentation pathway of 9. The mechanism involved the heterolytic cleavage of R group from the closed shell 9 leading to the formation of an ion molecule complex 10; the loss of R from 9 is associated with the transfer of an H from the R group to one of the oxygens (Scheme 2).

Similar to complex 6, the carbocation in the optimized complex 10 was hyperconjugated; as expected, the phthalic anhydride partner in 10 (Figure 10) has a closed five-membered ring structure. Similar to 6, complex 10 of DEHP showed an alkyl group that migrates to the adjacent carbocation center; this rearrangement produces a secondary carbocation that is more stable than the initial primary carbocation. The C1C2C3H1dihedral angle was 91, 91, and 85° for DBP, DnOP, and DEHP, respectively; therefore, the R' bond is shown perpendicular to the plane of the carbocation center (Scheme 2). Loss of an alkene from 10 leads to the formation of 1. Structures in these mechanisms were confirmed using labeled phthalated.[6]

The CID spectrum of $M^{\bullet+}$ from BBP showed a peak at *m/z* 206. This peak is characteristic for BBP and was observed in its full scan spectrum. Scheme 3 shows the mechanism proposed for the formation of *m/z* 206; this mechanism involves a H rearrangement to give a distonic ion followed by a neutral elimination of benzaldehyde from distonic $M^{\bullet+}$ (Scheme 3).[6] The resulting *m/z* 206 is stabilized by resonance. Similar mechanism for the initial H rearrangement was proposed for the elimination of CO_2 from methyl benzoate.[29] The proposed mechanism (Scheme 3) was energetically feasible (Table 1).

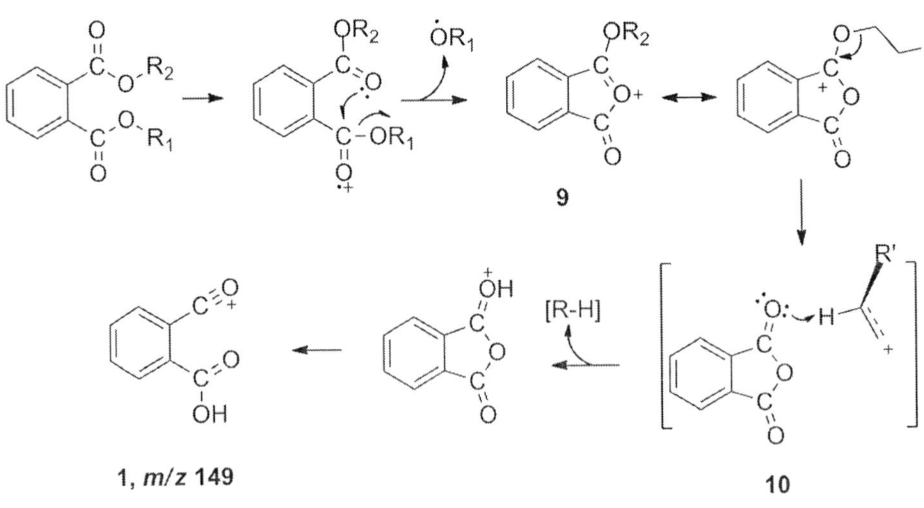

Scheme 2.

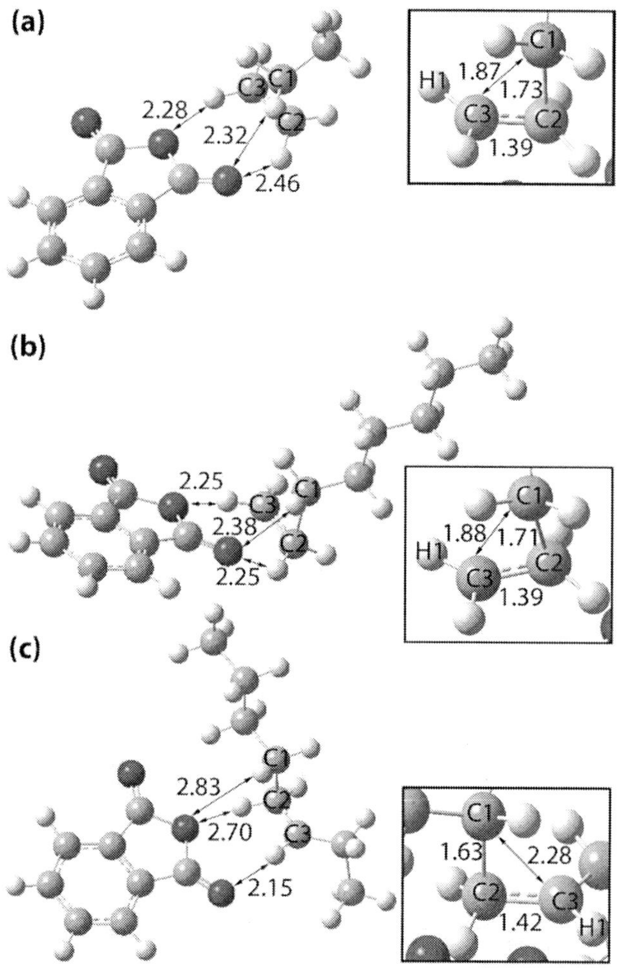

Figure 10. Optimized structure of hyperconjugated cation molecule complex 10 for: (a) DBP, (b) DnOP, and (c) DEHP.

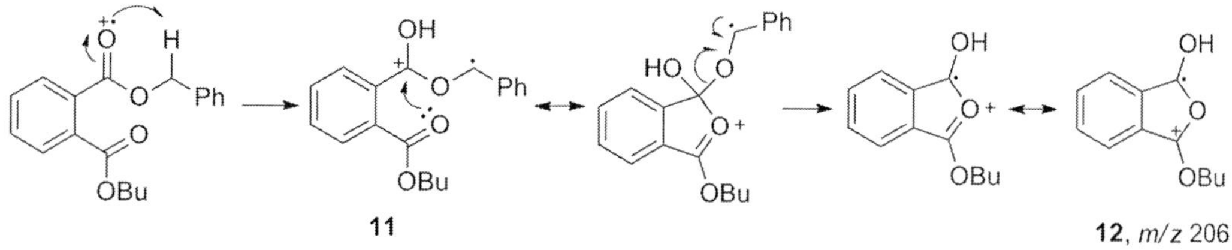

Scheme 3.

Fragmentation Reactions Under Positive Chemical Ionization MS

Positive chemical ionization (PCI) is a useful method used to determine the molecular weight of a compound. Recently, CID under PCI conditions was shown to provide important structural features of phthalates.[8] PCI precursor ion scanning spectra of phthalates (Figure

11) showed significant peaks of $[M+H]^+$, $[M+C_2H_5]^+$ and $[M+C_3H_5]^+$. Diagnostic ions associated with the fragmentation of these peaks were identified at *m/z* 149, 177 and 189 assigned as protonated phthalic anhydride (1) and an adduct of phthalic anhydride with $C_2H_5^+$ and $C_3H_5^+$, respectively. Targeting these ions by CID provided important structural information.[8] CID studies showed that each $[M+H]^+$, $[M+C_2H_5]^+$ and $[M+C_3H_5]^+$ leads to distinct fragmentation pathway to 1.

Two distinct pathways were proposed under PCI fragmentation pathways of phthalates: (a) $[M+C_2H_5]^+$ pathways through an adduct of phthalic anhydride with $C_2H_5^+$ (15), and (b) $[M+H]^+$ pathways through 13 and 14. Scheme 1 shows that 16 and 17 are associated with the $[M+C_2H_5]^+$ pathway and 8 is associated with the $[M+H]^+$ pathway. The first two steps of the mechanism were addition of $C_2H_5^+$ and elimination of [R-H] to give 16. Two routes were proposed for 15 to give 16 (Scheme 4).

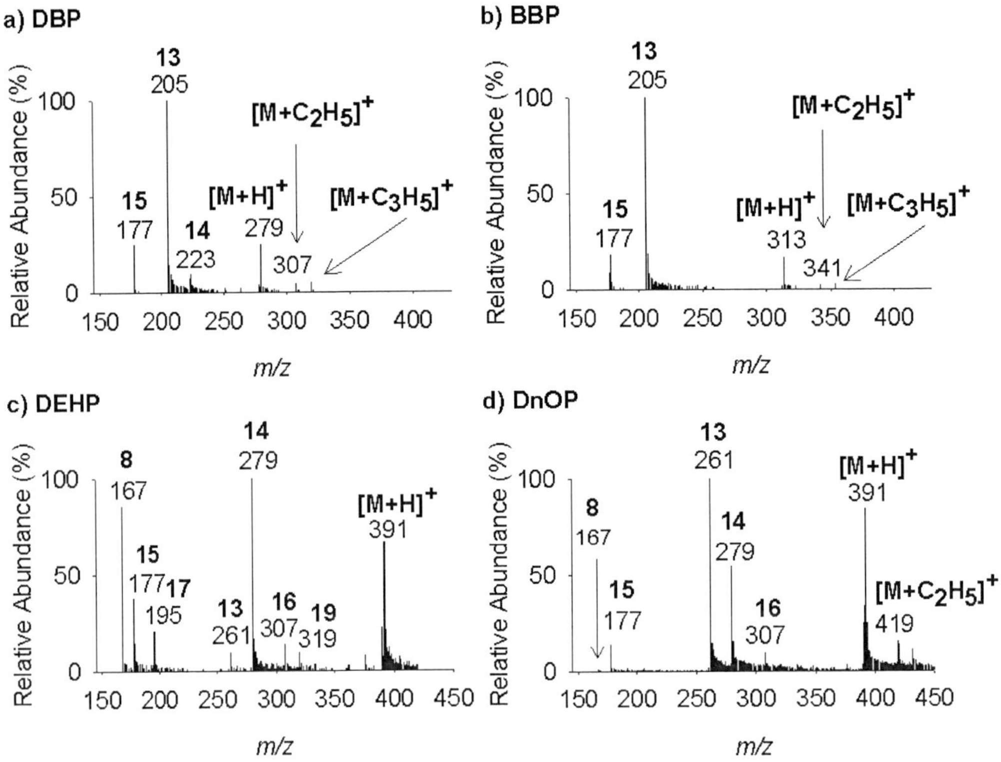

Figure 11. PCI precursor ion scan of the protonated phthalic anhydride (m/z 149).

Ion 15 was diagnostic for the $[M+C_2H_5]^+$ fragmentation, and it was observed in the PCI full scan spectra of the phthalates. Loss of C_2H_4 from 15 led to the protonated phthalic anhydride. For diethyl phthalate (DEP), a peak at *m/z* 177 was typically observed in the full scan spectrum that could result from neutral loss of CH_3CH_2OH from $[M+H]^+$. The product ion spectrum of *m/z* 177 from DEP was found to be quite different from the product ion spectra of *m/z* 177 from the rest of phthalates. A significant difference was that the DEP product ion spectrum had peaks at *m/z* 148 and 149, while the rest of the phthalates showed a peak of *m/z* 149 only. A distinct pathway leading to *m/z* 148 was proposed.[8]

The selected phthalates had product ion spectra of $[M+C_2H_5]^+$ with several common (Figure 12). The intensity of 15 (*m/z* 177 peak) even exceeds the intensity of 1. These spectra complemented well the precursor ion spectra for the proposed $[M+C_2H_5]^+$ fragmentation mechanism through 15 and 16. In Scheme 1, both 13 and 14 were associated in the fragmentation of $[M+H]^+$. Product ion scanning of $[M+H]^+$ was performed from 5 to 25 eV. At higher collision energies, the dominant peak was 1. The product ion spectra recorded at collision energy of 5 eV showed 13 in all spectra. Loss of ROH from $[M+H]^+$ gave 13 that eventually fragmented to 1. The spectrum from BBP showed both *m/z* 205 and *m/z* 239 corresponding to 13 with butyl and benzyl groups, respectively. DEHP was the only compound showing McLafferty rearrangement product 14. On the other hand, 14 was frequently observed in the precursor ion spectra of 1 (Figure 11).

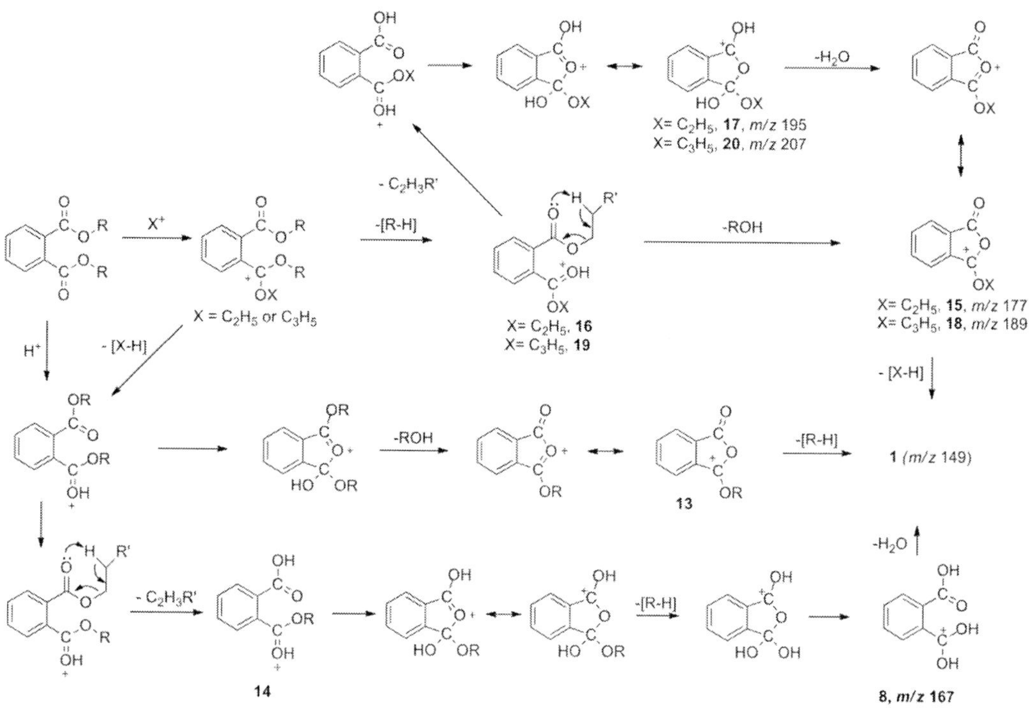

Scheme 4.

Dimethyl phthalate and diethyl phthalate gave simple PCI spectra that were dominated by loss of alkoxy group.[8] For DMP, the product ion spectrum showed *m/z* 163, corresponding to loss of an OCH_3 group. Further CID of *m/z* 163 gave *m/z* 135 and 77, corresponding to $[M-COOCH_3]^+$ and phenylium ion, respectively. A stepwise decomposition was proposed for DMP similar to a previous study using mass-analyzed ion kinetic energy mass spectrometry.[30] For DEP, the product ion spectra of either $[M+H]^+$ or $[M+C_2H_5]^+$ showed two major peaks at *m/z* 177 and 149, corresponding to $[M-OR]^+$ and 1, respectively. For DEP, precursors of *m/z* 149 were identified at *m/z* 177, 223, 251, and 263, corresponding to $[M-OR]^+$, $[M+H]^+$, $[M+C_2H_5]^+$ and $[M+C_3H_5]^+$, respectively. Fragmentation of DMP and DEP proceeded via loss of an alkoxy group from either $[M+H]^+$ or $[M+C_2H_5]^+$ that eventually gave 1.

$[M+C_3H_5]^+$ was a precursor ion of 1, and it was observed in the PCI full scan spectra. Prominent ions from $[M+C_3H_5]^+$ fragmentation were not observed in the precursor ion of *m/z* 149 spectra. The $[M+C_3H_5]^+$ pathway was studied by product ion scanning. Product ion spectra of $[M+C_3H_5]^+$ showed both 18 and 13 as common product ions for all phthalates. All of DBP, DEHP and DnOP spectra had 19, consistent with the mechanism in Scheme 4. McLafferty fragmentation pathway were slightly energetically favored over loss of alcohol pathway.[8]

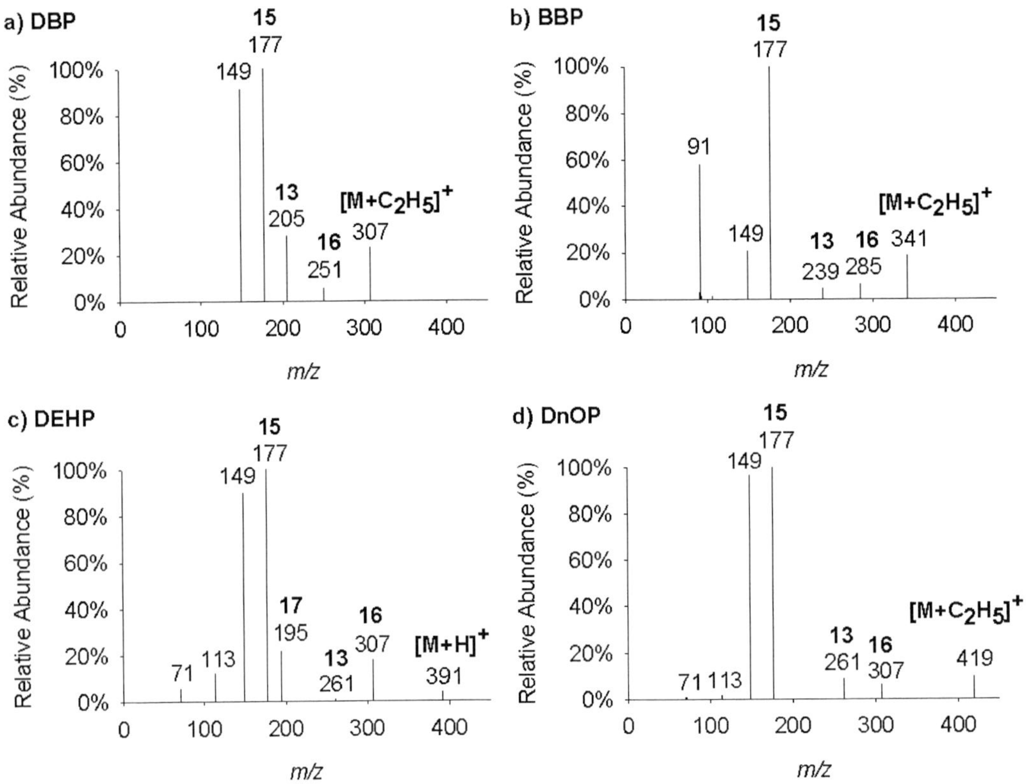

Figure 12. PCI product ion spectra of [M+C2H5]+.

Fragmentation Reactions Under Negative Chemical Ionization MS

Negative chemical ionization (NCI) is typically sensitive to certain compounds. There are limited studies focusing on phthalates by negative chemical ionization NCI-MS.[31] The NCI spectra show important structural features that can be used for identification purposes. The advantage of NCI-MS of phthalates is the observation of an intense $[M-R]^-$ peak;[2] this structural information allows discrimination of homologues phthalates. The use of O_2/Ar-NCI MS was shown to distinguish phthalates from terephthalates and isophthalates.[3] Formation of $[M-R]^-$ was attributed to substitution nucleophilic bimolecular (S_N2) mechanism by $O_2^{-\bullet}$ at the alkyl carbon of the ester group:

$$M + O_2^{-\bullet} \longrightarrow [M + O_2]^{-\bullet} \longrightarrow [M\text{-}R]^-$$

The negatively charged phthalic anhydride is typically observed in negative chemical ionization (NCI) mass spectrum of phthalates.[3,32] Optimization at B3LYP/6-311G(d,p) showed that the anhydride ring remains closed with a similar structure to that of the neutral anhydride.

Using low energy electrons, the fragmentation patterns of several phthalates were studied by electron-monochromator MS.[31] The phthalates were shown to give a peak at *m/z* 165 in the mass spectrum corresponding to $[M\text{-}R^1\text{-}R^2\text{+}H]^-$. The *m/z* 165 was assigned to a singly charged anion of phthalic acid. DMP did not show this type of fragmentation. Mazurkiewicz and Deinzer suggested that the lack of the peak corresponding to $[M\text{-}R^1\text{-}R^2\text{+}H]^-$ in DMP spectrum is due to lack of β-hydrogens.[31] A two step mechanism was proposed for the formation *m/z* 165; the first step was loss of the first alkyl group. An "intramolecular" bimolecular elimination (E2) mechanism was proposed for the second step. In this step, COO^- acts as a base to remove β-hydrogen from the alkyl chain of the second *ortho*-ester group.

CONCLUSION

Major advances in understanding fragmentation reactions mechanisms of phthalates have been made recently. These mechanisms allow the proper assignments of major MS peaks. PCI mode provides a wealth of structural information by using CID of $[M+H]^+$, $[M+C_2H_5]^+$ and $[M+C_3H_5]^+$. Fragmentation mechanisms under EI mode contributed significantly in modeling of ion-molecule complexes. In particular, computational studies have been crucial in the determination of ion structures and energetics of the fragmentation pathways. The most stable structure of the protonated phthalic anhydride has an open anhydride ring. These evidences allow the correct assignment of structures.

ACKNOWLEDGMENT

Natarajan Ravi acknowledges the Department of Energy (DOE) Grant No. DE-FG52-09NA29518.

REFERENCES

[1] Lovekamp-Swan, T.; Davis, B. J.: Mechanisms of Phthalate Ester Toxicity in the Female Reproductive System. *Environ, Health Perspect.* 2003, *11*, 139-145.

[2] Le pine, F.; Boismenu, D.; Milot, S.; Mamer, O. A.: Collision of Molecular Anions of Benzenedicarboxylic Esters with Oxygen in a Triple Quadrupole Mass Spectrometer. *J. Am. Soc. Mass Spectrom.* 1999, *10*, 1248–1252.

[3] Stemmler, E. A.; Diener, J. L.; Swift, J. A.: Gas-Phase Reactions of O2-• with Alkyl and Aryl Esters of Benzenedicarboxylic Acids. *J. Am. Soc. Mass Spectrom.* 1994, *5*, 990-1000.

[4] Flammang, R.; Dechamps, N.; Gerbaux, P.; Nam, P.-C.; Nguyen, M. T.: Unimolecular Chemistry of Dimethyl Isophthalate Radical Cations. *Int. J. Mass Spectrom. Ion Processes* 2008, *275*, 110-116.

[5] Yinon, J.: Mass Spectral Fragmentation Pathways in Phthalate Esters. A Tandem Mass Spectrometric Collision-induced Dissociation Study. *Org. Mass Spectrom.* 1988, *23*, 755-759.

[6] Jeilani, Y. A.; Cardelino, B. H.; Ibeanusi, V. M.: Density Functional Theory and Mass Spectrometry of Phthalate Fragmentations Mechanisms: Modeling Hyperconjugated Carbocation and Radical Cation Complexes with Neutral Molecules *J. Am. Soc. Mass Spectrom.* 2011, *22*, 1999-2010.

[7] Emery, E. M.: Mass Spectra of Aromatic Esters. *Anal. Chem.* 1960, *32*, 1495-1506.

[8] Jeilani, Y. A.; Cardelino, B. H.; Ibeanusi, V. M.: Positive Chemical Ionization Triple-Quadrupole Mass Spectrometry and ab initio Computational Studies of the Multi-Pathway Fragmentation of Phthalates. *J. Mass Spectrom.* 2010, *45*, 678-685.

[9] Feás, C. P.; Barciela-Alonso, M. C.; Bermejo-Barrera, P.: Direct LC–ES-MS/MS Determination of Phthalates in Physiological Saline Solutions. *J. Chromatogr., B* 2011, *879*, 231–235.

[10] Becke, D.: Density-Functional Exchange-Energy Approximation with Correct Asymptotic Behavior. *Phys. Rev. A* 1988, *38*, 3098-3100.

[11] Lee, C.; Yang, W.; Parr, R. G.: Development of the Colle-Salvetti Correlation-Energy Formula into a Functional of the Electron Density. *Phys. Rev. B* 1988, *37*, 785-789.

[12] Hehre, W. J.; Radom, L.; Schleyer, P. V. R.; Pople, J. A.: *ab Initio Molecular Orbital Theory*; Wiley: New York, 1986.

[13] Mclafferty, F. W.; Gohlke, R. S.: Mass Spectrometric Analysis Aromatic Acids and Esters. *Anal. Chem.* 1959, *31*, 2076-2082.

[14] Djerassi, C.; Fenselau, C.: Mass Spectrometry in Structural and Stereochemical Problems. LXXXVI. The Hydrogen-Transfer Reactions in Butyl Propionate, Benzoate, and Phthalate. *J. Am. Soc. Mass Spectrom.* 1965, *87*, 5756-5762.

[15] Hunt, D. F.; Shabanowitz, J.; Harvey, T. M.; Coates, M.: Scheme for the Direct Analysis of Organics in the Environment by Tandem Mass Spectrometry. *Anal. Chem.* 1985, *57*, 525-537.

[16] Zha, Q.; Hayes, R. H.; Nishimura, T.; Meisels, G. G.; Gross, M. L.: A Study of the Metastable Dissociations of Formate Esters: A Mclafferty Rearrangement to a Distonic Radical Cation. *J. Phys. Chem.* 1990, *94*, 1286-1290.

[17] Mclafferty, F. W.; Turecek, F.: *Interpretation of Mass Spectra*; 4th ed.; University Science Books: CA (USA), 1993.

[18] Tu, Y.-P.: Dissociative Protonation Sites: Reactive Centers in Protonated Molecules Leading to Fragmentation in Mass Spectrometry. *J. Org. Chem.* 2006, *71*, 5482-5488.

[19] Tobitaa, S.; Tajimab, S.; Ishiharab, Y.; Kojimab, M.; Shigihara, A.: Single and Double Hydrogen Atom Migrations in Substituted Alkyl Benzoates: a Study on the Substituent Effect Using MIKE Spectrometry. *Int. J. Mass Spectrom. Ion Processes* 1994, *132*, 129-136.

[20] McAdoo, D. J.; Hudson, C. E.; Skyiepa, M.; Broido, E.; Griffin, L. L.: Ion-Neutral Complexes as Intermediates in the Decompositions Of C5Hl0O2•+ Isomers. *J. Am. Chem. Soc.* 1987, *109*, 7648-7653.

[21] Morton, T. H.: The Reorientation Criterion and Positive Ion-Neutral Complexes. *Org. Mass Spectrom.* 1992, *27*, 353-368.

[22] Bernigaud, V.; Drenck, K.; Huber, B. A.; Hvelplund, P.; Jabot, T.; Kadhane, U.; Kirketerp, M. B. S.; Liu, B.; Lykkegaard, M. K.; Manil, B.; Nielsen, S. B.: Electron-Capture-Induced Dissociation of Protoporphyrin IX Ions. *J. Am. Soc. Mass Spectrom.* 2008, *19*, 809-813.

[23] Hudson, C. E.; McAdoo, D. J.; Griffin, L. L.: Preference for an Ion-Neutral Complex-Mediated Pathway over a Five-Membered-Ring H Shift in the Isomerization of $CH_3O^+HCH_2CH_2$. to $CH_3CH_2CH_2OH^+$. by Ab initio Theory. *J. Am. Soc. Mass Spectrom.* 1995, *6*, 1037-1046.

[24] Kingston, D. G. I.; Bursey, J. T.; Bursey, M. M.: Intramolecular Hydrogen Transfer in Mass Spectra. II. The Mclafferty Rearrangement and Related Reactions. *Chem Rev.* 1974, *74*, 215-242.

[25] McAdoo, D. J.: Ion-Neutral Complexes in Unimolecular Decompositions. *Mass Spectrom. Rev.* 1988, 363-393.

[26] Gerbaux, P.; Julien De Winter, J. D.; Flammang, R.; Nguyen, V. S.; Nguyen, M. T.: Metastable Dimethyl Phthalate Molecular Ions: Does the Loss of a Methoxyl Radical Proceed with or Without Anchimeric Assistance? *Int. J. Mass Spectrom. Ion Processes* 2010, *290*, 127-132.

[27] Siehl, H.-U.; Walter, H.: Conformational Dependence of the β-Deuterium Equilibrium Isotope Effect in Acyclic Carbocations. Suppression of Hyperconjugation by Restricted Rotation in the 2,4,4,6-Tetramethylheptyl Cation. *J. Chem. Soc., Chem. Commun.* 1985, 76-78.

[28] Vrcek, I. V.; Vrcek, V.; Siehl, H.-U.: Quantum Chemical Study of Degenerate Hydride Shifts in Acyclic Tertiary Carbocations. *J. Phys. Chem. A* 2002, *106*, 1604-1611.

[29] Dechamps, N.; Flammang, R.; Gerbaux, P.: Decarboxylation of Metastable Methyl Benzoate Molecular Ions. *J. Am. Soc. Mass Spectrom.* 2006, *17*, 807-814.

[30] Tajima, S.; Kojima, A.; Sugimura, T.; Nakajima, S.; Takahashi, Y.; Nibbering, N. M.: On the Chemistry Following Methoxy Migration in the Metastably Decomposing $(M-COOCH_3)^+$ Ions (*m/z* 135) from Dimethyl Phthalate, Isophthalate and Terephthalate. *Int. J. Mass Spectrom. Ion Processes* 2003, *228*, 891-899.

[31] Mazurkiewicz, P. H.; Deinzer, M. L.: Steric Effects on Gas-Phase Negative Ion Resonances of Phthalates. *J. Am. Chem. Soc.* 1999, *121*, 3421-3427.

[32] Harvan, D. J.; Hass, J. R.; Alboro, P. W.; Friesen, M. D.: Mass Spectrometry of Di-(2-Ethylhyexyl)phthalate Metabolites. *Biol. Mass Spectrom.* 1980, *7*, 242-246.

In: Phthalates
Editors: Gerardo L. Moretti and Drago Romano
ISBN: 978-1-62081-994-4

Chapter 7

THE EFFECTS OF DI(2-ETHYLHEXYL) PHTHALATE EXPOSURE ON BRAIN DEVELOPMENT

C. A. Smith and M. R. Holahan*
Department of Neuroscience, Carleton University, Ottawa, Canada

ABSTRACT

Phthalates are synthetic chemicals used primarily to increase the flexibility and the durability of plastic products, such as food storage containers, children's toys, and personal care products. Given the widespread use of these products, it is likely that exposure to phthalates is unavoidable in our daily lives. There has been increasing concern over the dangers of exposure to these compounds, especially in developing organisms as key organ systems, such as the brain, are undergoing rapid development at this time. Phthalates have anti-androgenic properties, meaning these chemicals interfere with the biological role of androgens (i.e., hormones critical for the masculinization of the brain) as well as suppress enzyme activity, mRNA expression, and protein levels important for brain masculinization and steroid synthesis. Interfering with androgens may directly or indirectly alter or prevent brain masculinization resulting in a more feminized version of the brain. A recent study has shown that phthalate exposure also affects estrogen synthesis in both males and females in addition to interfering with normal steroid homeostasis. Research investigating the effects of phthalates on brain development is limited. In a recent study, male rats were more vulnerable to the effects of phthalate exposure during development showing obvious structural deficits in the hippocampus (important for learning and memory), whereas brains of phthalate-exposed females were similar to control females. Researchers have also linked phthalate exposure to alterations in the dopaminergic pathway (important for movement and reward). The mechanisms of action underlying phthalate-induced changes in neurodevelopment have not been fully elucidated; however, it is likely that altered steroidogenesis may play a vital role. While research on neurodevelopmental effects of phthalate exposure is increasing, information is incomplete on the underlying neural and molecular mechanisms of phthalate toxicity in the brain. We recommend more extensive animal studies on phthalate exposure during vulnerable developmental periods when rapid structural and functional changes in the brain are taking place. Animal models will be

* Email: csmith@connect.carleton.ca

crucial in establishing a causal link between phthalate exposure and changes in neuro-circuitry and in understanding the neural and molecular mechanisms which contribute to phthalate-induced changes in the developing brain.

INTRODUCTION

Phthalates are industrial chemicals that are added to plastic products to increase their flexibility and durability (Lyche, et al., 2009). Phthalates have been measured in individuals of all ages, but researchers have reported infants and children have a greater exposure to these chemicals than adults (Koch, et al., 2004; Wormuth, et al., 2006). Researchers have linked phthalate exposure to a number of adverse neurodevelopmental and behavioural effects in rodents, including hyperactivity and alterations in neuro-circuitry (Ishido, et al., 2004; Smith et al., 2011). It is unclear whether exposure to phthalates in humans results in similar changes in neurobiology and behaviour.

Exposure to Di(2-Ethylhexyl) Phthalate (DEHP)

Phthalates are not chemically bound to plastic polymers so over time approximately 2-50% of phthalates easily migrate out of these products into food, saliva, blood, etc. contaminating those who are in close contact (Heudorf, et al., 2007; Schettler, 2006). Ingestion is the most common route of exposure for DEHP with approximately 90% occurring through oral consumption of food in adults (Clarke et al., 2003a, 2003b; Latini, 2005; Wormuth et al., 2006). Exposure to DEHP through food intake is much lower in formula-fed and breastfed infants with 44% and 60% of the estimated intake of DEHP occurring via this route, respectively (Clarke et al., 2003a, 2003b). The United States Environmental Protection Agency (EPA) has recommended a reference dose of 20 μg/kg of body weight for the human population and the European Union has recommended tolerable daily intake of 20 μg/kg of body weight and 48 μg/kg of body weight for newborns and adults, respectively (Koch et al., 2006).

Neurotoxicity of DEHP

The health impact of phthalates is related to the length of the ester side chain. Only transitional phthalates with ester side chains in the "ortho" position, like DEHP, have been identified as having a toxic potential in humans (Fabjan et al., 2006). It is possible that other phthalates may present some health risk to the human population; however, research on these phthalates is quite limited. According to the Federal Hazardous Substance Act (FHSA), DEHP is labelled a chronic toxicant with no acute toxicity to humans (CPSC, 2010). Chronic exposure can be defined as an exposure period longer than 365 days while acute exposure is an exposure period of less than 14 days (CPSC, 2010). The CPSC (2010) considers DEHP to be a known animal, but only a "possible" human, reproductive and developmental toxicant (CPSC, 2010).

In the last decade, a number of neurotoxic effects of DEHP exposure have been reported in the literature. Examinations into DEHP-induced neurotoxicology have been limited to animal studies and it is unknown whether these toxicants adversely affect human brains as well. Adverse neurodevelopment effects from DEHP exposure have been documented in both male and female animals; although males have been more widely studied and seen to be more sensitive to phthalate exposure.

Neuroanatomical Changes

Adult male and female rats, and adult male mice that were exposed to chronic oral doses of DEHP showed significant increases in brain weight at high doses (rats: 789-938.5 mg/kg/day; male mice: 1266.1 mg/kg/day) while only male mice showed a decrease in brain weight at chronic low doses of DEHP (19.2 mg/kg/day; David et al., 2000a; 2000b). The brain weight of female mice was unaffected by DEHP exposure. Changes in brain weight were not associated with morphological changes in the brains of affected animals (David et al., 2000a; 2000b). In contrast, Rhodes et al., (1986) observed no differences in brain weight in phthalate-exposed and control marmoset monkeys suggesting that changes in brain weight may be restricted to rodents.

Hippocampal-Connectivity

Non-genetic experiences in the womb and shortly after birth can radically alter the normal developmental progression of the brain and the behaviours it produces. If the brain is adversely affected by toxins, development may be compromised producing detrimental effects throughout the lifespan of the organism. As the brain develops, neurons become more and more intricately connected and with this increased connectivity, increasingly complex mental health functions emerge. The developing nervous system requires large amounts of pruning and remodeling in order to achieve its final, fully functioning adult configuration (Kantor & Kolodkin, 2003). These connectivity-based changes are determined by a combination of innate genetic factors and responses to external stimuli (see Sur & Rubenstein, 2005 for review) such as environmental toxins. These connectivity-based changes occur in spurts or sensitive periods that can be observed brain-wide or in localized brain regions (Fenoglio et al., 2006). Insults that occur during these sensitive periods often have a long-lasting, irreversible effect.

One localized brain region that shows widespread connectivity-based changes during developmental sensitive periods is the hippocampus. The hippocampus has long been studied for its prominent role in learning and memory (Nadel & Moscovitch, 2001). The anatomy of the hippocampus is classically divided into four principal regions: the dentate gyrus (DG), the hippocampus proper made up of CA1, CA2 and CA3 subregions, the subiculum and the entorhinal cortex (Amaral & Witter, 1989; see Figure 1). From a general perspective, these hippocampal fields are connected through unidirectional excitatory projections (Goodrich-Hunsaker et al., 2008). One set of projections, referred to as the mossy fibers, connect the DG region to the CA3 subfield. There is increasing evidence that mossy fibers in the adult rat may regulate mental health function (e.g., executive function) by remodeling or changing their synaptic connectivity (Holahan et al., 2006; Ramirez-Amaya et al., 2001; Ramirez-Amaya et al., 1999). During postnatal development, the mossy fibers show a pattern of connectivity-based changes characterized by widespread neuroanatomical adjustments suggestive of a sensitive period (Holahan et al., 2007; 2010).

Previous work has highlighted developmental differences in the Long-Evans hooded (LER) and Wistar (WR) rat strains in terms of hippocampal mossy fiber connectivity (Holahan et al., 2007). From 18 to 21 days after birth (postnatal day 18 - 21, abbreviated as p18 - p21; approximately 4 years old in humans), both strains show a rapidly changing and expanding mossy fiber projection throughout the entire CA3 region. By p24, these changes stabilize in the LER, whereas in WR, at p24, there is a less-pronounced mossy fiber connectivity pattern in CA3 (Holahan et al., 2007). Thus, although LER and WR show a growth spurt in the mossy fiber connectivity patterns between p18 and p21, the WR connectivity pattern appears to retract leaving minimal mossy fiber innervation that persists to adulthood (Holahan et al., 2006). It is worthwhile to note that the LER axonal distribution in the hippocampus more closely matches that of human and non-human primates (Seress, 1992).

The structural development described above has recently been correlated with the emergence of memory function (Keeley et al., 2010). Using the LER and WR rat strains, cognitive function was assessed in a water maze task – a task routinely used in the study of learning and memory in rodents (Maei et al., 2009). From p16 to p19, no overall differences were observed between strains when measuring memory performance. At p20, the performance between the two strains showed a striking difference; LER showed a dramatic improvement in performance compared to WR. This enhanced function in the LER persisted into adulthood (Holahan et al., 2006; Keeley et al., 2010).

These results parallel the earlier studies of LER and WR mossy fiber development indicating a sensitive developmental period between p18 and p24 for both neuroanatomical connectivity and behavioural-based memory performance.

Additional evidence exists to support the idea that this is indeed a sensitive developmental period: 1) maternal separation just prior to this period results in a significant reduction in mossy fiber axon connectivity and impairs spatial learning (Huot et al., 2002), 2) early-life experience given to rats from p16 to p20 results in improved spatial performance when compared to age-matched controls without early life experience (Carman & Mactutus, 2001; Keeley et al., 2010) and 3) daily administration of a neuronal toxin from p17 to p20 in LER reduces the widespread mossy fiber connectivity pattern as observed in controls (Holahan et al., 2007). These observations suggest that a sensitive developmental period in rats spans from p18 to p21 (approximately 4 years old in humans) thus allowing for the study of factors that influence both brain and behavioural development during this time.

Acute exposure to low doses of DEHP (10 mg/kg) during this sensitive development period (postnatal day 16 to 22) in Long Evans rats differentially affected hippocampal connectivity between genders (Smith et al., 2011). Male rats exposed to DEHP show a reduction in axonal staining in the stratum oriens (see Figure 2) as well as a reduction in cell density in the stratum pyramidale compared to saline-treated males, suggesting that DEHP exposure interfered with proper connectivity and development of this hippocampal region (Smith et al., 2011). DEHP-treated male rats also showed a reduction in the number of doublecortin positive cells in the dentate gyrus when compared to control males. The same markers in the hippocampus of female rats were similar following saline and DEHP-treatment suggestingthe hippocampus of male rats may be more vulnerable to DEHP exposure (Smith et al., 2011).

Neurophysiological and Chemical Changes

A number of alterations in the physiology and chemistry of the brain have been reported in young and adult rats following DEHP exposure. Adult rats exposed to DEHP or DEHP metabolites showed significantly reduced activity in membrane Na^+/K^+-ATPase that coincided with neuronal degeneration (Dhanya et al., 2003; 2004). Na^+/K^+-ATPase is an enzyme important for cation transportation (also referred to as the sodium-potassium pump) and when inhibited can lead to neuronal cell death (Xiao et al., 2002). Dhanya et al. hypothesized that the neuronal degeneration observed in these studies may be mediated by disruptions to ion homeostasis (such as increased intracellular levels of Ca^{+2}) caused by the inhibition of Na^+/K^+-ATPase activity. In another study, researchers showed that intracellular levels of Ca^{+2} were affected after DEHP exposure and levels increased quite rapidly (within minutes) in neurosecretory and pheochromocytoma cells (Tully et al., 2000).

A single low dose of dicyclohexyl phthalate (DCHP) in five-day old rat pups has been reported to lead to a number of alterations in the dopaminergic pathway, including the down-regulation of dopamine D4 receptors in the midbrain and striatum, and dopamine transporters in the midbrain, and reduced tyrosine hydroxylase labeling in the substantianigra and the ventral tegmental area (Ishido, et al., 2004).

Researchers also found an increase in TUNEL labelling (a process which detects the presence of DNA fragmentation found in cells in the last stage of apoptosis) in the substantianigra (Ishido, et al., 2004). Other neurochemical alterations due to DEHP exposure include supressed activity at γ-aminobutyric acid (GABA) type A receptors and increased activity at glycine receptors in the female gameotocytes of frogs (Yang et al., 2007).

Lipid Profiles

The fetal brain contains very high concentrations of lipids, second only to adipose tissue (Ballabriga, 1994). Lipids are transferred from the mother to the fetus via the placenta (Haggerty, 2004). Lipids play an essential role in normal neurodevelopment of the fetus, including neurogenesis and myelination (Innis, 2003) and thus, alterations in lipid homeostasis in the placenta may lead to altered lipid composition and abnormal brain development in the fetus.

Rat dams were orally exposed to DEHP via gavage from gestation day 0 to 19, inclusive (Xu et al., 2007). Fetal rat brains were extracted on gestation day 20 and the concentration of eleven lipid classes were analyzed using gas chromotography mass spectrometry. Fetal brains that were exposed to DEHP *in utero* showed a 33% and 54% decrease in the concentration of free cholesterol ester and sphingomyelin, respectively, when compared to controls (Xu et al., 2007).

DEHP exposure also led to decreases in the composition of mono- and polyunsaturated fatty acids. Finally, five lipid classes (cholesterol esters, diacylglycerol, phosphatidylserine, lysophosphatidylcholine, and sphingomyelin) showed reductions in docosahexaenoic acid composition and two lipid classes (cholesterol esters and lysophosphatidylcholine) had decreased arachidonic acid composition (Xu et al., 2007).

These results indicate that maternal exposure to DEHP leads to altered lipid profiles in the fetal rat brain which may produce downstream changes in brain development.

Neurotoxicity in Humans

Although research that examines DEHP exposure in animal (specifically rodent) neurodevelopment has identified a number of adverse neurological effects, it is unclear whether human exposure to DEHP results in similar changes in neurobiology. More recently, a number of behavioural consequences in humans have been correlated with urinary concentrations of phthalate metabolites (Cho, et al., 2010; Engel, et al., 2010; Kim, et al., 2009; Swan et al., 2010). Given what we know from animal studies, it is likely that these behavioural changes correlate with neurological changes indicating more elaborate human studies which examine both neurological and behavioural consequences need to be completed.

Behavioural Effects of DEHP

Researchers have examined the effect of DEHP exposure on a variety of behaviours in animals and more recently, in humans (Boberg et al., 2011; Engel, et al., 2010; Tanaka 2002; 2005; Swan et al., 2010).

There are some discrepancies between animal and human studies indicating that 1) more research is necessary to draw conclusions about the effects of DEHP exposure on behaviour and 2) there may be species-specific differences in how DEHP exposure affects humans and animals.

Learning and Memory

Animal studies have produced conflicting results in regards to tasks involving spatial learning (Boberg et al., 2011; Tanaka 2002; 2005). In one study, female rats exposed to DINP showed a dose-dependent decrease in path length and latency to locate a hidden platform in a water maze task compared to control females, suggesting a beneficial effect of phthalate exposure in females (Boberg et al., 2011).

There were no differences between DINP-exposed and control males. In the same study, DINP exposure did not alter task performance on a radial arm maze (another hippocampal dependent task) in male or female rats, suggesting changes in spatial learning may be dependent on the type of maze used (Boberg et al., 2011). In another group of studies, it was reported that DEHP exposure did not affect water T maze learning in male or female mice (Tanaka, 2002; 2005). The inconsistencies in findings between these studies may be accounted by the use of 1) different phthalates (DEHP verses DINP) and 2) difference species (rats verses mice).

In human studies, children exposed to higher concentrations of phthalates prenatally (based on urinary concentrations of phthalate metabolites in mothers during their third trimester of pregnancy) performed worse on working memory tasks at ages 4 and 9 (Engel et al., 2010). This finding in humans is surprising given that phthalate exposure in animal studies either had no effect or an improvement in cognitive performance (Boberg et al., 2011, Tanaka 2002; 2005).

In addition, because of the negative impact on hippocampal development (a brain region important for spatial learning and memory) after DEHP treatment in rodents (Smith et al., 2011), it would be of interest to thoroughly examine spatial learning and memory changes associated with DEHP exposure in humans. These data point to the importance of examining the experimental link between phthalate exposure and cognitive functioning.

Attention Deficit-Hyperactivity Disorder (ADHD)

ADHD is a developmental disorder that has been characterized by attention problems, hyperactivity and impulsivity (Zwi et al., 2000). Animal and human studies support the hypothesis that exposure to phthalates early in life may represent one underlying causal factor of this disorder. In animal studies, researchers have reported that a single dose of DCHP in male rats increased spontaneous motor activity by a function of 1.4 when compared to controls (Ishido et al., 2004).

DCHP-induced hyperactivity increased throughout the dark phase of the light-dark cycle and was dose-dependent - with rats exposed to higher doses of DCHP showing more pronounced increases in hyperactivity (Ishido et al., 2004). As mentioned earlier, DCHP exposure also leads to changes in dopaminergic pathway, including the down-regulation of dopamine D4 receptors and dopamine transporters, and reduced tyrosine hydroxylase (a precursor of dopamine) staining in rats (Ishido et al., 2004). Given that the neurochemical basis of ADHD is heavily centered on abnormalities in the dopaminergic functioning (see Tripp & Wickens, 2008 for review) it is quite likely that in some cases phthalate exposure early in life may lead to the development of ADHD.

In humans, researchers have shown strong associations between the concentrations of phthalate metabolites in the urine of school-age children (grade three to five) and symptoms of ADHD, where children with higher concentrations of phthalates were rated by their teachers as showing more symptoms of ADHD (Engel et al., 2010; Kim et al., 2009). While the connection between ADHD and phthalate exposure has not been directly approached through experimental research, it is possible that phthalate exposure may represent a potential risk factor for the pathophysiology of ADHD.

Other Behavioural Effects

Researchers have also reported positive correlations between concentrations of phthalate metabolites in the urine of school-age children (grade three to five) and Intelligence Quotient scores (Cho et al., 2010) and negative correlations between phthalate metabolites in boys and sex-typical play behaviours (Swan et al., 2010).

Possible Mechanisms of Toxicity

DEHP exposure appears to affect many aspects of neurodevelopment and behaviour, and likely acts through many molecular mechanisms, including disruptions in the endocrine system and aberrant apoptosis. It is possible that theses molecular mechanisms may be different for males and females and may explain why some studies have reported differential effects of DEHP-treatment across gender (Andrade et al., 2006; Boberg et al., 2011; David et al., 2000a; 2000b; Smith et al., 2011).

Endocrine Disruption

The endocrine system is important for maintaining a number of homeostatic mechanisms such as reproduction, metabolism, and other regulatory functions (Crisp et al., 1998). It is comprised of a number of glands (including the pituitary and adrenal glands) which release hormones directly into the bloodstream.Hormones are slow acting, but can have long-lasting effects on growth and development (Crisp et al., 1998).

Endocrine disrupting chemicals (EDC) can be defined as natural or synthetic compounds that mimic or interfere with the biosynthesis, metabolism or action of endogenous hormones and thus, interfere with homeostatic maintenance (Crisp et al., 1998). DEHP has been described as a synthetic toxicant with endocrine disrupting characteristics. Although the mechanism of action underlying DEHP neurotoxicity is not yet understood, it is likely that DEHP's action as an EDC may at least in part contribute to the neurodevelopmental and behavioural changes observed in animals and humans. In particular, the effects of DEHP exposure on androgen and estrogensignalling, and peroxisome proliferator-activated receptors (PPAR) will be discussed.

*Anti-androgeniceffects.*Androgens, such as testosterone, are critical for the masculinization of the brain. Directly or indirectly interfering with androgen levels can prevent brain masculinization resulting in a more feminized version of the brain. It is suggested that males may show an increased vulnerability to DEHP exposure, which may be explained through DEHP action as an anti-androgenic compound (i.e., compounds which interfere with the biological role of androgens).

The anti-androgenic properties of DEHP do not appear to involve the androgen receptor (AR). DEHP is not an AR antagonist (Gray et al., 1994; Kelce et al., 1994) and neither DEHP nor MEHP shows an affinity for AR (Parks et al., 2000; Takeuchi et al., 2005). Instead, the anti-androgen effects of DEHP may be related to the disruption of Sertoliand Leydigcell function during sensitive developmental periods. Treatment with DEHP during early developmental periods led to a 35 % decrease in the number of Sertoli cells (Dostal et al., 1988) and disrupted Leydig cell function (Jones et al., 1993; Parks et al., 2000). Sertoli and Leydig cells are essential for the synthesis of testosterone and thus, a reduction in the number or in the functioning of these cells can lead to altered testosterone levels.

A number of studies have reported decreased serum testosterone levels in animals following treatment with DEHP (Akingbemi et al., 2001; Borch et al., 2004; 2006; Ge et al., 2007; Noreiga et al., 2009; Parks et al., 2000). In one study, a 60-85% reduction in testosterone production was reported following DEHP treatment, such that testosterone levels in males were reduced to levels as seen in females (Parks et al., 2000). These finding support the idea that feminization of the male brain following DEHP treatment may be related to the suppression of testosterone production.

There is a slight discrepancy in the literature at what dose DEHP decreases testosterone production, with some studies reporting reductions at low (10 mg/kg) and high doses (200 mg/kg; Akingbemi et al., 2001; Borch et al., 2004; 2006; Ge et al., 2007), while others have demonstrated effects at only high doses (300 and 900 mg/kg; Noreiga et al., 2009). Additionally, it appears the timing of DEHP exposure is very important in determining how testosterone production will be affected. Akingbemi et al., (2001) have shown that maternal exposure to DEHP reduced testosterone production in male offspring at postnatal days 21 and 35, but at postnatal day 90, testosterone production was equivalent to control males. Postnatal treatment with DEHP for 14 days also reduced testosterone levels in males, but when animals were exposed for 28 days testosterone levels increased (Akingbemi et al., 2001). These data point to the importance of examining the effects of DEHP at different time points in development and at different doses.

The production of testosterone in Leydig cells is regulated by luteinizing hormone (LH) – a hormone released from the anterior pituitary gland. DEHP treatment can alter LH levels and thus, can indirectly affect testosterone production (Akingbemi et al., 2001). Timing also appears to be central in how LH levels will be altered following DEHP treatment. Maternal and acute postnatal (14 days) exposure to DEHP reduced LH levels in males, while longer postnatal exposure periods of 28 days led to increased LH levels in males (Akingbemi et al., 2001).In a more recent study, mRNA levels of LH were significantly up-regulated and LH receptors were down-regulated at low doses of DEHP following maternal and postnatal exposure in males and females (Pocar et al., 2012). These findings highlight the significance of exploring the connection between testosterone production and neurodevelopment in males to determine if changes in neuro-circuitry observed in DEHP-exposed males can be attributed to low levels of testosterone.

*Estrogenic effects.*There is strong evidence to support the idea that DEHP may lead to alterations in estrogensignalling pathways in animals. *In vitro,* exposure to DEHP (but not MEHP) led to weak estrogenic activity via human estrogen receptor α (ERα) but not human estrogen receptor β (ERβ) and in the presence of 17 β-estradiol, weak anti-estrogenic effects have been reported in ERα but not ERβ (Takeuchi et al., 2005). In studies conducted *in vivo*, the administration of DEHP did not induced estrogenic effects in ovariectomized rats (Zacharewski et al., 1998).

Decreased levels of circulating estrogen have been reported in adult female rodents and in cultured female rat granulosa cells following exposure to DEHP and MEHP, respectively, in a dose-dependent manner (Davis et al., 1994a; 1994b; Lovekamp-Swan et al., 2003). Estrogen levels in male rodents were not evaluated in these studies. Researchers have also demonstrated significant down-regulation of the progesterone receptor (an estrogen target) in males and females, and down-regulation of CYP17a1 (a gene important in the synthesis of steroid hormones) in females at low doses of DEHP, suggesting low levels of serum estrogen levels in DEHP exposed animals (Pocar et al., 2012).

Pocar et al. (2012) hypothesized that low levels of estrogen may have a negative effect on the hypothalamic-pituitary-gonadal axis and thus, can affect the secretion of gonadotropins, such as LH and follicle-stimulating hormone (FSH). As mentioned, LH is important in testosterone synthesis and LH mRNA levels are up-regulated and LH receptors are down-regulated in DEHP-exposed males and females (Pocar et al., 2012). FSH is important in promoting normal gonadal function and in normal controls. FSH release elevates intracellular levels of cyclic adenosine monophosphate (cAMP) – a protein important in steroidogenesis (Heindeland Chapin, 1989). Following treatment with MEHP, the ability of FSH to elevate cAMP was reduced by 40% (Heindeland Chapin, 1989). In another study, researchers found that FSH mRNA was up-regulated in males (but not females) after DEHP exposure and the FSH receptor was down-regulated in both males and females (Pocar et al., 2012).

The decrease in circulating estrogen levels in females may be related to the suppression of the aromatase enzyme (a key enzyme in the conversion of testosterone into estrogen and in the masculinization of the brain) by MEHP (Davis et al., 1994b).

Researchers have also shown decreased mRNA expression and protein levels in the aromatase enzyme following MEHP treatment in females (Lovekampand Davis, 2001) as well as a significant down-regulation in CYP19a1 (the gene which encodes aromatase enzyme) following DEHP in males and females (Pocar et al., 2012).

In the brain, *in utero* and lactational exposure to DEHP at low doses suppressed aromatase enzyme activity in the hypothalamic/preoptic area in one-day-old male rat pups, whereas higher doses produced the opposite effect (Andrade, et al., 2006). In one-day old females, DEHP exposure did not affect aromatase activity, suggesting phthalate exposure may differentially affect the brains of developing males and females. The suppression of this enzyme during early brain development in males (but not females) at low (and more ecologically relevant) doses may account for the development of a feminized male brain.

At postnatal day 22 (p22), females showed an increase in aromatase activity at all phthalate exposure doses, but in males only one of the lower doses led to an increase in activity (Andrade et al., 2006). These data indicate that 1) exposure to DEHP can indirectly effect brain masculinization by interfering with the aromatization of androgens into estrogens, 2) the effects of DEHP exposure on the male and the female brain are distinctly different throughout development and 3) brain development in females may also be affected by DEHP exposure.

*Peroxisome proliferation.*Alterations in the lipid profiles of fetal rat brains following maternal DEHP may be mediated by abnormal activation of peroxisome proliferator-activated nuclear hormone receptor family (PPAR) by DEHP and its metabolites (Xu et al., 2005; 2006).PPAR receptors play an important role in the regulation of lipid/essential fatty acid transporters and enzymes as well as in the maintenance of lipid/essential fatty acid homeostasis (Lemberger et al., 1996). DEHP and its metabolites (MEHP and 2-ethylhexanoic acid) can bind readily to PPAR receptors– acting as an agonist (Xu et al., 2005; 2006).

Exposure to DEHP, MEHP and 2-ethylhexanoic acid increased the uptake, transport, and accumulation of essential fatty acids in rat placental cells (Xu et al., 2005; 2006).

Researchers have also indicated a possible role of PPARs in the synthesis of steroid hormones, such as estrogens and thus, alterations in PPAR functioning may lead to disruptions in estrogen synthesis (Lovekamp-Swan et al., 2003). These finding suggest an important role for PPARs in not only maintaining lipid/essential fatty acid homeostasis, but in steroidogenesis and thus, PPAR-dysfunction may be associated with abnormal neurodevelopment.

Apoptosis

Apoptosis is a normal activity-dependent process that occurs routinely in our brains. As mentioned previously, acute exposure to DEHP postnatally led to a reduction in mature neurons in the CA3 region of juvenile rat hippocampi (Smith et al., 2011). It is possible that the decreased cell density in this region was related to enhanced apoptosis from DEHP exposure. Researchers have demonstrated that MEHP induced apoptosis in testicular germ cells via the release of cytochrome c and activation of caspase-3, and through the FAS receptor pathway – all important proteins in apoptosis (Dalgaarrd et al., 2001; Kasahara et al., 2002; Richburg et al., 1999; 2000). In another study, up to 53.4% of cells showed a dose-dependent MEHP-induced cell death compared to controls (Yokoyama et al., 2003). Levels of pro-apoptotic protein, BAX, and anti-apoptotic protein, Bcl-2 were increased and decreased, respectively, following MEHP treatment (Yokoyama et al., 2003). Caspase-3 in cells was increased in a dose-dependent manner with increasing concentrations of MEHP, which was reduced following the administration of PPAR gamma antagoinists suggest an alternative role for PPAR in DEHP mediated affects (Yokoyama et al., 2003).

Conclusion

Research investigating the effects of phthalates on brain development has been somewhat limited until recent years. It is apparent phthalate exposure adversely affects many aspects of neurodevelopment in animals, particularly rodents.

The effects on human brains have not yet been evaluated, suggesting a strong need for these types of studies in the future. It is clear that phthalate exposure leads to behavioural changes in the human population. It will be important to expand on these findings and evaluate how these changes in behaviour arecorrelated with brain structure and function through more extensive neurobehaviouralanimal studies and imaging studies in humans. Thorough investigations which explore gender differences in neurodevelopment and behaviour after phthalate exposure are needed.

It is clear that both males and females are affected by phthalate-exposure, but the neural and molecular mechanisms that underlie neurotoxicity have not been fully elucidated. Since similar phthalate treatment results in differential effects across the genders, it is likely that phthalates alter neurodevelopment via different mechanisms in males and females. Researchers have indicated endocrine dysfunction and abnormal apoptosis may be significant contributors to phthalate-induced neurotoxicity.We recommend more comprehensive experimental research studies that examine steroid hormone levels and factors which alter steroidogenesis in the brain following phthalate exposure and how these changes are associated with brain structure and function. Animal models will be crucial in establishing a causal link between phthalate exposure and changes in neuro-circuitry and in understanding the neural and molecular mechanisms which contribute to phthalate-induced changes in the developing brain.

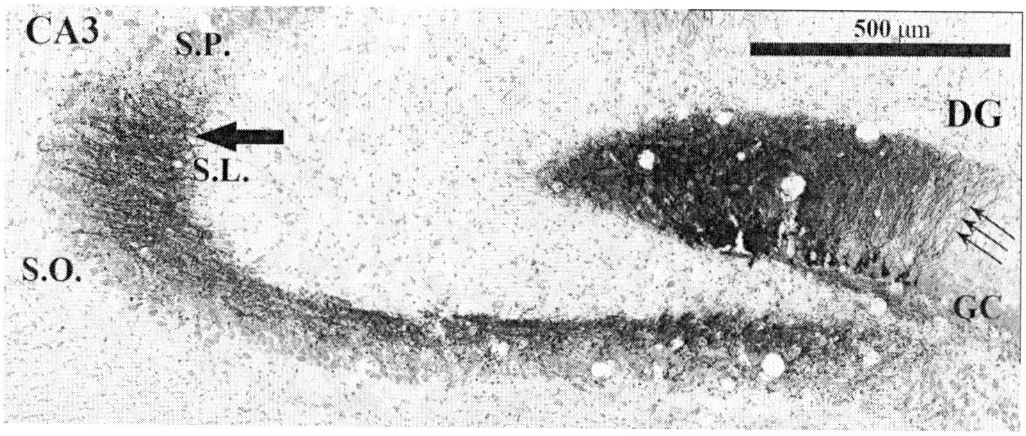

Figure 1. Hippocampal connectivity.Biocytin-labeling of granule cells (GC) in the dentate gyrus (DG).Information is transmitted through the axons of GCs and terminates on pyramidal neurons in the stratum pyramidale (SP) of the CA3 to form the mossy-fibre pathway. SO: *stratum oriens*; SL: *stratum lucidum*.

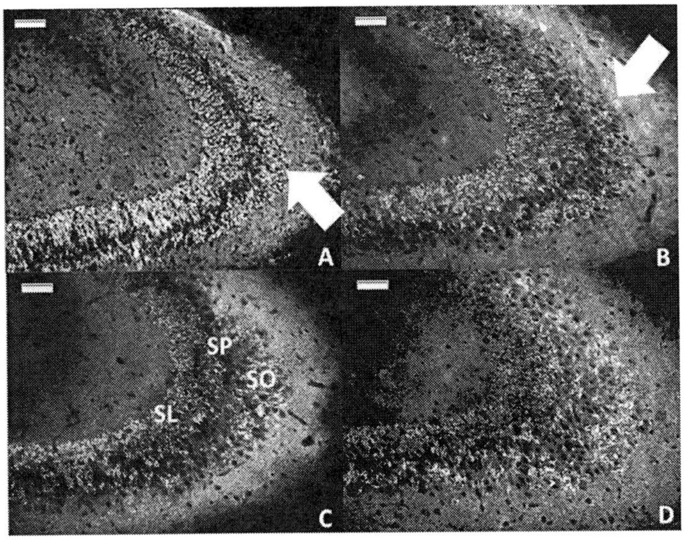

Figure 2A-D. Synaptophysin staining in the CA3 region.Representative photomicrographs of synaptophysin staining in the CA3 (20x) of saline-treated males (A) and females (C) and phthalate-treated males (B) and females (D). Downward arrows show that males exposed to phthalate showed a significant reduction in the synaptophysin stain in the stratum oriens compared to saline-treated males. SO: *stratum* oriens; SP: *stratum pyramidale*; SL: *stratum* lucidum; DEHP: phthalate.

REFERENCES

Akingbemi, B.T., Youker, R.T., Sottas, C.M., Ge, R., Katz, E., Klinefelter, G.R. et al., (2001).Modulation of rat Leydig cell steroidogenic function by di(2-ethylhexyl)phthalate. *Biol. Reprod.* 65:1252-1259.

Amaral, D. G., and Witter, M. P. (1989). The three-dimensional organization of the hippocampal formation: a reivew of anatomical data. *Neuroscience, 31*(3), 571-591.

Andrade, A.J.M., Grande, S.W., Talsness, C.E., Grote, K. and Chahoud, I. (2006). A dose response study following *in utero* and lactational exposure to di-(2-ethylhexyl)-phthalate (DEHP): Non-monotonic dose–response and low dose effects on rat brain aromatase activity. *Toxicology* 227:185-192.

Ballabriga A (1994) Essential fatty acids and human tissue com-position.An overview.*Acta Paediatr. Suppl.* 402:6–8.

Borch, J., Ladefoged, O., and Vinggaard, A.M. (2004).Steroidogenesis in fetal male rats is reduced by DEHP and DINP, but endocrine effects of DEHP are not modulated by DEHA in fetal, prepubertal and adult male rats. *Reprod. Toxicol.* 18, 53–61.

Borch, J., Metzdorff, S.B., Vinggaard, A.M., Brokken, L. andDalgaard, M. (2006).Mechanisms underlying the anti-androgenic effects of diethylhexyl phthalate in fetal rat testis. *Toxicology* 223:144-155.

Boberg J., Christiansen, S., Axelstad, M., Kledal, T.S., Vinggaard, A.M., Dalgaard, M. et al., (2011). Reproductive and behavioral effects of diisononyl phthalate (DINP) in perinatally exposed rats. *Reprod. Toxicol* 31:200-209.

Carman, H. M., and Mactutus, C. F. (2001). Ontogeny of spatial navigation in rats: a role for response requirements? *Behav. Neurosci, 115*(4), 870-879.

Cho, S-C., Bhang, S-Y., Hong, Y-C., Shin, M-S., Kim, B.N., Kim, J.W. et al., (2010).Relationship between environmental phthalate exposure and the intelligence of school-aged children. *Environ. Health Perspect* 118:1027-1032.

Clark, K., Cousins, I. T., and Mackay, D. (2003a).Assessment of critical exposure pathways.In The handbook of environmental chemistry ed. Staples, C. A., pp. 227–262. New York: Springer-Verlag.

Clark, K., Cousins, I. T., Mackay, D., and Yamada, K. (2003b).Observed concentrations in the environment. In The handbook of environ-mental chemistry, ed. Staples, C. A., pp. 125–177. New York; Springer-Verlag.

Crisp, T.M., Clegg, E.D., Cooper, R.L., Wood, W.P., Anderson, D.G., Baetcke, K.P., et al., (1998).Environmental endocrine disruption: An effects assessment and analysis. *Environ. Health Perspect.* 106: 11–56.

CPSC. (2010). Toxicity review of di(2-ethylhexyl)phthalate (DEHP). Bethesda, MD: Consumer Product Safety Commission, http://www.cpsc.gov/about/cpsia/toxicity DEHP.pdf.

Dalgaard, M., Nellemann, C., Lam, H.R., Sorensen, I.K., Ladefoged, O. (2001). The acute effects of mono(2-ethylhexyl) phthalate (MEHP) on testes of prepubertalWistar rats. *Toxicol.Lett.* 122: 69-79.

David, R.M., Moore, M.R., Finney, D.C., and D. Guest. 2000a. Chronic toxicity of di(2-ethylhexyl)phthalate in rats. *Toxicological Sciences*. 55: 433-443.

David, R.M., Moore, M.R., Finney, D.C., and D. Guest. 2000b. Chronic toxicity of di(2-ethylhexyl)phthalate in mice. *Toxicological Sciences*. 58: 377-385.

Davis, B.J., Maronpot, R.R. and Heindel, J.J. (1994a). Di-(2-ethylhexyl) phthalate suppresses estradiol and ovulation in cycling rats. *Toxicol. Appl. Pharmacol.,* 128:216–223.

Davis, B.J., Weaver, R., Gaines, L.J.andHeindel, J.J. (1994b). Mono-(2-ethylhexyl) phtha late suppresses estradiol production independent of FSH-cAMP stimulation in rat granulosa cells. *Toxicol. Appl. Pharmacol.* 128(2):224-8.

Dhanya C. R., Gayathri N. S., Mithra K., Nair K. V. and Kurup P. A. (2004) Vitamin E prevents deleterious effects of di (2-ethyl hexyl) phthalate, a plasticizer used in PVC blood storage bags. *Indian J. Exp. Biol.* 42, 871-5.

Dhanya CR, Indu AR, Deepadevi KV, Kurup PA (2003) Inhibition of membrane Na(+)-K+ ATPase of the brain, liver and RBC in rats administered di(2-ethyl hexyl) phthalate (DE-HP) a plasticizer used in polyvinyl chloride (PVC) blood storage bags. *Indian J. Exp. Biol.* 41:814–820.

Dostal, L.A., Chapin, R.E., Stefanski, S.A., Harris, M.W., and B.A. Schwetz. (1988). Testicular toxicity and reduced Sertoli cell numbers in neonatal rats by di(2-ethylhexyl) phthalate and the recovery of fertility as adults. *Toxicol. Appl. Pharmacol.* 95: 104-121.

Engel, S.M., Miodovnik, A., Canfield, R.L., Zhu, C., Silva, M.J., Calafat, A.M. et al., (2010). Prenatal phthalate exposure is associated with childhood behaviour and executive functioning. *Environ Health Perspect* 118:565-571.

Fabjan, E., Hulzebos, E., Mennes, W., and Piersma, A. H. (2006).A category approach for reproductive effects of phthalates. *Crit. Rev.Toxicol.* 36:695–726.

Fenoglio, K. A., Brunson, K. L., and Baram, T. Z. (2006). Hippocampal neuroplasticity induced by early-life stress: functional and molecular aspects. *Frontiers in Neuroendocrinology, 27*(2), 180-192.

Gray, L.E., Ostby, J., and Kelce, W.R. (1994). Developmental effects of an environmental antiandrogen: the fungicide vinclozolin alters sex differentiation of the male rats. *Toxibol. Appl. Phatmacol.* 129:46-52.

Ge, R. S., Chen, G. R., Dong, Q., Akingbemi, B., Sottas, C. M., Santos, M., Sealfon, S. C., Bernard, D. J., and Hardy, M. P. (2007). Biphasic effects of postnatal exposure to diethylhexylphthalate on the timing of puberty in male rats. *J. Androl.*28, 513–520.

Goodrich-Hunsaker, N. J., Hunsaker, M. R., and Kesner, R. P. (2008). The interactions and dissociations of the dorsal hippocampus subregions: how the dentate gyrus, CA3, and CA1 process spatial information. *Behav. Neurosci., 122*(1), 16-26.

Haggarty P (2004) Effect of placental function on fatty acid requirements during pregnancy.*Eur. J. Clin. Nutr.* 58:1559–1570.

Heindel, J.J. and Chapin, R.E. (1989). Inhibition of FSH-stimulated cAMP accumulation by mono(2-ethylhexyl) phthalate in primary rat Sertoli cell cultures. *Toxicol. Appl. Pharmacol.* 97:377-385.

Heudorf, U., Mersch-Sundermann, V., and Angerer, J. (2007). Phthalates: Toxicology and exposure. *Int. J. Hyg. Environ. Health* 210:623–634.

Holahan, M. R., Honegger, K. S., and Routtenberg, A. (2007). Expansion and retraction of hippocampal mossy fibers during postweaning development: strain-specific effects of NMDA receptor blockade. [Research Support, N.I.H., Extramural]. *Hippocampus, 17*(1), 58-67. doi: 10.1002/hipo.20242.

Holahan, M. R., Honegger, K. S., and Routtenberg, A. (2010). Ectopic growth of hippocampal mossy fibers in a mutated GAP-43 transgenic mouse with impaired spatial memory retention. [Research Support, N.I.H., ExtramuralResearch Support, Non-U.S. Gov't]. *Hippocampus, 20*(1), 58-64. doi: 10.1002/hipo.20635.

Holahan, M. R., Rekart, J. L., Sandoval, J., and Routtenberg, A. (2006). Spatial learning induces presynaptic structural remodeling in the hippocampal mossy fiber system of two rat strains. [Research Support, N.I.H., ExtramuralResearch Support, U.S. Gov't, Non-P.H.S.]. *Hippocampus, 16*(6), 560-570. doi: 10.1002/hipo.20185.

Huot, R. L., Plotsky, P. M., Lenox, R. H., and McNamara, R. K. (2002). Neonatal maternal separation reduces hippocampal mossy fiber density in adult Long Evans rats. *Brain Res., 950*(1-2), 52-63.

Innis SM (2003) Perinatal biochemistry and physiology of long-chain polyunsaturated fatty acids. *J. Pediatr* 143:S1–S8.

Ishido, M., Masuo, Y., Sayato-Suzuki, J., Oka, S., Niki, E. and Morita, M. (2004). Dicyclohexylphthalate causes hyperactivity in the rat concomitantly with impairment of tyrosine hydroxylase immunoreactivity. *J. Neurochem*. 91:69-76.

Jones, H.B., Garside, D.A., Liu, R. and Roberts, J.C. (1993). The influence of phthalate esters on Leydig cell structure and function in vitro and in vivo.*Exp. Mol. Pathol.* 58:179–193.

Kantor, D. B., and Kolodkin, A. L. (2003). Curbing the excesses of youth: molecular insights into axonal pruning. *Neuron, 38*(6), 849-852.

Kasahara, E., Sato, E.F., Miyoshi, M., Konaka, R., Hiramoto, K., Sasaki, J.(2002).Role of oxidative stress in germ cell apoptosis induced by di(2-ethylhexyl)phthalate. *Biochem. J.* 365, 849_/856.

Keeley, R. J., Wartman, B. C., Hausler, A. N., and Holahan, M. R. (2010). Effect of juvenile pretraining on adolescent structural hippocampal attributes as a substrate for enhanced spatial performance. *Learn Mem, 17*(7), 344-354. doi: 10.1101/lm.1849910.

Kelce, W.R., Monosson, E., Gamesik, M.P., Laws, S. and Gray, L.E., Jr. (1994). Environmental hormone disruptors: evidence that vinclozolin developmental toxicity is mediated by antuandorgenicmetaboloties. *Toxicol. Appl. Pharmacol.* 126:276-285.

Kim, B-N., Cho, S.C., Kim, Y., Shin, M-S., Yoo, H-J., Kim, J-W., Yang, Y.H., Kim, H.W., Bhang, S-Y. and Hong, Y-C. (2009). Phthalates exposure and attention-deficit/hyperactivity disorder in school-age children. *Environ. Health Perspect* 66:958-963.

Koch, H.M., Drexler, H. and Angerer, J. (2004). Internal exposure of nursery-school children and their parents and teachers to di(2-ethylhexyl)phthalate (DEHP). *Int. J. Hyg. Environ Health* 207:15-22.

Koch, H. M., Preuss, R., and Angerer, J. (2006).Di(2-ethylhexyl)phthalate (DEHP): Human metabolism and internal exposure—An update and latest results. *Int. J. Androl.* 29:155–165.

Latini, G. (2005). Monitoring phthalate exposure in humans.*Clin. Chim. Acta* 361:20–29.

Lemberger, T., Braissant, O., Juge-Aubry, C., Keller, H., Saladin, R., Staels, B., et al., (1996). PPAR tissue distribution and interactions with other hormone-signaling pathways. *Ann. NY Acad. Sci.* 804:231–251.

Lovekamp, T.N. and Davis, B.J. (2001). Mono-(2-ethylhexyl) phthalate suppresses aromatase transcript levels and estradiol production in cultured rat granulosa cells. *Toxicol. Appl. Pharmacol.* 172: 217-224.

Lovekamp-Swan, T., Jetten, A.M. and Davis, B.J. (2003).Dual activation of PPARalpha and PPARgamma by mono-(2-ethylhexyl) phthalate in rat ovarian granulosa cells. *Mol. Cell. Endocrinol.* 201: 133-141.

Lyche, J.L., Gutleb, A.C., Berginan, A., Eriksen, G.S., Murk, A.J., Ropstad, E., Saunders, M. and Skaare, J.U. (2009).Reproductive and developmental toxicity of phthalates. *J. Toxicol. Environ. Health B. Crit. Rev.* 12:225-249.

Maei, H. R., Zaslavsky, K., Teixeira, C. M., and Frankland, P. W. (2009). What is the Most Sensitive Measure of Water Maze Probe Test Performance? *Front Integr Neurosci, 3*, 4.

Nadel, L., and Moscovitch, M. (2001). The hippocampal complex and long-term memory revisited. *Trends in Cognitive Sciences, 5*(6), 228-230.

Noriega, N., Howdeshell, K.L., Furr, J., Lambright, C.R., Wilson, V.S. and Gray, L.E., Jr. (2009). Pubertal administration of DEHP delays puberty, supresses testosterone production and inhibits reproductive tract development in male Sprqgue-Dawley and Long-Evans rats. *Toxicol. Sci.* 111:163-178.

Parks, L.G., Ostby, J.S., Lambright, C.R., Abbott, B.D., Klinefelter, G.R., Barlow, N.J. et al., (2000). The plasticizer diethylhexylphthalate induces malformations by decreasing fetal testosterone synthesis during sexual differentiation in the male rat. *Toxicol. Sci.* 58:339–349.

Pocar, P., Fiandanese, N., Secchi, C., Berrini, A., Fischer, B., Schmidt, J.S. et al., (2012). Exposure to Di(2-ethyl-hexyl) phthalate (DEHP *in Utero* and Lactation causes long-term pituitary=gonagal axis disruption inmale and female mouse offspring. *Endocrinol.* 153:937-948.

Ramirez-Amaya, V., Balderas, I., Sandoval, J., Escobar, M. L., and Bermudez-Rattoni, F. (2001). Spatial long-term memory is related to mossy fiber synaptogenesis. *J. Neurosci, 21*(18), 7340-7348.

Ramirez-Amaya, V., Escobar, M. L., Chao, V., and Bermudez-Rattoni, F. (1999). Synaptogenesis of mossy fibers induced by spatial water maze overtraining. *Hippocampus, 9*(6), 631-636.

Rhodes, C., Orton, T.C., and I.S. Pratt. (1986). Compartive pharmaocokinetics and subacute toxicity of di(2-ethylhexyl)phthalate in rats and marmosets: Extrapolation of effects in rodents to man. *Environmental Health Perspectives*. 65: 299-308.

Richburg, J.H., Nanez, A. and Gao, H. (1999). Participation of the Fas-signaling system in the initiation of germ cell apoptosis in young rat testes after exposure to mono-(2-ethylhexyl) phthalate. *Toxicol. Appl. Pharmacol.* 160, 271_/278.

Richburg, J.H., Nanez, A., Williams, L.R., Embree, M.E. and Boekelheide, K. (2000).Sensitivity of testicular germ cells to toxicant-induced apoptosis in gld mice that express a nonfunctional form of Fas ligand.Endocrinology 141, 787_/793.

Schettler, T. (2006).Human exposure to phthalates via consumer products. *Int. J. Androl.* 29:134-139.

Seress, L. (1992). Morphological variability and developmental aspects of monkey and human granule cells: differences between the rodent and primate dentate gyrus. In C. E. Ribak, C. M. Gall and I. Mody (Eds.), *The Dentate Gyrus an its Role in Seizures* (pp. 3-28). San Diego: *Elsevier Science Publishers.*

Smith, C.A., MacDonald, A., and Holahan, M.R. (2011).Acute postnatal exposure to di(2-ethylhexyl) phthalate adversely impacts hippocampal development in the male rat. *Neuroscience*. 193, 100-108.

Sur, M., and Rubenstein, J. L. (2005). Patterning and plasticity of the cerebral cortex. *Science, 310*(5749), 805-810.

Swan, S.H., Liu, F., Hines, M., Kruse, R.L., Wang, C., Redmon, J.B. et al., (2010). Prenatal phthalate exposure and reduced masculine play in boys. *Inter. J. Andr*. 33:259-269.

Takeuchi, S., Iida, M., Kobayashi, S., Jin, K., Matsuda, T. and Kojima, H. (2005). Differential effects of phthalate esters on trandscriptional activities via human estrogen receptors a and b, and androgen receptor. *Toxicology*, 210: 223-233.

Tanaka, T. (2002). Reproductive and neurobehavioural toxicity study of bis(2-ethylhexyl) phthalate (DEHP) administered to mice in the diet. *Food Chem. Toxicol*. 40:1499-1506.

Tanaka, T. (2005). Reproductive and neurobehavioural toxicity study of bis(2-ethylhexyl) phthalate (DEHP) in a cross-mating toxicity study of mice. *Food Chem. Toxicol.* 43:581-589.

Tripp, G. and Wickens, J.R. (2008). Research review: dopamine transfer deficit: a neurobiological theory of altered reinforcement mechanisms in ADHD. *J. Child Psychol. Psychiatry*, 49: 691-704.

Tully K., Kupfer D., and A.M. Dopico. (2000). A plasticizer released from IV drip chambers elevates calcium levels in neurosecretory terminals. *Toxicol. Appl. Pharmacol.* 168: 183-188.

Wormuth, M., Scheringer, M., Vollenweider, M. and Hungerbühler, K. (2006) What are the Sources of Exposure to Eight Frequently Used Phthalic Acid Esters in Europeans? *Risk Anal.* 26:803–824.

Xiao A. Y., Wei L., Xia S., Rothman S. and Yu S. P. (2002) Ionic mechanism of ouabain-induced concurrent apoptosis and necrosis in individual cultured cortical neurons. *J. Neurosci.* 22, 1350-62.

Xu, Y., Cook, T.J., and G.T. Knipp.(2005). Effects of di-(2-ethylhexyl)-phthalate (DEHP) and its metabolites on fatty acid homeostasis regulating proteins in rat placental HRP-1 trophoblast cells.*Toxicological Sciences*. 84: 287-300.

Xu, Y., Agrawal, S., Cook, T.J., and G. T. Knipp. (2006). Di-(2-ethylhexyl) phthalate affects lipid profiling in fetal rat brain upon maternal exposure. *Archives of Toxicology*. 81(1): 57-62.

Xu, Y., Agrawal, S., Cook, T.J. and Knipp, G.T. (2007). Di-(2-cthylhcxyl)-phthalatc affccts lipid profiling in fetal rat brain upon maternal exposure. *Reprod. Toxicol*. 81:57-62.

Yang, L., Milutinovic, P.S., Brosnan, R.J., Eger II, E.I., and J.M. Sonner. (2007). The plasticizer di(2-ethylhexyl) phthalate modulates γ-aminobutyric acid type A and glycine receptor function. *Anesthesia and An algesia*. 105(2): 393-396.

Yokoyama, Y., Okubo, T., Kano, I., Sato, S. and Kano, K. (2003). Induction of apoptosis by mom(2-ethylhexyl) phthalate (MEHP) in U937 cells. *Toxicol.Lett*. 144: 371-381.

Zacharewski, T.R., Meek, M.D., Clemons, J.H., Wu, Z.F., Fielden, M.R., and J.B. Matthews. (1998). Examination of the in vitro and in vivo estrogenic activities of eight commercial phthalate, esters. *Toxicological Sciences*. 46(2): 282-293.

Zwi, M., Ramchandani, P. and Joughin, C. (2000).Evidence and belief in *ADHD. BMJ*. 321(7267): 975–6.

In: Phthalates
Editors: Gerardo L. Moretti and Drago Romano
ISBN: 978-1-62081-994-4

Chapter 8

PHTHALATES: CHEMICAL PROPERTIES, IMPACTS ON HEALTH AND THE ENVIRONMENT

Anderson Joel Martino-Andrade[1] and Ibrahim Chahoud[2,*]
[1]Universidade Federal do Paraná, Departamento de Fisiologia, Curitiba, Brazil
[2]Charité Universitätsmedizin Berlin, Campus Charité Mitte, Institute of Clinical Pharmacology and Toxicology, Department of Toxicology, Berlin, Germany

ABSTRACT

Phthalate esters comprise a family of common industrial compounds, dialkyl- or alkyl/aryl esters of phthalic acid, which are used as plasticizers to impart flexibility to polyvinylchloride (PVC) plastics, as well as additives in many chemical and pharmaceutical formulations. They are present in a wide range of consumer products including building materials, food packaging, clothing, toys, medical devices, cosmetics, lubricant oils and solvents. Phthalates are not chemically bound to PVC and can leach out of plastics, which contributes to their ubiquitous presence in the environment. The industrial application as well as toxicity profile of phthalates is largely dependent upon the characteristics of the alkyl/aryl side chains of the ester molecule. Although phthalates usually display low acute toxicity, they have been focus of significant concern because of the evidence of reproductive adverse effects in experimental animals. In general, phthalates with medium- (e.g. dibutyl phthalate [DBP]) or branched long-side chains (e.g. di(2-ehtylhexyl phthalate [DEHP]) are considered reproductive toxicants while those with short- (diethyl phthalate) or linear long-side chains (di-n-octyl phthalate) are less active. Results of transgeneration studies in rats indicate that the reproductive system of developing animals is particularly vulnerable to active phthalates, which are able to reduce endogenous hormonal production by the fetal testis. *In utero* exposure of rats to certain phthalates results in a spectrum of effects that are remarkably similar to reproductive disorders that comprise the human testicular dysgenesis syndrome,

* Corresponding author: Prof. Dr. Ibrahim Chahoud Charité Universitätsmedizin Berlin, Campus Charité Mitte, Institute of Clinical Pharmacology and Toxicology - Department of Toxicology Charitéplatz 1, Berlin - Germany; Tel: ++49 30 450525551; Fax: ++49 30 450525962; E-mail address: ibrahim.chahoud@charite.de.

including cryptorchidism, hypospadias and low sperm counts. In addition, there are also evidences of endocrine disruption and reproductive effects in wildlife, including alterations in aquatic organisms. In the present review, we describe the reproductive effects and possible mode of action of major phthalates on laboratory animals, findings in wildlife and humans, as well as the relevance of these data to understanding the risks for human health and the environment.

Keywords: Phthalates; reproductive and developmental toxicity; testicular dysgenesis syndrome

1. Introduction

Phthalates are a group of chemicals that have been used for more than 80 years for several industrial applications. Chemically, they are dialkyl- or alkyl/aryl esters of 1,2-benzenedicarboxylic acid (phthalic acid). The properties and applications, as well as the toxicological profile of the different phthalate ester molecules, are largely dependent upon the alcohol that makes up the alkyl side chain (from methanol up to tridecanol, either straight or branched chain) (Lorz et al., 2002; Wittassek et al., 2011). Figure 1 displays the structure of several common phthalate esters.

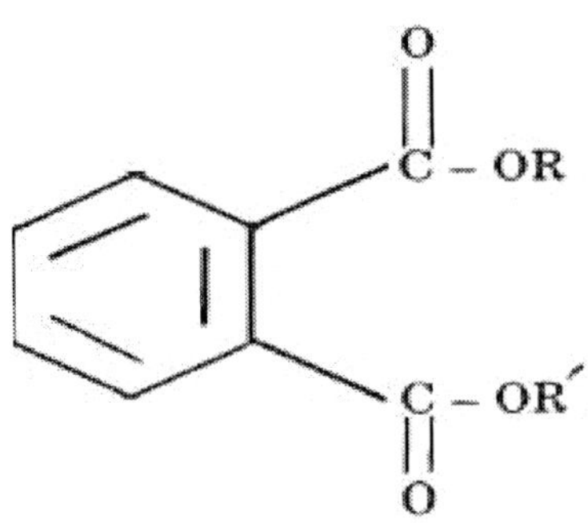

Phthalate esters

R	R'	Phthalate
- H	- H	Phthalic acid
- CH_3	- CH_3	Dimethyl phthalate
- $(CH_2)_3CH_3$	- $(CH_2)_3CH_3$	Di-n-butyl phthalate
- $(CH_2)_2(CH_3)_2$	- $(CH_2)_2(CH_3)_2$	Diisobutyl phthalate
- $(CH_2)_4CH_3$	- $(CH_2)_4CH_3$	Di-n-pentyl phthalate
- $(CH_2)_7CH_3$	- $(CH_2)_7CH_3$	Di-n-octyl phthalate
- $CH_2CH(CH_2)_3CH_3$	- $CH_2CH(CH_2)_3CH_3$	Di(2-ethylhexyl) phthalate
- $(CH_2)_3CH_3$	- $CH_2C_6H_5$	Butyl benzyl phthalate
- $(CH_2)_6CH(CH_3)_2$	- $(CH_2)_6CH(CH_3)_2$	Diisononyl phthalate
- $(CH_2)_7CH(CH_3)_2$	- $(CH_2)_7CH(CH_3)_2$	Diisodecyl phthalate

Figure 1. Structure of several common phthalates. Adapted from Kluwe, 1982.

Phthalates are the most widely used plasticizers, imparting flexibility to polyvinyl chloride (PVC) based plastics used in the automotive industry, clothing, toys, building materials, cables and wires, flooring, and food contact products. Long-chain phthalates (C8-C9), such as di(2-ethylhexyl) phthalate (DEHP), diisononyl phthalate (DiNP) and diisodecyl phthalate (DiDP) are the main phthalates used as PVC plasticizers. Short chain phthalates, including dimethyl phthalate (DMP), diethyl phthalate (DEP) and dibutyl- (DBP) and diisobutyl (DiBP) phthalate are often used in non PVC applications, mainly as solvents, fixatives and dye carriers in cosmetics, perfumes, nail polish, sealants, formulations of pesticides, adhesives, paints and enteric-coated tablets (Lorz et al., 2002; Wittassek et al, 2011).

In recent years, several restrictions have been applied to the use of certain phthalates, mainly due to the potential endocrine disruptive properties of some of these compounds in rats and the consequent induction of reproductive adverse effects, particularly in the developing fetus and pubertal animals. Currently, there are specific regulations in Europe which restrict the use of reproductive toxic phthalates in toys, cosmetics and food contact materials. In 2005, the European Union approved the ban of reproductive toxic phthalates in all toys and childcare articles (EC, 2005).

2. Human Exposure

Phthalates can be released into the environment by migration, evaporation, leaching and abrasion of and from products containing these compounds (Petersen & Breindahl, 2000; Bosnir et al., 2003; Wittassek et al., 2011). They are rapidly metabolized in the human body, but the widespread contact with phthalate containing products contributes to a relatively constant exposure of the general population (Kavlock et al., 2001; Wittassek et al., 2011). Studies revealed the presence of DEHP and other phthalates in several media including food, water, infant formula, house dust and air (Petersen & Breindahl, 2000; Becker et al., 2004; Mortensen et al., 2005). The major pathway of human exposure is through ingestion of contaminated food and water but other routes including inhalation and dermal contact may also contribute to overall exposure (Kavlock et al., 2002; Wittassek et al., 2011). The ubiquitous exposure of the general human population has been demonstrated in several biomonitoring studies in the United States of America and Europe, particularly in Germany (Koch et al., 2003, 2006; Silva et al., 2004). The use of specific phthalate metabolites generated in the human body (oxidized secondary urinary metabolites), which are not subject to external contamination, has allowed reliable assessment of internal exposure for several phthalates. There are overall agreement in the urinary concentrations of phthalate metabolites in the general US and German populations, with some minor specific differences (Wittassek et al., 2011). The urinary levels of mono-n-butyl phthalate, and mono-iso butyl phthalate, the main DBP and DiBP metabolites, respectively, are higher in Germany, whereas the concentration of diethyl and butylbenzyl phthalate metabolites are higher in the USA. According to Wittassek et al. 2011, different patterns in the formulation of personal care products, for instance a preferred usage of DEP compared to DBP in the USA, may account for the observed differences in exposure scenarios. On the other hand, higher concentrations of diisononyl phthalate metabolites have been found in the US population. In German study populations, the highest daily intakes have been calculated for DBP and DEHP, although continuously decreasing estimates have been found over the years, which are probably related to the shift in the European phthalate market that has taken place in the last years (Wittassek et al., 2011). In a study by Koch *et al.* 2006 the estimated median and 95th percentile of daily DEHP intake for a German population (n=85) was 5.6 and 21 μg/kg body weight/day, respectively. For children aged 3-14 years, these values were significantly higher - 7.7 and 25 μg DEHP/kg body weight/day (n=254). Other studies have also indicated that children have higher phthalate metabolite concentrations than adolescents and adults (reviewed by Wittassek et al., 2011). These data are particularly important as experimental studies using laboratory animals showed that developing organisms are more vulnerable to changes in the endocrine system and the consequent adverse effects in the development and function of the reproductive system. Special exposure scenarios are additional causes of concern. Critically ill patients and neonates hospitalized in intensive care units can be exposed to significantly higher doses of phthalates that migrate from medical devices such as blood bags, catheters, and nasogastric and intravenous tubes (Koch *et al.* 2006). According to the results published by Koch et al. (2006), the median and 95th percentile estimate exposure of 45 neonates undergoing different medical procedures was 0.042 and 1.780 mg DEHP/kg bw/day, respectively, and the maximum calculated intake was 2.3 mg DEHP/kg bw/day.

3. Evidence of Adverse Effects from Laboratory Studies

The male reproductive tract has been long recognized as a main target for phthalates. The first finding of phthalate-induced testicular injury in experimental animals was reported by Shaffer *et al.* in 1945. Treatment of adult male rats with high doses of certain phthalates (e.g., 2000 mg DEHP/kg/day) results in rapid and severe changes in the testis (Gray & Gangolli 1986; Dostal *et al.* 1988). The Sertoli cell is believed to be one of the main cellular targets of active phthalates, leading to secondary alterations in germ cells (Gray & Gangolli 1986). There is also experimental evidence showing that phthalates can target Leydig cells and induce multiple hormonal disturbances (Akingbemi *et al.* 2004; Lin *et al.* 2008).

Despite the evidence of male reproductive toxicity of phthalates in adult animals, it was not until the end of the 1990's that transgeneration studies revealed the potential of some phthalates to disrupt the pre-natal development of the male reproductive system (Mylchreest *et al.* 1998, 1999; Gray et al, 1999, 2000; Nagao et al., 2000). Reproductive tract malformations induced during the perinatal period were reported for a number of phthalate diesters, including dibutyl- (Mylchreest *et al.* 1998; 1999; Fisher et al., 2003), diiso butyl- (Saillenfait et al., 2008), butyl benzyl- (Gray *et al.* 2000; Nagao *et al.* 2000), dipentyl- (Hannas et al., 2011) dihexyl- (Saillenfait et al., 2009), di(2-ethylhexyl)- (Gray *et al.* 2000; Moore et al., 2001; Andrade *et al.* 2006a) and diisononyl phthalate (Gray *et al.* 2000). Male rat offspring exposed *in utero* or both *in utero* and during lactation to high doses of selected phthalates (e.g. 500 mg DBP/kg/day) display reproductive tract abnormalities compatible with disruption of androgen-dependent development and impaired testicular function. The phenotypic alterations manifested in male offspring include cryptorchidism, hypospadias, atrophy or agenesis of sex accessory organs, testicular lesions (e.g. small fluid-filled testes), reduced daily sperm production, delayed preputial separation, permanent retention of nipples, and decreased (feminized) anogenital distance (Mylchreest *et al.* 1999; Gray *et al.* 2000; Andrade *et al.* 2006a, 2006b).

Active phthalates can act as "antiandrogens" and disrupt the development of androgen-dependent structures mainly by suppressing fetal testicular testosterone biosynthesis (Parks *et al.* 2000; Wilson *et al.* 2004; Howdeshell *et al.* 2008). This effect is mediated by changes in gene expression of enzymes and proteins involved in androgen production by fetal Leydig cells, including the steroidogenic acute regulatory protein (StAR), which participates in the transport of cholesterol to the inner mitochondrial membrane, the step in the steroidogenic pathway that is considered to be rate limiting (Shultz *et al.* 2001; Lehmann *et al.* 2004). The expression of another product of the fetal Leydig cell, the insulin-like factor 3 (Insl3), has also been shown to be reduced in phthalate-exposed animals (Wilson *et al.* 2004), an effect that could explain the incidence abdominal cryptorchidism following phthalate exposure, as Insl3 is involved in the initial stages of testicular descent into the scrotum (Wilson *et al.* 2004; Foster 2006).

In utero phthalate exposure results in testicular histological alterations that are seen early in the fetal testis, with the presence of dysgenetic areas characterized by malformed seminiferous cords containing multinucleated gonocytes and aggregates of Leydig cells (Barlow *et al.* 2003; Fisher *et al.* 2003). Adult rats exposed *in utero* display testes with reduced germ cell differentiation, Sertoli cell-only (SCO) tubules, Leydig cell aggregates and multinucleated giant germ cells (Gray *et al.* 2000; Fisher *et al.* 2003; Andrade *et al.* 2006b). Reductions in Sertoli cell number have also been reported in neonate rats treated with phthalates (Dostal *et al.* 1988; Li *et al.* 2000; Auharek et al, 2010), but this appears to be a transient effect, as no changes in the number of Sertoli cells are observed later in life (Dostal *et al.* 1988; Andrade *et al.* 2006b). However, according to Auharek et al. (2010), the compensatory proliferation of Sertoli cells may be attenuated by postnatal blockage of androgen action by maternal administration of phthalates or flutamide during lactation.

4. Structure-activity Relationship

The spectrum of phthalate effects seems to be structure dependent, and highly related to the length and ramification of the alkyl chain (Saillenfait 2006, 2011). Phthalates that have 4 to 6 carbons in the linear portion of the alkyl side chains, such as DBP, dipentyl phthalate (DPeP) and DEHP, show the greatest potency as developmental and reproductive toxicants. On the other hand, short methyl and ethyl phthalates (DMP, DEP), and phthalates with much longer side chains (e.g. di-n-octyl-, diisononyl- and diisodecyl phthalates) demonstrate low or no developmental reproductive toxicity in rodents (Saillenfait et al., 2011). Di-n-octyl phthalate (DnOP) is the linear isomer (C8 backbone) of DEHP and the absence of changes in markers of androgenic status following in utero exposure to doses up to 1000 mg DnOP/kg/day, such as anogenital distance in male fetuses, is in strong contrast with alterations obtained after gestational exposure to DEHP.

5. Human Relevance

The effects induced in rats following perinatal phthalate exposure are remarkably similar to the disorders seen in the human testicular dysgenesis syndrome (TDS), a condition comprising cryptochidism, hypospadias, low sperm counts and testicular cancer. It has been hypothesized that these disorders have a common origin in fetal life and that the clinical expression of these symptoms may vary with the severity of the syndrome. Accordingly, less severe manifestations would result in impaired spermatogenesis while other symptoms like testicular germ cell cancer may be present in more severely affected men (Skakkebaek *et al.* 2001). Similarly to the human TDS disorders, the effects induced by phthalates in rats constitute a continuum of response with the most severe manifestations and highest incidence of reproductive tract malformations observed at high doses (Foster 2006). It is believed that in both humans and rats a disturbance of fetal Leydig and Sertoli cell functions plays a major role in the induction of TDS-like symptoms (Fisher *et al.* 2003). Although there is no animal model for human testicular germ cell cancer, rats exposed in utero to active phthalates show alterations in the germ cell population, such as the presence of multinucleated gonocytes in fetal/neonatal testis. In humans, the persistence of gonocytes in the adult testis is related to the development of carcinoma in situ (CIS), the known precursor of testicular germ cell cancer (Fisher et al., 2003; Ferrara et al., 2006).

Although the adverse effects induced by active phthalate in rats are observed at doses well above the estimated human exposure levels, the similarities between the human TDS and the rat phthalate syndrome have raised concerns on possible associations between phthalate human exposure and reproductive disorders. Moreover, the possibility of cumulative effects resulting from multiple exposures is also a cause of concern, as reproductive toxic phthalates can cause a similar spectrum of effects and molecular changes in rats (Howdeshell et al., 2008; Martino-Andrade et al., 2009). Recently, a number of studies have shown that the *in vivo* effects of phthalates and other anti-androgenic compounds can be additive (Howdeshell et al.2008, Rider et al., 2008, 2010).

In addition, results from epidemiological studies indicate associations of phthalate exposure with a number of human reproductive disorders, including poor semen quality in adult males (Duty et al, 2003, 2004; Hauser et al., 2006), short pregnancy duration (Latini et al., 2003), endometriosis (Cobellis et al., 2003; Reddy et al., 2006), premature breast development in girls (Colon et al., 2000) and reduced anogenital distance in male infants (Swan et al., 2005). However, most of these studies have limitations that do not allow us to establish clear conclusions on human health risks (Matsumoto et al, 2008). In addition, it is likely that no single factor is the cause of reproductive abnormalities such as those seen in the human TDS. In fact, high human phthalate exposures may be associated with other factors or lifestyle practices that are the actual causative agents (Sharpe, 2005). For instance, some studies in human populations have suggested associations between monoethyl phthalate (MEP), a metabolite of diethyl phthalate (DEP), and low sperm quality (Duty et al, 2003, 2004; Hauser et al., 2006). However, DEP belongs to the group of phthalates that cause no reproductive effects either in adult or developing laboratory animals. Additional studies using larger human populations are needed, before we can determine the real risks of phthalates to human reproductive health.

5.1. Human Relevance and Chemical Disposition

Although most reproductive effects have been described in rats, phthalates can induce testicular injury in several other species including mice (Lamb *et al.* 1987), guinea pigs (Gray *et al.* 1982), and ferrets (Lake *et al.* 1976). However, some species, such as hamsters and nonhuman primates, seem to be less sensitive than rats (Gray *et al.* 1982; Foster *et al.* 1983; Kurata *et al.* 1998), and part of this variability may be attributed to differences in phthalate bioavailability (Foster *et al.* 1983; Ito *et al.* 2005). Accordingly, it has been argued that humans and nonhuman primates are less susceptible to the effects of phthalates, due to the lower conversion of parent compounds into active monoester metabolites, which are considered the proximate toxicants (Mckee *et al.* 2004). In a study by Kurata *et al.* (1998), no changes in testis weight or histopathology were observed in adult marmosets administered orally at 100, 500 or 2500 mg DEHP/kg/day for 13 weeks. Absence of testicular effects was also reported in young adult cynomolgus monkeys and juvenile marmosets exposed to high DEHP doses (500-2500 mg/kg/day) (Pugh *et al.* 2000; Tomonari *et al.* 2006). However, nonhuman primates have not been extensively evaluated during fetal and neonatal periods, which represent developmental windows especially susceptible to exogenous insults. A work by Hallmark *et al.* (2007) indicates that neonatal marmosets treated orally with 500 mg/kg/day di-butyl phthalate (DBP) respond similarly to rats in relation to changes in testosterone production and Leydig cell alterations, although marmosets seem to be able to reverse the suppression of testosterone production more efficiently than rats.

5.2. Human Relevance and Mechanism of Action

The suppression of testosterone biosynthesis is a key factor for the induction of several phthalate effects. It is interesting to note that the fetal mice testis seems to be resistant to the androgen insufficiency induced by active phthalates, although other effects, such as the presence of multinucleated gonocytes, are still present (Gaido et al., 2007). A central question for the assessment of potential risks of phthalates to humans is whether the response of the fetal human testis is more similar to the rat or mice testis. Recent studies using testis xenographs in nude (immune-compromised) castrated mice demonstrated that the human fetal testis is less susceptible to the suppression of testosterone production induced by di-butyl phthalate (DBP) when compared to the rat fetal testis (Mitchell et al., 2012). These results are in accordance with a previous *in vivo* study using marmoset (nonhuman primate) that showed absence of any masculinization impairment following gestational exposure to mono-butyl phthalate, the active monoester metabolite of DBP (Mitchell et al., 2008). However, other phthalate-mediated effects such as abnormalities in germ cell development have been observed in both human and rat testes and the mechanisms underlying such effects are currently under investigation (Saffarini et al., 2011; Mitchell et al., 2012).

In addition, it is important to note that most hormonal alterations associated with exposure to active phthalates in laboratory animals appear to be secondary to maldevelopment (dysgenesis) of the testis. Although several target genes involved in the development and function of fetal Leydig and Sertoli cells have been identified so far (Lehmann et al., 2004; Liu et al., 2005), the mechanism by which phthalates alter the expression of these genes are currently unknown.

6. Evidence of Adverse Effects in the Environment

The assessment of phthalate effects in the aquatic environment is challenged by the low water solubility of these compounds, especially phthalates with medium and long side chains. Even so, some studies have addressed the effects of phthalates in fish and other aquatic organisms. Studies on the ability of phthalates to induce the synthesis of vitellogenin - a biomarker of estrogenic activity - have lead to contradictory results (reviewed by Oehlmann et al., 2009). However, vitellogenin induction in male and reproductive effects on female fish appear to be restricted to high (unrealistic) concentrations of phthalates.

On the other hand and in contrast to mammals, fish appear to bioaccumulate phthalates in their bodies. In salmon (Salmo salar), exposure to 1500 mg DEHP/kg in the food during early life for four months resulted in a small incidence of intersex (Norman et al. 2007).

DBP has also been shown to alter the sexual differentiation in amphibians, delaying gonadal development in male Rana rugosa tadpoles and altering the spermatogenesis process in eight-month-old X. laevis after exposure during larval development at 1– 10 mg DBP/L (Reviewed by Oehlmann et al., 2009).

Conclusion

The widespread use of phthalates in mass consumer products and the consequent potential for high exposure scenarios have raised concerns on possible adverse impacts to human health and the environment. The resemblance of rat phthalate effects with common human male reproductive disorders, the results of recent epidemiological studies showing possible associations between exposure and human effects, and the potential for cumulative effects have raised concerns of a possible link between phthalate exposure and human disease. However, additional mechanistic studies are needed to further evaluate the relevance of *in vivo* laboratory studies to humans. In addition, biomonotoring and epidemiological studies using larger human samples, with particular emphasis on vulnerable populations are also of critical interest.

References

Akingbemi, B. T., Ge, R., Klinefelter, G. R., Zirkin, B. R., Hardy, M. P., Phthalate-induced Leydig cell hyperplasia is associated with multiple endocrine disturbances, *Proc. Natl. Acad. Sci. USA* 2004, *101*, 775-780.

Andrade, A. J., Grande, S. W., Talsness, C. E., Gericke, C., Grote, K., Golombiewski, A., Sterner-Kock, A. & Chahoud, I. A dose response study following *in utero* and lactational exposure to di-(2-ethylhexyl) phthalate (DEHP): Reproductive effects on adult male offspring rats. *Toxicology*, 2006b, *228*, 85-97.

Andrade, A. J., Grande, S. W., Talsness, C. E., Grote, K., Golombiewski, A., Sterner-Kock, A. and Chahoud, I. A dose-response study following *in utero* and lactational exposure to di-(2-ethylhexyl) phthalate (DEHP): Effects on androgenic status, developmental landmarks and testicular histology in male offspring rats. *Toxicology*, 2006a, *225*, 64-74.

Auharek S. A., de Franca L. R., McKinnell C., Jobling M. S., Scott H. M., Sharpe R. M. Prenatal plus postnatal exposure *Endocrinology*. 2010, 151, 2868-75.

Barlow N. J. and Foster P. M. Pathogenesis of male reproductive tract lesions from gestation through adulthood following in utero exposure to Di(n-butyl) phthalate. *Toxicol. Pathol.*, 2003, *31*, 397-410.

Becker K., Seiwert M., Angerer J., Heger W., Koch H. M., Nagorka R., Rosskamp E., Schlüter C., Seifert B., Ullrich D. DEHP metabolites in urine of children and DEHP in house dust, *Int. J. Hyg. Environ. Health* 2004, 207, 409-417.

Bosnir J., Puntaric D., Skes I., Klaric M., Simic S., Zoric I. Migration of phthalates from plastic products to model solutions, *Coll.Antropol.* 2003, 27, 23-30.

Cobellis L., Latini G., De Felice C., Razzi S., Paris I., Ruggieri F., Mazzeo P., Petraglia F. High plasma concentrations of di-(2-ethylhexyl)-phthalate in women with endometriosis, *Hum. Reprod.* 2003, 18, 1512-1515.

Colon, I., Caro, D., Bourdony, C. J., Rosario, O., Identification of phthalate esters in the serum of young Puerto Rican girls with premature breast development, *Environ. Health Perspect.* 2000, *108*, 895-900.

Dostal, L. A., Chapin, R. E., Stefanski, S. A., Harris, M. W., Schwetz, B. A., Testicular toxicity and reduced Sertoli cell numbers in neonatal rats by di(2-ethylhexyl)phthalate and the recovery of fertility as adults, *Toxicol. Appl. Pharmacol.* 1988, *95*, 104-121.

Duty S. M., Silva M. J., Barr D. B., Brock J. W., Ryan L., Chen Z., Herrick R. F., Christiani D. C., Hauser R.Phthalate exposure and human semen parameters, *Epidemiology* 2003, *14*, 269-277.

Duty S. M., Calafat A. M., Silva M. J., Brock J. W., Ryan L., Chen Z., Overstreet J., Hauser R. The relationship between environmental exposure to phthalates and computer-aided sperm analysis motion parameters. *J. Androl.* 2004, *25*, 293-302.

European Commission (EC), Restrictions on the marketing and use of certain dangerous substances and preparations (phthalates in toys and childcare articles) DIRECTIVE 2005/84/EC. *Official Journal of the European Union* 2005, *L344*, 40-43.

Ferrara D., Hallmark N., Scott H., Brown R., McKinnell C., Mahood I. K., Sharpe R. M. Acute and long-term effects of in utero *Endocrinology*. 2006, 147, 5352-62.

Fisher, J. S., Macpherson, S., Marchetti, N., Sharpe, R. M., Human 'testicular dysgenesis syndrome': a possible model using in-utero exposure of the rat to dibutyl phthalate, *Hum. Reprod.* 2003, *18*, 1383-1394.

Foster, P. M., Cook, M. W., Thomas, L. V., Walters, D. G., Gangolli, S. D., Differences in urinary metabolic profile from di-n-butyl phthalate-treated rats and hamsters. A possible explanation for species differences in susceptibility to testicular atrophy, *Drug Metab. Dispos.* 1983, *11*, 59-61.

Foster, P. M., Disruption of reproductive development in male rat offspring following in utero exposure to phthalate esters, *Int. J. Androl.* 2006, *29*, 140-147.

Gaido K. W., Hensley J. B., Liu D., Wallace D. G., Borghoff S., Johnson K. J., Hall S. J., Boekelheide K. Fetal mouse phthalate exposure *Toxicol. Sci.* 2007, 97, 491-503.

Gray, T. J., Rowland, I. R., Foster, P. M., Gangolli, S. D., Species differences in the testicular toxicity of phthalate esters, *Toxicol. Lett.* 1982, *11*, 141-147.

Gray, T. J., Gangolli, S. D., Aspects of the testicular toxicity of phthalate esters, *Environ. Health Perspect.* 1986, *65*, 229-235.

Gray, L. E., Jr., Wolf, C., Lambright, C., Mann, P. *et al.*, Administration of potentially antiandrogenic pesticides (procymidone, linuron, iprodione, chlozolinate, p,p'-DDE, and ketoconazole) and toxic substances (dibutyl- and diethylhexyl phthalate, PCB 169, and ethane dimethane sulphonate) during sexual differentiation produces diverse profiles of reproductive malformations in the male rat, *Toxicol. Ind. Health* 1999, *15*, 94-118.

Gray, L. E., Ostby, J., Furr, J., Price, M., Veeramachanemi, D. N. & Parks, L. Perinatal exposure to the phthalates DEHP, BBP, and DINP, but not DEP, DMP, or DOTP, alters sexual differentiation of the male rat. *Toxicol. Sci.*, 2000, *58*, 350-365.

Hallmark, N., Walker, M., McKinnell, C., Mahood, I. K., Scott, H., Bayne, R., Coutts, S., Anderson, R. A., Greig, I., Morris, K. and Sharpe, R. M. Effects of monobutyl- and di-(n-butyl) phthalate in vitro on steroidogenesis and leydig cell aggregation in fetal testis explants from the rat: comparison with effects in vivo in the fetal rat and neonatal marmoset and in vitro in the human. *Environ. Health Perspect*, 2007, 115, 390-396.

Hannas B. R., Furr J., Lambright C. S., Wilson V. S., Foster P. M., Gray LE Jr. Dipentyl phthalate dosing *Toxicol Sci.*, 2011, 120, 184-93.

Hauser, R., Meeker, J. D., Duty, S., Silva M. J., Calafat, A. M., Altered Semen Quality in Relation to Urinary Concentrations of Phthalate Monoester and Oxidative Metabolites, *Epidemiology* 2006, 17, 682-691.

Howdeshell, K. L., Furr, J., Lambright, C. R., Rider, C. V., Wilson, V. S. & Gray, L. E. Jr. Cumulative effects of dibutyl phthalate and diethylhexyl phthalate on male rat reproductive tract development: altered fetal steroid hormones and genes. *Toxicol. Sci.*, 2007, *99*, 190-202.

Howdeshell KL, Furr J, Lambright CR, Wilson VS, Ryan BC, Gray LE Jr. Gestational and lactational exposure *Toxicol Sci,* 2008, 102, 371-82.

Ito Y., Yokota, H., Wang, R., Yamanoshita, O., Ichihara, G., Wang, H., Kurata, Y., Takagi, K. and Nakajima, T. Species differences in the metabolism of di(2-ethylhexyl) phthalate (DEHP) in several organs of mice, rats, and marmosets. *Arch. Toxicol.*, 2005, *79*, 147-154.

Kavlock R., Boekelheide K., Chapin R., Cunningham M., Faustman E., Foster P., Golub M., Henderson R., Hinberg I., Little R., Seed J., Shea K., Tabacova S., Tyl R.,Williams P., Zacharewski T. NTP Center for the Evaluation of Risks to Human Reproduction: phthalates expert panel report on the reproductive and developmental toxicity of di(2-ethylhexyl) phthalate, *Reprod. Toxicol.* 2002, 16, 529-653.

Kluwe W. M. Overview of phthalate ester *Environ Health Perspect.*, 1982, 45, 3-9.

Koch, H. M., Drexler, H., Angerer, J., An estimation of the daily intake of di(2-ethylhexyl)phthalate (DEHP) and other phthalates in the general population, *Int. J. Hyg. Environ. Health* 2003, *206*, 77-83.

Koch, H. M., Preuss, R., and Angerer, J. Di(2-ethylhexyl)phthalate (DEHP): human metabolism and internal exposure-- an update and latest results. *Int J Androl*, 2006, *29*, 155-165.

Kurata, Y., Kidachi, F., Yokoyama, M., Toyota, N., Tsuchitani, M. & Katoh, M. Subchronic toxicity of Di(2-ethylhexyl)phthalate in common marmosets: lack of hepatic peroxisome proliferation, testicular atrophy, or pancreatic acinar cell hyperplasia. *Toxicol. Sci.*, 1998, *42*, 49-56.

Lake, B. G., Brantom, P. G., Gangolli, S. D., Butterworth, K. R., Grasso, P., Studies on the effects of orally administered Di-(2-ethylhexyl) phthalate in the ferret, *Toxicology* 1976, *6*, 341-356.

Lamb, J. C., Chapin, R. E., Teague, J., Lawton, A. D., Reel, J. R., Reproductive effects of four phthalic acid esters in the mouse, *Toxicol. Appl. Pharmacol.* 1987, *88*, 255-269.

Latini G., De Felice C., Presta G., Del Vecchio A., Paris I., Ruggieri F., Mazzeo P. *In utero* exposure to di-(2-ethylhexyl) phthalate and duration of human pregnancy. *Environ. Health Perspect.*, 2003, *111*, 1783-5.

Lehmann, K. P., Phillips, S., Sar, M., Foster, P. M., Gaido, K. W., Dose-dependent alterations in gene expression and testosterone synthesis in the fetal testes of male rats exposed to di (n-butyl) phthalate, *Toxicol. Sci.* 2004, *81*, 60-68.

Li L. H., Jester, W. F., Jr., Laslett, A. L., Orth, J. M., A single dose of Di-(2-ethylhexyl) phthalate in neonatal rats alters gonocytes, reduces sertoli cell proliferation, and decreases cyclin D2 expression, *Toxicol. Appl. Pharmacol.* 2000, *166*, 222-229.

Lin, H., Ge, R. S., Chen, G. R., Hu, G.X., Dong, L., Lian, Q. Q., Hardy, D. O., Sottas, C. M., Li, X. K. & Hardy, M. P. Involvement of testicular growth factors in fetal Leydig cell aggregation after exposure to phthalate in utero. *Proc. Natl. Acad. Sci. U S A.*, 2008, 105, 7218-22.

Liu, K., Lehmann, K. P., Sar, M., Young, S. S., Gaido, K. W., Gene expression profiling following in utero exposure to phthalate esters reveals new gene targets in the etiology of testicular dysgenesis, *Biol. Reprod.* 2005, *73*, 180-192.

Lorz, P. M., Towae, F. K., Enke, W., Jaeckh, R., Bhargava, N., Phthalic acid and derivatives, in: Wiley-VCH (Ed.), *Ullmann's Encyclopedia of Industrial Chemistry Release 2006*, 7th Edition Online, Wiley-VCH, Weinheim 2002, pp. 1-36.

Martino-Andrade, A. J., Morais, R. N., Botelho, G. G., Muller, G., Grande S. W., Carpentieri, G. B., Leão, G. M., Dalsenter, P. R. Coadministration of active phthalates Int. *J. Androl.* 2009, 32, 704-712.

Matsumoto, M., Hirata-Koizumi, M., Ema, M., Potential adverse effects *Regul. Toxicol. Pharmacol.* 2008, 50, 37-49.

Mckee, R. H., Butala, J. H., David, R. M. And Gans, G. NTP center for the evaluation of risks to human reproduction reports on phthalates: addressing the data gaps. *Reprod. Toxicol.*, 2004, *18*, 1-22.

Mitchell R. T., Cowan G., Morris K. D., Anderson R. A., Fraser H. M., Mckenzie K. J., Wallace W. H., Kelnar C. J., Saunders P. T., Sharpe R. M. Germ cell differentiation in the marmoset (Callithrix jacchus) during fetal and neonatal life closely parallels that in the human. *Hum. Reprod.*, 2008, *23*, 2755–2765.

Mitchell R. T., Childs A. J., Anderson R. A., van den Driesche S., Saunders P. T., McKinnell C., Wallace W. H., Kelnar C. J., Sharpe R. M. Do Phthalates Affect Steroidogenesis by the Human Fetal Testis? Exposure of Human Fetal Testis Xenografts to Di-n-ButylPhthalate. *J Clin Endocrinol Metab.*, 2012 (in press).

Moore, R. W., Rudy, T. A., Lin, T. M., Ko, K., Peterson, R. E., Abnormalities of sexual development in male rats with in utero and lactational exposure to the antiandrogenic plasticizer Di(2-ethylhexyl) phthalate, *Environ. Health Perspect.* 2001, *109*, 229-237.

Mortensen, G. K., Main, K. M., Andersson, A. M., Leffers, H., Skakkebaek, N. E., Determination of phthalate monoesters in human milk, consumer milk, and infant formula by tandem mass spectrometry (LC-MS-MS), *Anal. Bioanal. Chem.* 2005, 382, 1084-1092.

Mylchreest, E., Cattley, R. C., Foster, P. M., Male reproductive tract malformations in rats following gestational and lactational exposure to Di(n-butyl) phthalate: an antiandrogenic mechanism?, *Toxicol. Sci.* 1998, *43*, 47-60.

Mylchreest, E., Wallace, D. G., Cattley, R. C., Foster, P. M., Dose-dependent alterations in androgen-regulated male reproductive development in rats exposed to Di(n-butyl) phthalate during late gestation, *Toxicol. Sci.* 2000, *55*, 143-151.

Nagao, T., Ohta, R., Marumo, H., Shindo, T., Yoshimura, S. and Ono, H. Effect of butyl benzyl phthalate in Sprague-Dawley rats after gavage administration: a two-generation reproductive study. *Reprod. Toxicol.*, 2000, *14*, 513-532.

Norman A., Börjeson H., David F., Tienpont B., Norrgren L. Studies of uptake, elimination, and late effects in Atlantic salmon (Salmo salar) dietary exposed to Di-2-ethylhexyl phthalate (DEHP) during early life. *Arch Environ Contam Toxicol.*, 2007, 52, 235-42.

Oehlmann J., Schulte-Oehlmann U., Kloas W., Jagnytsch O., Lutz I., Kusk K. O., Wollenberger L., Santos E. M., Paull G. C., Van Look K. J., Tyler C. R. A critical analysis *Philos Trans R Soc Lond B Biol Sci.*, 2009, 364, 2047-62.

Parks, L. G., Ostby, J. S., Lambright, C. R., Abbott, B. D., Klinefelter, G. R., Barlow, N. J. and Gray, L. E. The plasticizer diethylhexyl phthalate induces malformations by decreasing fetal testosterone synthesis during sexual differentiation in the male rat. *Toxicol. Sci.*, 2000, *5*, 339-349.

Petersen, J. H., Breindahl, T., Plasticizers in total diet samples, baby food and infant formulae, *Food Addit. Contam.* 2000, *17*, 133-141.

Pugh, G., Jr., Isenberg, J. S., Kamendulis, L. M., Ackley, D. C., Clare, L. J., Brown, R., Lington, A. W., Smith, J. H. & Klauning, J. E. Effects of di-isononyl phthalate, di-2-ethylhexyl phthalate, and clofibrate in cynomolgus monkeys. *Toxicol. Sci.*, 2000, *56*, 181-188.

Reddy, B. S., Rozati, R., Reddy, B. V., Raman, N. V., Association of phthalate esters with endometriosis in Indian women, BJOG. 2006, 113, 515-520. *Reprod Toxicol.* 2009, 28, 468-76.

Rider C, V,, Furr J,, Wilson V, S., Gray L. E. Jr. A mixture of seven antiandrogens induces reproductive malformations in rats. *Int J Androl.*, 2008, 31, 249-62.

Rider C. V., Furr J. R., Wilson V. S., Gray L. E. Jr. Cumulative effects of in utero *Int J Androl.*, 2010, 33, 443-62.

Saffarini C. M., Heger N. E., Yamasaki H., Liu T., Hall S. J., Boekelheide K. Induction and Persistence of Abnormal Testicular Germ Cells Following Gestational Exposure to Di-(n-butyl) Phthalate in p53 *J Androl.* 2011 (in press).

Saillenfait A. M., Sabaté J. P., Gallissot F. Developmental toxic effect *Toxicol Lett.* 2006, 165, 39-46.

Saillenfait A. M., Sabaté J. P., Gallissot F. Diisobutyl phthalate impairs the androgen *Reprod Toxicol.* 2008, 26, 107-15.

Saillenfait A. M., Sabaté J. P., Gallissot F. Effects of in utero *Reprod. Toxicol.*, 2009, 28, 468-76.

Saillenfait A. M., Roudot A. C., Gallissot F., Sabaté J. P. Prenatal developmental toxicity *Reprod Toxicol.* 2011, 32, 268-76.

Shaffer, C. B., Carpenter, C. P., Smyth, H. R., Jr., Acute and subacute toxicity of di-(2-ethylhexyl) phthalate with note upon its metabolism, *J. Ind. Hyg. Toxicol.* 1945, *27*, 130-135.

Sharpe, R. M., Phthalate exposure *Environ. Health Perspect.* 2005, 113, A504-5.

Shultz, V. D., Phillips, S., Sar, M., Foster, P. M., Gaido, K. W. Altered gene profiles in fetal rat testes after in utero exposure to di(n-butyl) phthalate. *Toxicol. Sci.,* 2001, *64*, 233-242.

Silva, M. J., Barr, D. B., Reidy, J. A., Malek, N. A., Hodge, C. C., Caudill, S. P., Brock, J. W., Needham, L. L., and Calafat, A. M. Urinary levels of seven phthalate metabolites in the U.S. population from the National Health and Nutrition Examination Survey (NHANES) 1999-2000. *Environ. Health Perspect.*, 2004, *112*, 331-338.

Skakkebaek, N. E., Rajpert-De Meyts, E., Main, K. M., Testicular dysgenesis syndrome: an increasingly common developmental disorder with environmental aspects, *Hum. Reprod.* 2001, *16*, 972-978.

Swan S. H., Main K. M., Liu F., Stewart S. L., Kruse R. L., Calafat A. M., Mao C. S., Redmon J. B., Ternand C. L., Sullivan S., Teague J. L.; Study for Future Families Research Team. Decrease in anogenital distance among male infants with prenatal phthalate exposure. *Environ Health Perspect.* 2005. 113, 1056-61.

Tomonari, Y., Kurata, Y., David, R. M., Gans, G. Kawasuso, T., Katoh, M. Effect of di(2-ethylhexyl) phthalate (DEHP) on genital organs from juvenile common marmosets: I. Morphological and biochemical investigation in 65-week toxicity study, *J. Toxicol. Environ. Health Part A*, 2006, *69*, 1651-1672.

Wilson, V. S., Lambright, C., Furr, J., Ostby, J., Wood, C., Held, G. & Gray, L. E. Phthalate ester-induced gubernacular lesions are associated with reduced insl3 gene expression in the fetal rat testis. *Toxicol. Lett.*, 2004, *146*, 207-215.

In: Phthalates
Editors: Gerardo L. Moretti and Drago Romano

ISBN: 978-1-62081-994-4

Chapter 9

PREVENTION OF PHATALTES MIGRATION FROM FLEXIBLE PVC PRODUCTS

***Veronica Ambrogi*[1], *Cosimo Carfagna*[1,2], *Marianna Pannico*[2*] *and Paola Persico*[2]**

[1]Department of Materials and Production Engineering,
University of Naples "Federico II", Piazzale Tecchio, Napoli, Italy
[2]Institute of Chemistry and Technology of Polymers,
National Research Council of Italy, Pozzuoli NA, Italy

ABSTRACT

Di-(2-ethylhexyl) phthalate (DEHP) is an ester of phthalic acid, which is mainly used as plasticizer in polyvinyl chloride (PVC) to increase its flexibility, elongation and processability. The main drawback, related to its low molecular weight, is the leaching out from flexible PVC (pPVC), which affects the polymer properties limiting its employ. Moreover, the use of phthalate-based plasticizers is being questioned worldwide because of their potential toxicity to humans and environment. Migration of such harmful substances from PVC can be avoided by substituting the unsafe plasticizer or by preventing its mass transfer from polymer matrix.

In this communication, promising alternatives to reduce phthalate mass transfer are discussed in detail: the chemical cross-linking of flexible PVC and the use of chlorinated polyethylene (CPE) as physical barrier against DEHP leaching. Cross-linked flexible PVC samples were obtained using a difunctional amine, namely isophoron diamine (IPDA), as cross-linking agent while two different CPE resins were employed for the preparation of new CPE/pPVC compounds. The effect of cross-linking and the efficiency of CPE in preventing plasticizer leaching were evaluated through migration tests and supported by tribological analysis. Thermal, mechanical and dynamic-mechanical properties were also studied.

[*] E-mail: marianna.pannico@ictp.cnr.it.

INTRODUCTION

Esters of phthalic acid are commercially very important low molecular weight compounds widely used in industrial applications. The largest market for these esters is as plasticizing agents for PVC, since they are able to decrease the interaction forces of adjacent polymer chains, lowering the glass transition temperature, promoting chain mobility and thus material flexibility. Phthalates plasticizers account for up to 40% of the final weight of the material and the wide variety of resin/plasticizer ratios permits the production of plastics that range from very hard to very flexible. DEHP is the most commonly used plasticizer in flexible PVC applications, its physical and chemical properties as well as its analytical chemistry, biological degradation, effects on human and ecological health, exposure and risk assessment are well documented [1]. Since phthalates are not chemically bounded to PVC, they can migrate or evaporate to atmosphere, foodstuff and other materials in contact, being harmful to health and environment [2, 3]. As a consequence, phthalate esters have been subjected to considerable regulatory scrutiny world-wide covering all aspects of their production, transportation, use, and disposal.

Plasticizer agents different from phthalates have been suggested as a possible solution to the toxicological and environmental problems. Among them, alkyl esters of adipic and citric acids have been considered as adequate due to their good technical performance, low toxicity and fast biodegradation [4].

Moreover, the loss of plasticizer is one of the most dominant adverse factors contributing to aging of flexible PVC products, making it useless for many applications [5, 6].

In this frame, several efforts are made by academic and industrial researchers in order to limit the plasticizer migration. A recent promising alternative to reduce mass transfer is based on covalent linkage of the additive to the polymer chains, keeping the PVC glass transition temperatures in the range of 0 °C [7]. However, the scaling up of the procedure for PVC modification could be difficult due to the use of solvents, which are cost effective and not safe from an industrial point of view.

A significant inhibition of plasticizer migration may also be achieved by blending PVC with a polymeric plasticizer [8-10]. Necessarily such plasticizers have a reduced ability to migrate compared to conventional non-polymeric ones. Nevertheless, the use of polymeric plasticizers causes the increase in tensile strength and decrease in percentage elongation at break as a result of reduced chain mobility and flexibility due to the high plasticizer viscosity.

In this communication, two approaches against DEHP migration are discussed and compared, taking care to minimize PVC properties alteration and to save easy industrial scale-up.

A specific tool aimed at reduction of plasticizer migration is the chemical cross-linking of PVC. Formation of more chemical bonds reduces the free volume and segmental mobility, thus affecting plasticizer diffusion through the matrix [11, 12]. In a previous work, authors [13] have already presented the effects of isophoron diamine (IPDA) used as cross-linking agent for flexible PVC dry blends.

The novel approach herein considered is the addition of CPE to flexible PVC formulations, and it is worth noticing that since now any similar studies have been reported in the literature.

CPE represents an important class of commercial resin mainly used as compatibilizer in polymer blends to improve their mechanical properties [14-16] for extruded, calendered, solution cast and injection-moulded parts and goods.

Usually it is a recommended additive for rigid PVC since it raises its impact strength and fire resistance, reduces brittle point, and improves the electrical properties [17, 18]. High molecular weight CPE is nonvolatile and does not migrate in blend with PVC, thus the rigid compositions exhibit a considerably higher service life.

To the best of our knowledge CPE has never been compounded with flexible PVC.

In this communication CPE was used as a tool to create a physical barrier against DEHP migration. Since it is well known that CPE/PVC compounds properties are influenced by the chlorine percentages in polyethylene and the components ratio in the formulation [19, 20], various amount of two CPEs, having different chlorine content, were selected for preparation of flexible PVC dry blends.

This approach was considered after preliminary tests on DEHP/CPE affinity. CPE was added to a large amount of DEHP (used as "solvent") obtaining a slurry viscous fluid. Differently, when rigid PVC was added to DEHP, the formation of a gelled system was observed. This result suggested a singular molecular interaction between DEHP and CPE, that was expected to improve its ability to retain the plasticizer.

Thermal, tensile and dynamic-mechanical analysis were performed in order to study the effect of IPDA cross-linking and CPE blending on the PVC properties. Moreover, tribological and migration tests were carried out on rigid PVC after being held in contact with cross-linked pPVC and CPEs/pPVC blends to test the effectiveness of these two strategies in preventing plasticizer migration.

EXPERIMENTAL PART

Materials

The PVC resin used for preparation of flexible samples was PVC Lacovyl® S7015 (PVC), k-value=70 (Arkema, France). The rigid PVC pellets were BENVICVIC® IH007W025AA (uPVC), Shore D=72 (Solvay Benvic, Italy). (Ca + Zn)-based powder Reapak B-NT/7060 (Ca + Zn 0.5-0.8 phr) (Ca/Zn- Reagens, Italy) and epoxy soybean oil (ESBO- Shangai Yanan Oil and Grease Co.) were selected as heat stabilizer and co-thermal stabilizer, respectively. Low-molecular weight commercial plasticizer Di-(2-ethylhexyl) phthalate (DEHP) DIPLAST® 0 (purity > 95%) was received from Lonza S.p.A., Italy. The cross-linker isophoron diamine (IPDA) was purchased from Sigma-Aldrich and used without further purification. Two types of CPE resins with different chlorine content were supplied by R.G. CHEM S.R.L., Italy, CPE6135A (35-37 wt.% chlorine) and CPE252 (24-26 wt.% chlorine).

Sample Preparation

Two flexible PVC-based systems were prepared:

- cross-linked pPVC using 1.2 wt.% (C1), 2.2 wt.% (C2) and 3.2 wt.% (C3) of IPDA;
- CPE/pPVC compounds using 7wt.% (A7) and 15 wt.% (A15) of CPE6135A and 7wt.% (B7) and 15 wt.% (B15) of CPE252.

All the formulations, reported in detail in table 1, have been studied in comparison with the reference pPVC obtained by dry-mixing of PVC resin (100 phr), DEHP (70 phr), Ca/Zn (1 phr) and ESBO (2 phr) as reported elsewhere [13].

Cross-linked pPVC systems were obtained by adding IPDA to pPVC dry blend which was then roll-milled in a plasticorder (Brabender *model PLE 67152*) at 40 rpm, T = 90°C for 15 min.

For the preparation of CPE/pPVC blends, CPE6135A and CPE252 were added to pPVC dry blend in different amount and then roll-milled at 40 rpm, T = 120 °C for 7 min.

The flexible sheets were obtained using a hot press (lab-scale Collin P 200 E) at 150°C and 200 bar for 7 min. Rigid PVC sheets were prepared by press-molding uPVC at 190 °C, 200 bar for 7 min.

Migration Tests

Small discs of pPVC-based samples (diameter d = 5.0 ± 0.1 cm) with approximately 1 mm thickness were placed between two uPVC discs (d = 5.0 ± 0.1 cm) with 2 mm thickness and kept under pressure (5.0 kg) in an oven at 60°C between two glassy plates for 32 days.

Plasticizer migration was determined by monitoring the extent of migration in terms of weight increase of rigid PVC. The specimens removed from the oven at different times were immediately weighted. The extent of migration is defined as:

$$Migration\ extent = \frac{\Delta P}{P_0} \qquad (1)$$

where P_0 = the weight of rigid part at time zero; ΔP = the weight difference $P - P_0$, where P is the weight of the rigid part at the selected times.

Table 1. pPVC-based formulations

	Samples						
	C1	**C2**	**C3**	**A7**	**A15**	**B7**	**B15**
IPDA (wt.%)	1.2	2.2	3.2	-	-	-	-
CPE6135A (wt.%)	-	-	-	7	15	-	-
CPE252 (wt.%)	-	-	-	-	-	7	15

Thermogravimetric Analysis(TGA)

Thermogravimetric analysis (TGA) was carried out at 10°C/min heating rate from 25°C to 600 °C under nitrogen flow using a Q5000 (TA Instruments) thermobalance.

Tensile Tests

The Young's modulus (E), ultimate tensile strength (σ_b), and elongation at break (ε_b) were determined from a traction test by an universal dynamometer INSTRON mod.5564 at room temperature using 1 KN load cell, at a crosshead speed of 10 mm/min. The dog bone shaped mini tensile bars of 2 mm thickness, 4 mm width and 28 mm length were cut from compression moulded sheets. For each sample a batch of five specimens was tested and the average values were reported.

Dynamic-Mechanical Analysis

Dynamic Mechanical Thermal Analysis (DMTA) measurements were carried out using a Triton Technology mod. Tritec 2000 testing machine. Tests were performed in single cantilever bending mode, using a constant frequency of 1 Hz and an oscillation amplitude of 0,01 mm. Samples were heated from -80°C to 40 °C at 5°C/min heating rate. The glass transition temperature (Tg) was taken as the peak temperature of tanδ curve.

Tribological Analysis

Friction tests were performed using a Nanoevea pin-on-disk tribometer from Micro Photonics. The pin used was steel ball (SS302) made by Salem Specialty Ball with 0.32 mm diameter. The normal force applied was 5.0 N. Tribological tests were performed on rigid PVC samples after migration tests. The friction tests lasted 30 minutes at the sliding velocity of 3000 revolution numbers per minute; the radius of the frictional track on the disk was 2.0 mm. Preliminary experiments at different sliding rates (not reported) were performed in order to find the test conditions able to reduce the points scattering in the friction trace.

Results and Discussion

Migration Tests

Migration tests were performed for 32 days in order to evaluate the effect of cross-linking and the efficiency of CPE in preventing plasticizer leaching. The extent of migration has been related to the weight increase of rigid PVC in contact with pPVC-based samples which was calculated according to equation (1).

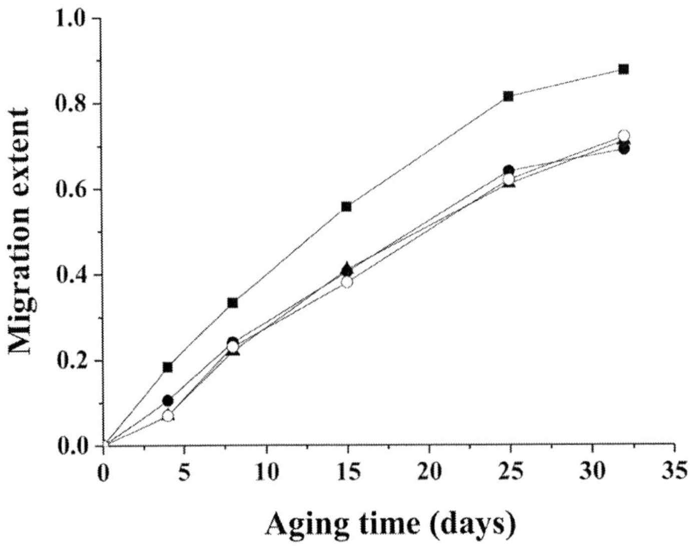

Figure 1. Migration extent of DEHP for standard pPVC (-■-), C1 (-●-), C2 (-▲-), C3 (-○-).

Figure 1 reports the migration extent versus time determined for uPVC sheets kept in contact with standard pPVC, C1, C2 and C3.

The cross-linked materials showed comparable abilities to retain the plasticizer; the amount of DEHP migrating turned out to be independent of the cross-linker loading. This behavior was explained in terms of comparable cross-linking density values characterizing the three samples [13].

After 32 days the migration extent of DEHP for cross-linked samples was about 18 % lower than that of standard pPVC.

The migration extent of DEHP versus time for CPE/pPVC compounds containing 7 wt.% and 15 wt.% of CPE resins are displayed in Figure 2a and 2b, respectively.

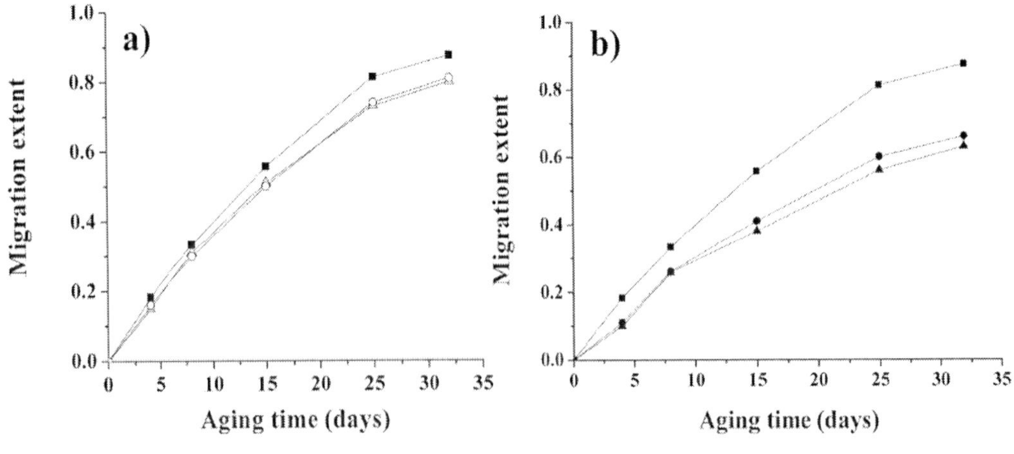

Figure 2. Migration extent of DEHP for: a) standard pPVC (-■-), A7 (-○-), B7 (-△-); b) standard pPVC (-■-), A15 (-●-), B15 (-▲-).

Results showed the CPEs ability to retain the plasticizer: when CPE was added at 7 wt.% the plasticizer migration was moderately prevented, at higher CPE content, i.e. 15 wt.%, the DEHP migration was hindered more efficiently.

As a matter of fact, after 32 days, the migration extent of DEHP for A15 and B15 was about 23 % lower than that of standard pPVC. This behavior was independent of the chlorine content.

Migration results proved that pPVC chemical cross linking and the addition of CPE to pPVC could be efficient tools to overcome the disadvantage related to the plasticizer migration. The chemical approach led to the formation of more intermolecular bonds reducing the free volume and segmental mobility of the matrix; the physical approach provided the occurrence of strong interactions between CPE and DEHP, hampering its release.

More precisely, the effectiveness of IPDA, at any percentage, in preventing phthalates leaching was almost comparable to CPEs/pPVC blends containing 15 wt.% of CPE.

Thermal Stability

Thermal stability of pPVC-based samples was evaluated by thermogravimetric analysis. Thermogravimetric (TG) and differential (DTG) curves of flexible cross-linked samples are shown and compared with standard pPVC in Figure 3.

In table 2 are listed: the temperature at which the original weight of the sample decreased by 5 wt.% ($T_{5\%}$) and the temperature at which dehydrochlorination (DHCl) occurs (T_{max}).

The thermal degradation process proceeded in two basic stages. Within the range of 200–350°C the loss of DEHP and PVC dehydrochlorination (DHCl) occurred separately [21]. These two distinct phenomena were clearly observed in the case of standard pPVC for which the loss of DEHP and the DHCl took place at T = 280°C and T_{max} = 291°C, respectively. On the other hand, C1, C2 and C3 exhibited one broad degradation peak, centred at lower temperatures, indicating simultaneous occurrence of the two phenomena.

The cross-linking reaction promoted by IPDA involved the loss of two HCl molecules for each amine unit, which were responsible for the enhanced catalytic effect on PVC dehydrochlorination [22] resulting in its shifting to lower temperatures and polymer yellowing.

In the second degradative step above 400 °C, which corresponds to scission of PVC covalent bonds, all the samples exhibited the same behavior.

Thermal stability of CPE resins was also evaluated. The thermograms of CPE6135A and CPE252 in comparison with PVC resin are reported in Figure 4 along with DTG (inset) degradation curves. TG and DTG (inset) degradation curves of standard pPVC and its blends with CPE6135A and CPE252 at 15wt.% are shown in Figure 5.

Similarly to PVC, DHCl was also the dominant reaction in the first step of CPE thermal degradation [23].

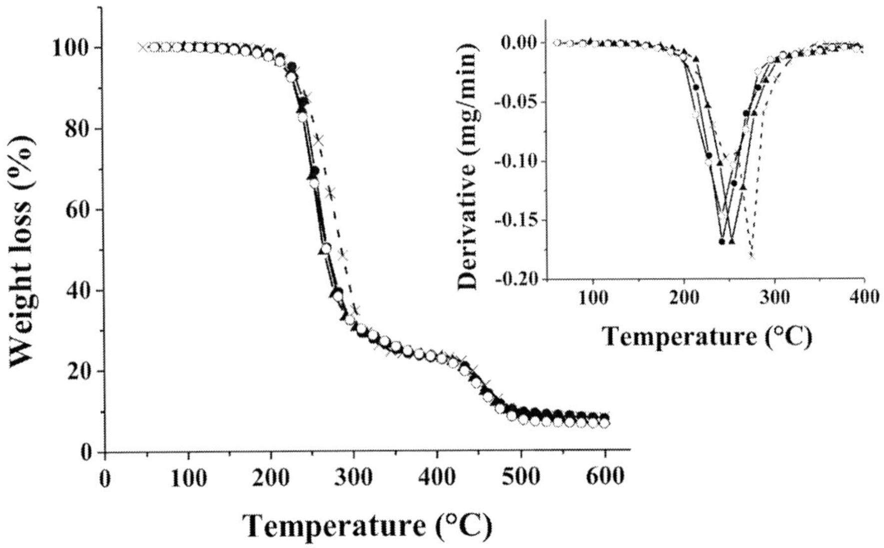

Figure 3. TG and DTG (inset) curves of standard pPVC (-x-),C1 (-●-), C2 (-▲-), C3 (-○-).

Table 2. $T_{5\%}$ and T_{max} of standard pPVC and cross-linked samples

Sample	$T_{5\%}$/°C	T_{max}/°C
pPVC	229	291
C1	224	260
C2	219	255
C3	217	270

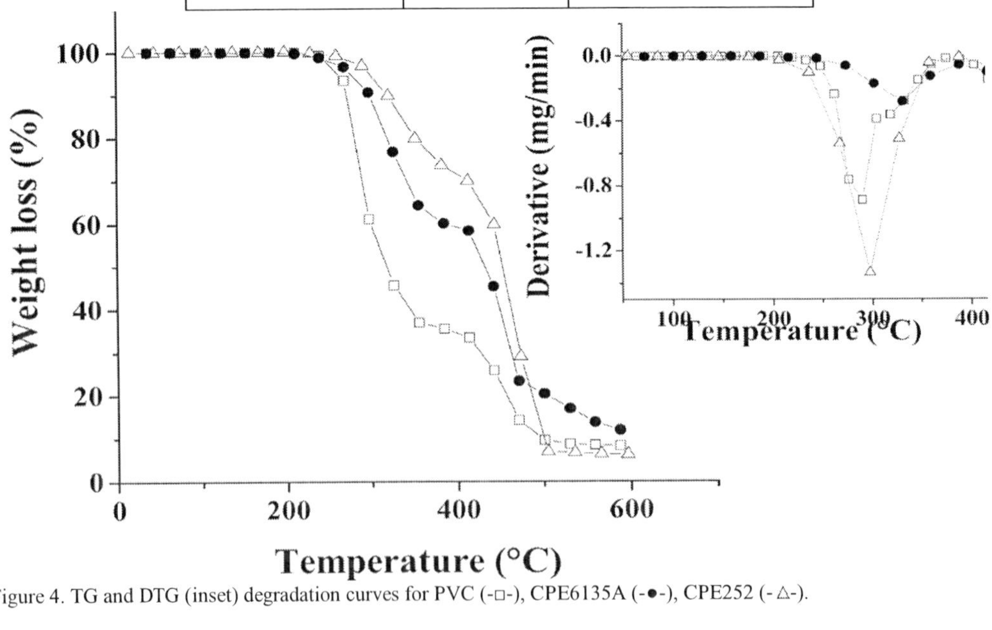

Figure 4. TG and DTG (inset) degradation curves for PVC (-□-), CPE6135A (-●-), CPE252 (-△-).

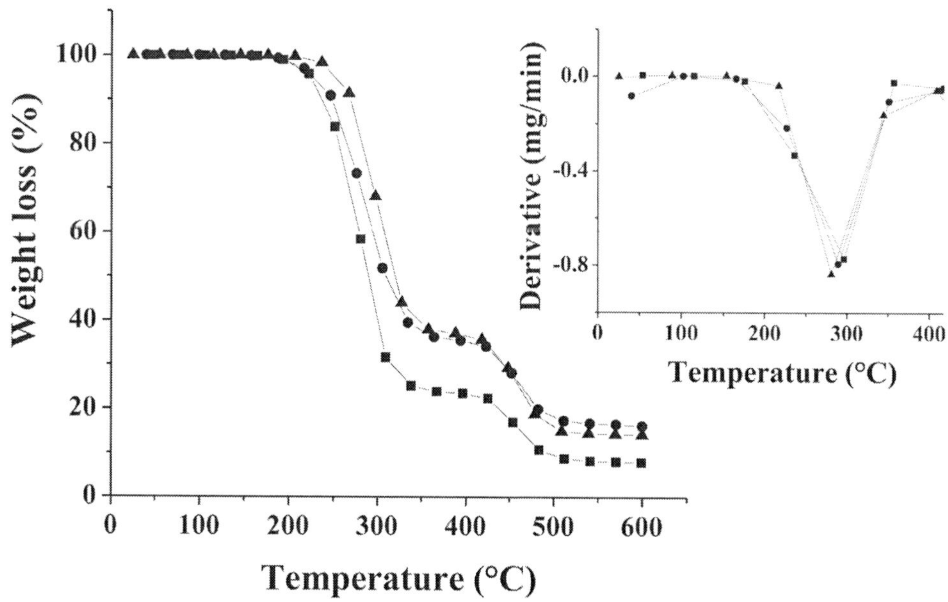

Figure 5. TG and DTG (inset) degradation curves of pPVC(-■-), A15 (-●-), B15 (-▲-).

In contrast to PVC, CPE had better thermal stability: the onset temperature ($T_{5\%}$) was increased and also T_{max} was located at slightly higher values (see table 3).

Moreover, the lower was the chlorine content in CPE, the higher its thermal stability due to its chemical structure more similar to polyethylene characterized by a higher amount of saturated backbone [18].

Table 3. $T_{5\%}$ and T_{max} of PVC, CPE6135, CPE252 resins and A7, A15, B7, B15

	Sample						
	pPVC	**CPE6135A**	**CPE252**	**A7**	**A15**	**B7**	**B15**
$T_{5\%}$/°C	265	286	300	227	229	258	259
T_{max}/°C	284	330	343	297	298	297	297

The degradation of CPE/pPVC blends also occurred through two steps: first DHCl, of both CPE and PVC, and then the degradation of polyene sequences took place.

When CPEs were added to pPVC, T_{max} values were shifted to slightly higher temperatures compared to standard pPVC. It is well established that blending may influence the thermal stability of individual polymers [24] and, since CPE was demonstrated to be thermally more stable than PVC resin, pPVC thermal stability was improved to some extent by the addition of CPEs, regardless of their concentration.

Moreover, compared to cross-linked samples, the DHCl was shifted to higher temperature (about 30 °C) in the case of CPEs/pPVC blends. Once more, the second degradation step occurring at temperature above 400 °C was not influenced by the addition of CPE.

Mechanical and Dynamic-Mechanical Properties

The dynamic-mechanical and tensile properties of cross-linked pPVC and CPEs/pPVC blends were measured and reported in table 4. It can be observed that the T_gs values of C1, C2 and C3 were only few degrees lower than that of standard pPVC, indicating the almost negligible effect of chemical cross-linking on the pPVC glass transition temperature.

As far as the tensile properties concerns, the Young's modulus values of C1, C2, and C3 were comparable to that of pPVC. At the same time, the elongation at break increased with IPDA content. This can be explained considering that the low concentrations of cross-linker used led to the formation of a lightly cross-linked network, in which the polymer chains were still able to slip without breaking the specimen [21]. Finally, ultimate tensile strength values seem to be not significantly influenced by the presence of cross-linked portions.

As concerning CPEs/pPVC blends, CPE resins were expected to have a stiffening effect on pPVC, due to their high molecular weight. The higher the CPE content, the higher were the T_g values of the blends.

Moreover, T_g values of samples containing the same amount of CPEs were higher when CPE6135A was used. This finding can be explained in terms of CPE chlorine content. Probably, with increasing the chlorine content, the polar interactions between the chlorine groups of PVC and CPE became more significant, reducing the chains mobility, thus increasing the T_g of the blends [25].

In terms of tensile properties, CPEs/pPVC blends underwent break at lower stresses and elongations than standard pPVC. Moreover, both stress and strain at break slightly decreased with increasing CPEs content. This finding was in agreement with DMA results confirming the CPE hardening effect on pPVC compounds.

Table 4. Mechanical and dynamic-mechanical results

Sample	Tg (°C)	E (MPa)	ε_b(%)	σ_b (MPa)
pPVC	11	6.7 ± 0.5	289.8 ± 11.1	9.40 ± 1.8
C1	5.5	6.0 ± 0.4	304.9 ± 14.0	10.4 ± 1.3
C2	3.5	6.0 ± 0.3	319.9 ± 11.5	11.4 ± 2.0
C3	3.1	4.9 ± 0.4	311.2 ± 11.2	9.40 ± 1.1
A7	21	7,2 ± 0.1	288.7 ± 19.60	7.30 ± 0.95
A15	27	7,8 ± 0.2	204.3 ± 8.61	5.66 ± 0.34
B7	17	6,6 ± 0.09	242.3 ± 13.84	7.20 ± 0.44
B15	19	7,2 ± 0.1	228.7 ±15.70	4.92 ± 0.36

In summary, the mechanical performances of chemically treated pPVC are encouraging since they were not appreciably affected by cross-linking. Differently, the addition of CPE caused a small reduction of elongation and stress at break values.

Tribological Analysis

Friction is a measure of the sliding resistance of a surface in contact with another. The friction expresses the "slip", which depends basically on the chemical nature of the material and the presence of additives which may promote it (sleep agents) or hinder it (anti-sleep agents). The materials friction behavior is influenced by the microstructure of the surface layer. The plasticizer motion towards the external layer should increase the adhesion between the pin and the specimen surface, leading to increased friction coefficient values. On the contrary, it should be noted decreased friction values.

Tribological tests were performed on both as-obtained uPVC and on uPVC after migration tests, with the aim of evaluating the effect that the chemical and physical modifications, operated on pPVC, had on DEHP migration.

In Figure 6 dynamic friction results for uPVC and standard pPVC as a function of the number of revolutions are shown.

The different surface layer features of as-obtained uPVC and standard pPVC were well evidenced by the comparison of the two friction profiles: uPVC was characterized by a friction averaged value of ≈ 0.15, which fell in the range between 0.1 and 0.5, typically exhibited by rigid polymers [26] whereas, pPVC was characterized by a higher value (≈ 0.5).

Tribological traces of uPVC held in contact for 32 days, at 60 °C, with C1, C2 and C3 (Figure 7) and with CPEs/pPVC (Figure 8a-8b) are reported. For comparison, as-obtained uPVC trace was also showed.

When pPVC-based samples were put in contact with uPVC in forced conditions, the plasticizer migrated toward the surface of rigid specimen, altering its tribological behavior. The cross-linking of pPVC was able to efficiently limit the DEHP migration. In fact, uPVC in contact with standard pPVC exhibited the highest friction values: a large amount of DEHP moved towards the interface increasing the adhesion between the relative sliding parts.

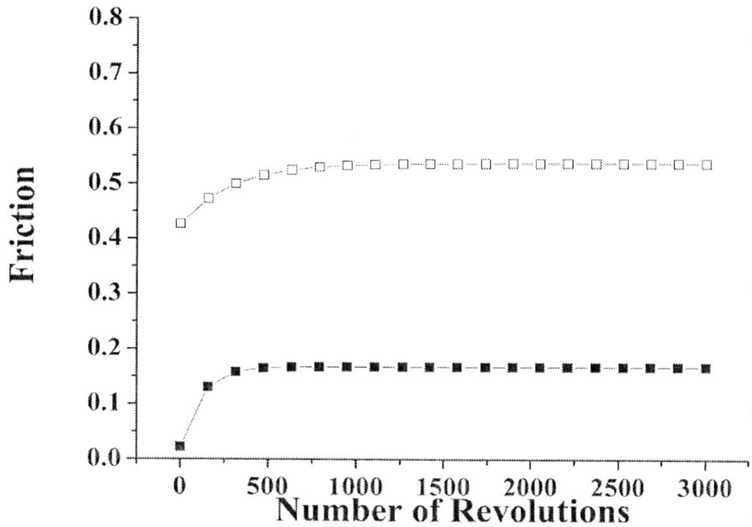

Figure 6. Friction of as-obtained uPVC (-■-) and standard pPVC (-□-).

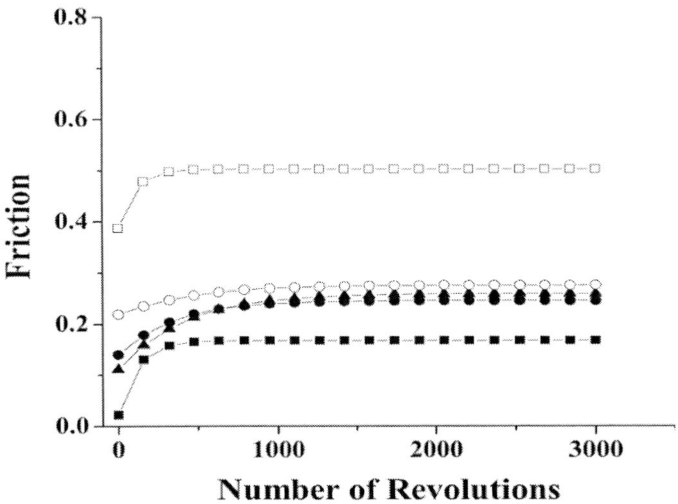

Figure 7. Friction of as-obtained uPVC (-■-) and uPVC after being in contact for 32 days with standard pPVC (-□-), C1 (-●-), C2 (-▲-) and C3 (-○-).

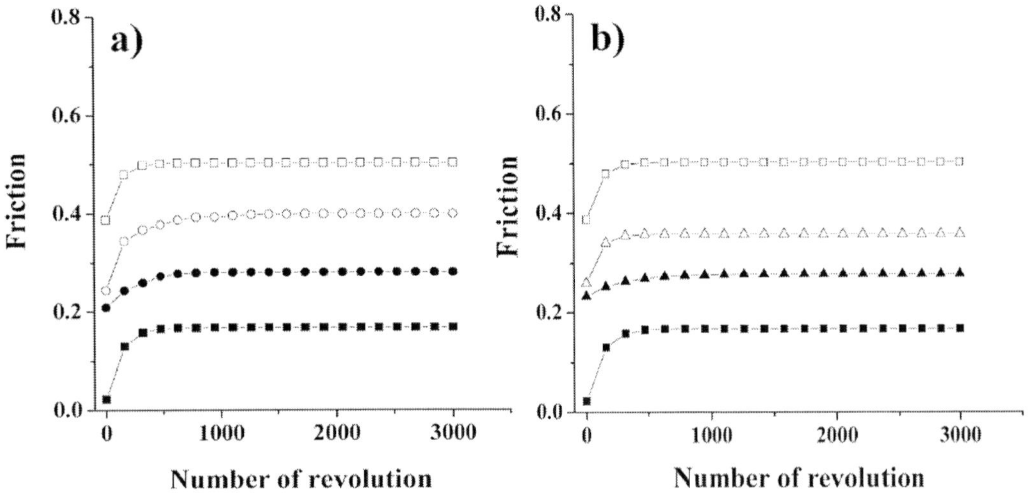

Figure 8. Friction of: a) as-obtained uPVC (-■-) and uPVC kept in contact with standard pPVC (-□-), A7 (-○-), A15.(-●-); b) as-obtained uPVC (-■-) and uPVC kept in contact with standard pPVC (-□-), B7 (- -), B15 (-▲-).

Rigid PVC in contact with cross-linked samples showed comparable friction behavior, relatively close to the as-obtained uPVC. This finding was in agreement with the migration test and validated again the effectiveness of cross-linking in partially preventing plasticizer migration.

The effectiveness of CPEs in hindering the plasticizer migration was also confirmed by the tribological tests. Independently of the chlorine content, the friction coefficient values measured on uPVC after contact with CPEs/pPVC blends, decreased with the addition of CPE, approaching the values of the as-obtained uPVC when 15 wt.% of CPE was contained in pPVC. This finding was in agreement with migration results, since the DEHP diffusivity was hindered especially when 15 wt.% of CPEs was used, as a further evidence of the effectiveness of CPE as a physical barrier against DEHP migration.

CONCLUSION

In this communication, the extent of plasticizer migration resulted hindered by both chemical cross-linking, independently of the amount of IPDA used, as well as by CPE blending, independently of chlorine content. The best results were obtained when 15 wt.% of CPE was mixed with pPVC. This trend was confirmed by friction tests.

Although IPDA was able to reduce plasticizer migration, without significantly affecting the mechanical properties of flexible PVC, it caused degradation of the polymer which in turn became yellow. The use of IPDA requires careful selection of processing times and temperatures in order to achieve polymer chains cross-linking and limiting detriment of PVC thermal stability. On the contrary, since CPE blending improves pPVC thermal properties, industrial scale-up will be facilitated due to mild processing conditions.

Hence, the chemical pPVC cross-linking and the CPE blending with pPVC can be considered possible available strategies in limiting the plasticizer leaching.

REFERENCES

[1] Latini, G.; De Felice, C.; Verrotti, A. *Reprod. Toxicol.* 2004, *19*, 27-33.

[2] Di Gennaro, P.; Collina, E.; Franzetti, A.; Lasagni, M.; Luridiana, A.; Pitea, D.; Bestetti, G. *Environ. Sci. Technol.* 2005, *39*, 325-330.

[3] Mackintosh, C. E.; Maldonado, J. A.; Ikonomou, M. G.; Gobas, F. C. *Environ. Sci. Technol.* 2006, *40*, 3481-3488.

[4] Persico, P.; Ambrogi, V.; Acierno, D.; Carfagna, C. *J. Vinyl Addit. Tech.* 2009, *15*, 139-146.

[5] Gumargalieva, K. Z.; Ivanov, V. B.; Zaikov, G.E.; Moiseev, J. V.; Pokholok, T. V. *Polym. Degrad. Stab.* 1996, *52*, 73-79.

[6] Marín, M. L.; López, J.; Sánchez, A.; Vilaplana, J.; Jiménez, A. *Bull. Environ. Contam. Toxicol.* 1998, *60*, 68-73.

[7] Navarro, R.; Perrino, M. P.; Tardajos, M. G.; Reinecke, H. *Macromolecules* 2010, *43*, 2377-2381.

[8] Choi, J.; Kwak, S. Y. *Environ. Sci. Tech.* 2007, *41*, 3763-3768.

[9] Lindstrom, A.; Hakkarainen, M. *J. Appl. Polym. Sci.* 2007, *104*, 2458-2467.

[10] Lindstrom, A.; Hakkarainen, M. *J. Appl. Polym. Sci.* 2006, *100*, 2180-2188.

[11] Romero Tendero, P. M.; Jimenez, A.; Greco, A.; Maffezzoli, A. *Eur. Polym. J.* 2006, *42*, 961-969.

[12] Lakshmi, S.; Jayakrishnan, S. *Polymer* 1998, *39*, 151–157.

[13] Ambrogi, V.; Brostow, W.; Carfagna, C.; Pannico, M.; Persico, P. *Polym. Eng. Sci.* 2012, *52*, 211–217.

[14] Gao, J.; Li, D.; Wang, D.; Yang, L. *Eur. Polym. J.* 2000, *36*, 2517-2522.

[15] Kabdi, S. A.; Belhaneche-Bensemra, N. *J. App. Polym. Sci.* 2008, *110*, 1750–1755.

[16] Eastwood, E. A.; Dadmu, M.D. *Polymer.* 2002, *43*, 6707–6717.
[17] Maksimov, R. D.; Zicans, J.; Ivanova, T.; Negreeva, S. N.; Plume, E. *Mech. Comp. Mat.* 2002, *38*, 141-144.
[18] Varma, A. J.; Deshpade, S. V.; Kondapalli, P. *Polym. Degrad. Stab.* 1999, *63*, 1–3.
[19] Siegmann, A.; Hiltner, A. *Polym. Eng. Sci.* 1984, *24*, 869-876.
[20] Doube,′ C. P.; Walsh, D. J. *Polymer* 1979, *20*,1115-1120.
[21] Hidalgo, M.; Beltràn, M. I.; Reinecke, H.; Mijangos, C. *J. Appl. Polym. Sci.* 1998, *70*, 865-872.
[22] Li, B. *Polym. Degrad. Stab.* 2000, *68*, 197-204.
[23] Stoeva, St.; Gjurova, K.; Zagorcheva, M. *Polym. Degrad. Stab.* 2000, *76*, 117-28.
[24] McNeil, I. C.; Gorman, J. G. *Polym. Degrad. Stab.* 1991, *33*, 263-267.
[25] Vlaev, L.; Stoeva, S.; Georgieva, V. *J. Therm. Anal. Calorim.* 2005, *81*, 329–332.
[26] Brostow, W.; Datashvili, T.; Huang, B. *Polym. Eng. Sci.* 2008, *48*, 292-296.

In: Phthalates
Editors: Gerardo L. Moretti and Drago Romano ISBN: 978-1-62081-994-4

Chapter 10

PHTHALATES: TOXICOGENOMICS AND HEALTH EFFECTS

Sher Singh*
National Taiwan Normal University, Taiwan
Steven Shoei-Lung Li†
Kaohsiung Medical University, Taiwan

ABSTRACT

Phthalates are a group of similar phthalic acid esters and are widely used as plasticizers to soften and increase the flexibility in polyvinyl chloride plastics. Since phthalates are not covalently bound to the plastic matrix, they can easily leach into the surrounding environment. There is sufficient evidence in rodents that phthalate exposure causes developmental and reproductive toxicities. We have recently analyzed the interactions between 16 phthalates and genes/proteins in the Comparative Toxicogenomic Database (CTD; http://ctd.mdibl.org), and a total of 445 interactions between the five most frequently curated phthalates (DEHP/MEHP and DBP/BBP/MBP) and 249 unique genes/proteins were found. The GeneOntology, pathways and networks of these 249 unique genes/proteins were fully analyzed using the computer software programs from MetaCore analytical Suite, Ingenuity Pathways Analysis and Pathway Studio. The pathways and networks of the top 34 genes/proteins were found to be very similar to those of the 249 unique genes/proteins. Thus, the top 34 genes/proteins may serve as molecular biomarkers to assay toxicities of phthalate exposure. The top three phthalate toxicity categories were found to be cardiotoxicity, hepatotoxicity and nephrotoxicity, and the top 20 diseases included cardiovascular, liver, urologic, endocrine and genital diseases. Recently, rat neonatal cardiomyocytes treated with clinically relevant DEHP concentrations were independently shown to exhibit global changes in expression of genes associated with cell electrical activity, calcium handling, adhesion and microtubular transport, and these changes may explain the arrythmogenic effects of phthalates on these cardiomyocytes.

---

* E-mail: sher@ntnu.edu.tw.
† E-mail: lissl@kmu.edu.tw.

ABBREVIATIONS

BBP, butylbenzyl phthalate;
CTD, Comparative Toxicogenomics Database;
DBP, Dibutyl phthalate;
DEHP, Diethylhexyl phthalate;
GO, GeneOntology;
IPA, Ingenuity Pathways Analysis;
MBP, monobutyl phthalate;
MEHP, Mono-(2-ethylhexyl) phthalate;
OMIN, Online Mendelian Inheritance in Man;
PAEs, diesters of phthalic acid; PVC, polyvinyl chloride.

INTRODUCTION

Phthalates (PAEs) are a group of similar diesters of phthalic acid, and are widely used as plasticizers to soften and increase the flexibility in polyvinyl chloride (PVC) plastics. Since phthalates are not covalently bound to the PVC plastics, they can easily leach into the surrounding environment [1]. Human exposure to phthalates mainly occurs through foods, because of the use of PVC in wrapping materials and food processing [2]. When ingested through food contamination, diethylhexyl phthalate (DEHP) is converted by intestinal lipases to mono-(2-ethylhexyl) phthalate (MEHP), which is then preferentially absorbed. Dibutyl phthalate (DBP) is used as a component of latex adhesives. It is also used in cosmetics and other personal care products, as a plasticizer in cellulose plastics, and as a solvent for dyes [3]. Monobutyl phthalate (MBP) is the embryotoxic metabolite of DBP and butylbenzyl phthalate (BBP). Phthalates are known to play critical roles in the etiology of many human diseases [4]. The impacts of phthalate exposure on human health have been extensively reviewed and reported by the National Toxicology Program- Center for the Evaluation of Risks to Human Reproduction [5]. There is sufficient evidence in rodents, but not yet in humans, that phthalate exposure causes developmental and reproductive toxicities.

The Comparative Toxicogenomics Database (CTD; http://ctd.mdibl.org) has been established to analyze the impact of environmental chemicals on human health [6]. Biocurators at the CTD manually curate toxicogenomic data, including over 116,000 interactions between 3,900 chemicals and 13,300 genes/proteins from 270 species, 5,900 gene/protein-disease direct relationships, and 2,500 chemical-disease direct relationships. In order to understand the potential adverse health effects of the most abundantly used DEHP/MEHP and DBP/BBP/MBP, the curated interactions between these five phthalates and genes/proteins were downloaded from the CTD, and GeneOntology (GO), pathways and networks of 249 phthalate-interacting genes/proteins were fully analyzed using the computer software programs from MetaCore analytical Suite, Ingenuity Pathways Analysis and Pathway Studio, and human diseases were inferred by the phthalate- gene/protein- disease relationships [7].

Phthalate-Gene/Protein Interactions

A total of 539 phthalate-gene/protein interactions (excluding no effects) were found for 16 phthalates from the CTD as July 2, 2009. Since DEHP is *in vivo* metabolized quantitatively to its monoester MEHP, the DEHP and MEHP were thus combined as a group. Likewise, DBP, BBP and MBP were pooled together, because MBP is embrotoxic metabolite of DBP and BBP. The five most frequently curated chemicals DEHP/MEHP and DBP/BBP/MBP exhibited 256 and 189 interactions, respectively, and their sum of 445 interactions accounted for 83% of all 539 interactions (Table 1). Only DEHP/MEHP and DBP/BBP/MBP groups were further analyzed in details. DEHP/MEHP and DBP/BBP/MBP interacted with 183 and 112 unique genes/proteins, respectively, and 46 of them were in common (Figure 1). Thus, there were 249 unique genes/proteins exhibiting a total of 445 interactions with DEHP/MEHP and DBP/BBP/MBP, and the top 34 unique genes/proteins with three and more interactions are summarized in Table 2. It is of interest that DEHP/MEHP was found to have at least five interactions with all three PPAR isotypes (PPARA, PPARG & PPARD), while DBP/BBP/MBP interacted only one each with these three genes/proteins. DBP/BBP/MBP was found to have eleven and nine interactions with ESR1 and AR, but DEHP/MEHP had only two and none, respectively. It may also be noted that ESR1 and VEGFA genes/proteins had nine and six interactions, respectively, with DEHP/MEHP and DBP/BBP/MBP in humans, but none in rodents.

Table 1. Curated interactions between 16 phthalates and genes/proteins

Phthalates	Phthalate-Gene Interactions
DEHP/MEHP[a]	256[c]
DBP/BBP/MBP[b]	189[c]
phthalic acid	17
di-n-pentyl phthalate	15
dicyclohexyl phthalate	12
diallyl phthalate	10
diethyl phthalate	10
diisodecyl phthalate	7
di-n-hexyl phthalate	6
diisobutyl phthalate	6
di-n-octyl phthalate	5
diisononyl phthalate	4
diheptyl phthalate	2
16 Phthalates Sum	**539[c]**
[a] DEHP, Diethylhexyl Phthalate	157
MEHP, mono-(2-ethylhexyl)phthalate	99
[b] DBP, Dibutyl Phthalate	150
BBP, butylbenzyl phthalate	33
MBP, monobutyl phthalate	6
[c] (256+189)/539 = 83%.	

This table is adapted with permission from Singh and Li (2011) [7].

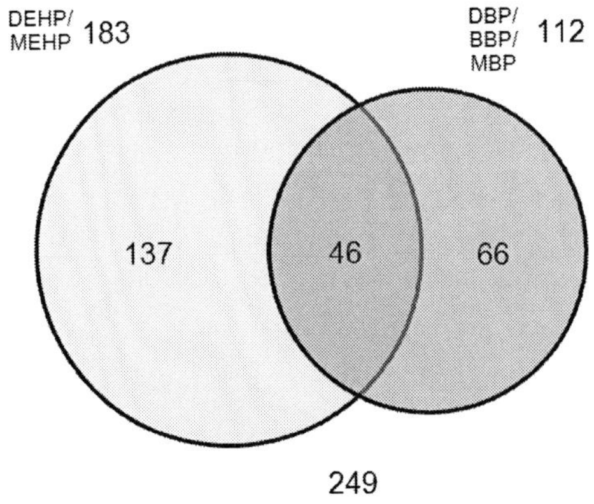

Figure 1.

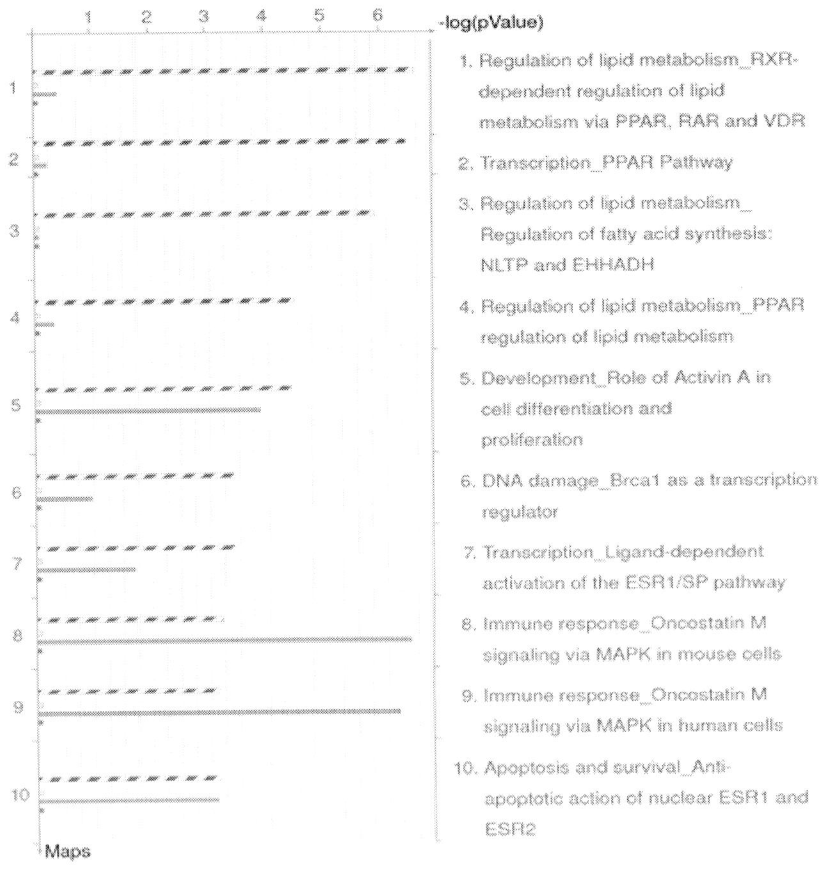

Figure 2. Comparison of GeneGo pathway maps and process networks of 249 and top 34 unique genes/proteins.

The top 10 GeneGo canonical pathway maps of top 34 genes/proteins and the rest 215 genes/proteins are indicated by blue/while strips and brown band, respectively. The degree of "relevance"to different GeneGo ontology categories is defined by p-value, so that the lower random p-value gets higher priority.

This figure is adapted with permission from Singh and Li (2011) [7].

Gene Ontology, Pathways and Networks

The GO molecular functions of 249 unique phthalate-interacting genes/proteins were analyzed by both MetaCore and Pathway Studio, and more than half of their top 20 functions were involved with different receptor activities. It should be noted that many of the GeneGo pathways and GeneGo process networks of the top 34 unique genes/proteins with three and more curated phthalate interactions were found to be very similar to those of 249 unique genes/proteins (Figure 2), suggesting that this subset of 34 unique genes/proteins may represent the overall biology of 249 genes/proteins responding to phthalate exposure. It is of interest that top four GeneGo pathway maps were involved in regulation of lipid metabolism, and PPAR pathway. It may be further noted that three PPARs complexing with retinoid X receptor (RXRA) played important roles in many normal and pathological biological processes such as atherosclerosis and hepatocarcinogenesis (Figure 3).

Table 2. 34 most frequently curated phthalate-interacting genes/proteins

Gene Symbols	DEHP/ MEHP	DBP/ BBP/ MBP	Sum	Human	Rat	Mouse	Others	Gene Names
PPARA	15	1	16	5	4	6	1	Peroxisome proliferator-activated receptor alpha
PPARG	9	1	10	3	5	1	1	Peroxisome proliferator-activated receptor gamma
PPARD	5	1	6	2	4	-	-	Peroxisome proliferator-activated receptor delta
NR1I2	5	4	9	4	2	-	3	Nuclear receptor subfamily 1, group I, member 2
STAR	4	8	12	-	6	2	4	Steroidogenic acute regulatory protein
HSD3B1	4	2	6	-	3	3	-	Hydroxy-delta-5-steroid dehydrogenase, 3 beta- and steroid delta-isomerase 1
INSL3	3	6	9	-	6	-	3	Insulin-like 3 (Leydig cell)
NR0B1	3	3	6	-	6	-	-	Nuclear receptor subfamily 0, group B, member 1
ESR1	2	11	13	9	-	-	4	Estrogen receptor 1
CYP11A1	2	7	9	1	4	-	4	Cytochrome P450, family 11, subfamily A, polypeptide 1
CYP17A1	2	5	7	-	4	-	3	Cytochrome P450, family 17, subfamily A, polypeptide 1
CYP1B1	2	2	4	4	-	-	-	Cytochrome P450, family 1, subfamily B, polypeptide 1
CYP4A1	2	2	4	-	3	1	-	Cytochrome P450, family 4, subfamily a, polypeptide 1
TSPO	2	2	4	-	4	-	-	Translocator protein (18kDa)
EGR1	2	1	3	-	3	-	-	Early growth response 1
MYC	2	1	3	1	1	1	-	V-myc myelocytomatosis viral oncogene homolog (avian)

Table 2. (Continued)

Gene Symbols	DEHP/ MEHP	DBP/ BBP/ MBP	Sum	Human	Rat	Mouse	Others	Gene Names
SCARB1	1	5	6	-	3	-	3	Scavenger receptor class B, member 1
VEGFA	1	5	6	6	-	-	-	Vascular endothelial growth factor A
ESR2	1	4	5	4	-	-	1	Estrogen receptor 2 (ER beta)
GRB14	1	3	4	-	4	-	-	Growth factor receptor-bound protein 14
LHCGR	1	3	4	-	3	-	1	Luteinizing hormone/choriogonadotropin receptor
AHR	1	2	3	-	-	3	-	Aryl hydrocarbon receptor
HNF1A	1	2	3	-	3	-	-	HNF1 homeobox A
INSIG1	1	2	3	-	3	-	-	Insulin induced gene 1
LDLR	1	2	3	-	3	-	-	Low density lipoprotein receptor
SVS5	1	2	3	-	3	-	-	Seminal vesicle secretory protein 5
FABP3	4	-		-	4	-	-	Fatty acid binding protein 3, muscle and heart (mammary-derived growth inhibitor)
SLC27A1	4	-	4	-	4	-	-	Solute carrier family 27 (fatty acid transporter), member 1
CELSR2	3	-	3	1	-	2	-	Cadherin, EGF LAG seven-pass G-type receptor 2 (flamingo homolog, Drosophila)
CYP19A1	3	-	3	-	3	-	-	Cytochrome P450, family 19, subfamily A, polypeptide 1
MT1	3	-	3	-	-	3	-	Metallothionein 1
MT2	3	-	3	-	-	3	-	Metallothionein 2
AR	-	9	9	3	3	1	2	Androgen receptor
SLC5A5	-	3	3	2	1	-	-	Solute carrier family 5 (sodium iodide symporter), member 5
34 SUM	**94**	**99**	**193**	**45**	**92**	**26**	**30**	
249 SUM	**256**	**189**	**445**	**55**	**301**	**40**	**49**	

This table is adapted with permission from Singh and Li (2011) [7].

Toxicity and Disease Associations

The toxicities of 249 unique genes/proteins interacting with DEHP/MEHP and DBP/BBP/MBP were analyzed using IPA Tox Functions, and their top three categories were cardiotoxicity, hepatotoxicity and nephrotoxicity (Table 3). The pathways and networks of top three toxicities of MEHP and DBP were further analyzed by Pathway Studio (Figure 4). The curated interactions of DEHP/MEHP and DBP/BBP/MBP with 249 unique genes/proteins in the CTD have inferred relationships with 1,437 GeneGo diseases identified using MetaCore by biomarkers, and the top 20 diseases are listed in Table 4. The top 20 diseases included cardiovascular diseases, liver diseases, urologic diseases, endocrine diseases, and genital diseases.

Table 3. Top three toxicity categories analyzed by IPA Tox Functions

Name	p-value
Cardiotoxicity:	
Heart Failure	8.16E-14
Tachycardia	1.32E-12
Cardiac Infarction	1.48E-08
Cardiac Congestive Cardiac Failure	4.20E-08
Cardiac Arrhythmia	5.64E-08
Hepatotoxicity:	
Liver Necrosis/Cell Death	3.54E-09
Liver Steatosis	1.28E-07
Liver Hepatomegaly	3.35E-07
Liver Hepatitis	4.51E-07
Liver Hyperplasia/Hyperproliferation	6.86E-07
Nephrotoxicity:	
Kidney Failure	3.31E-10
Renal Nephritis	6.14E-05
Renal Proliferation	1.15E-04
Renal Hydronephrosis	2.42E-04
Renal Necrosis/Cell Death	3.52E-04

This table is modified with permission from Singh and Li (2011) [7].

Table 4. Top 20 GeneGo diseases (by Biomarkers)

#	Disease Name	p-value
1	Disorders of Environmental Origin	2.755E-34
2	Cardiovascular Diseases	5.884E-32
3	Endocrine System Diseases	4.198E-30
4	Urologic Diseases	5.400E-28
5	Vascular Diseases	7.040E-28
6	Adenocarcinoma	8.261E-27
7	Endocrine Gland Neoplasms	2.822E-26
8	Liver Diseases	5.730E-26
9	Wounds and Injuries	1.019E-25
10	Body Weight	2.299E-25
11	Liver Neoplasms	1.708E-24
12	Urologic and Male Genital Diseases	8.423E-24
13	Autoimmune Diseases	2.106E-23
14	Pathological Conditions	3.817E-23
15	Mental Disorders	4.066E-23
16	Urogenital Diseases	4.328E-23
17	Prostatic Diseases	5.636E-23
18	Genital Diseases, Male	6.712E-23
19	Urogenital Neoplasms	6.728E-23
20	Endometrial Neoplasms	1.236E-22

This table is modified permission from Singh and Li (2011) [7].

Discussion

Phthalates (DEHP/MEHP and DBP/BBP/MBP) were found to interact with all three PPAR isotypes PPARA, PPARD and PPARG (Table 2). These three PPARs play broad roles in lipid and carbohydrate metabolism (Figure 3), and they are involved in metabolic disorders such as atherosclerosis, hyperlipidemia, and obesity [8]. Several urinary phthalate metabolites were also reported to be associated with key features of the metabolic syndrome (body mass index, abdominal adiposity and insulin resistance) [9, 10]. These disorders are consistent with the finding of the cardiotoxicity, including cardiovascular diseases, among top disorders caused by phthalate exposure (Table 4). The developmental effects of DBP/BBP/MBP depend on the embryonic age at the time of exposure and exposure duration. These effects are mainly on mesodermal structures, including cardiovascular and urogenital malformation. Recently, rat neonatal cardiomyocytes treated with clinically relevant DEHP concentrations were shown to exhibit global changes in expression of genes associated with cell electrical activity, calcium handling, adhesion and microtubular transport, and these changes may explain the arrythmogenic effects of phthalates on these cardiomyocytes [11]. The cardiomyocyte differentiation of murine embryonic stem cells was also reported to be inhibited by MBP [12].

Peroxisome proliferation is often seen in liver and kidney, and sustained peroxisome proliferation phenomenon can lead to carcinogenesis. The activation of PPARA, which was the most frequently curated gene/protein, was responsible for the DEHP/MEHP toxicity, especially, liver tumors in rodents [13]. This mechanism may be less relevant for humans because of lower expression of PPARA in the human liver compared to that seen in the rodent liver. However, PPARA-deficient mice did develop hepatocarcinogenesis under DEHP exposure [14], and this was due to activation of constitutive androstane receptor [15, 16]. DEHP/MEHP was also reported to have markered nephrotoxic effects on cultured kidney epithelial cells [17].

DEHP/MEHP causes animal toxicity in many organ systems [18, 19]. Studies at low doses of exposure, which appear to be of most concern for humans, have demonstrated subtle reproductive toxicity in male rodents [20]. Since fetal and neonatal stages seem to be the critical periods for exposure to DEHP/MEHP, a similar toxicity (for example, impaired fertility) in humans might be dormant and might not become manifest for many years, until adulthood is reached. In humans, population studies have established that dysmorphic disorders of the genital tract, observed in male infants, are significantly associated with prenatal exposure to phthalates. These effects likely result from the antiandrogenic and estrogeno-mimetic activities of DEHP/MEHP [21].

DBP/BBP/MBP, but not DEHP/MEHP, was found to have nine curated interactions with AR gene/protein (Table 2). However, these phthalates do not interact directly with the androgen receptor, but rather disrupt testosterone synthesis in the rat fetal testis [22]. DBP/BBP/MBP is known to affect male fertility and cause testicular atrophy in young adult rodents. DBP/BBP/MBP also has 7profound and permanent effects on the male reproductive development if exposure occurs during the critical periods of sexual differentiation (i.e. late in the gestation) [23]. The phenotypic alterations observed in male offspring rats exposed during the perinatal period have remarkable similarities with common human reproductive disorders, including cryptorchidism, hypospadias and low sperm counts [24].

The widespread exposure of the general human population has been demonstrated in several recent biomonitoring studies in the USA and Europe [2, 25, 26]. The potential public health risks associated with phthalate exposure include not only metabolic disorders, carcinogenesis, and reproductive toxicity but also endocrine disruption [27-29]. Estrogen receptor 1 (ESR1) was found to interact with both DEHP/MEHP and DBP/BBP/MBP in humans, but not in rodents (Table 2). The widely used phthalates are known to interfere with endocrine systems, including the estrogen and thyroid hormones [30-32]. Endocrine diseases and thyroid diseases were among the top 20 inferred diseases (Table 5). The vascular endothelial growth factor A (VEGFA) was also found to interact with DEHP/MEHP and DBP/BBP/MBP in humans, but not in rodents. DEHP has recently shown to exert species-specific metabolic consequence of PPARA activation [33].

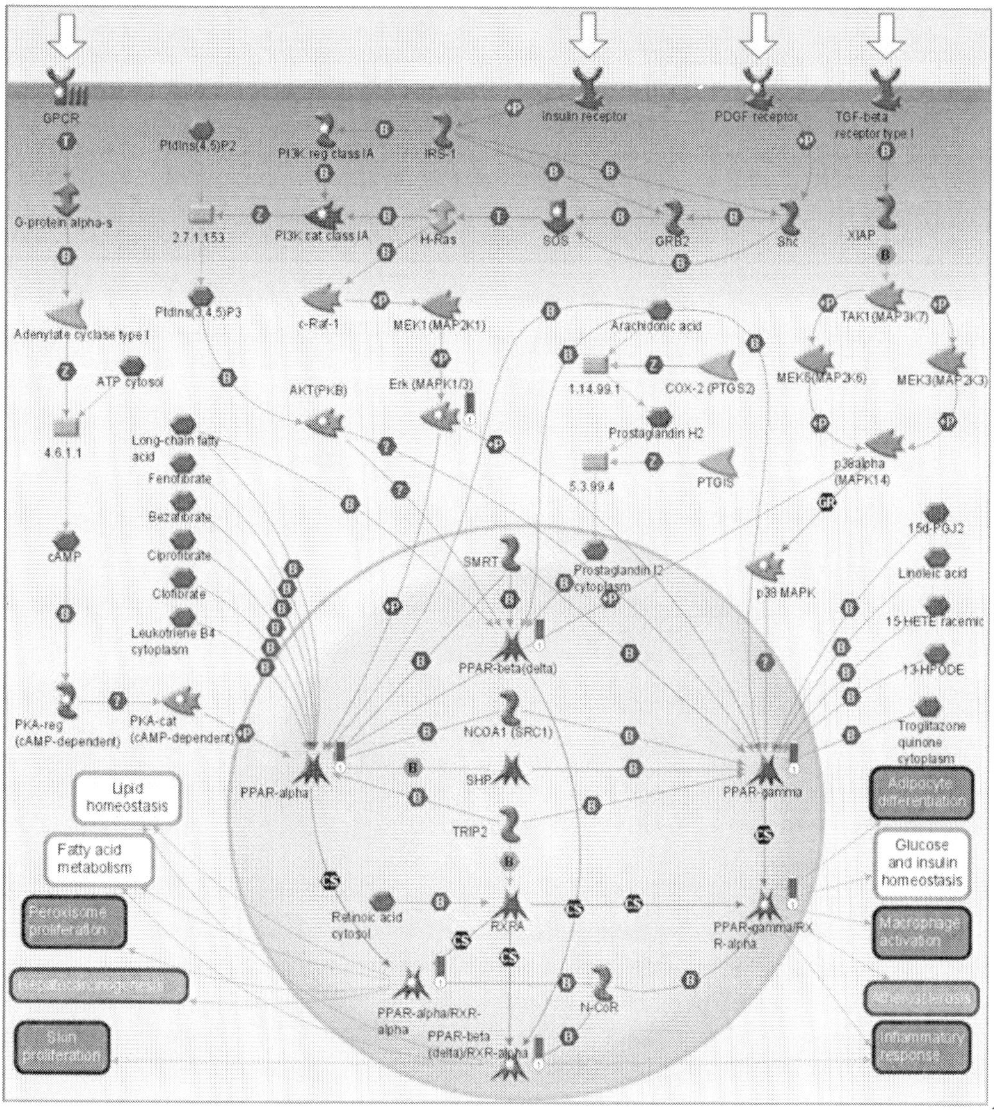

A

Figure 3. (Continued)

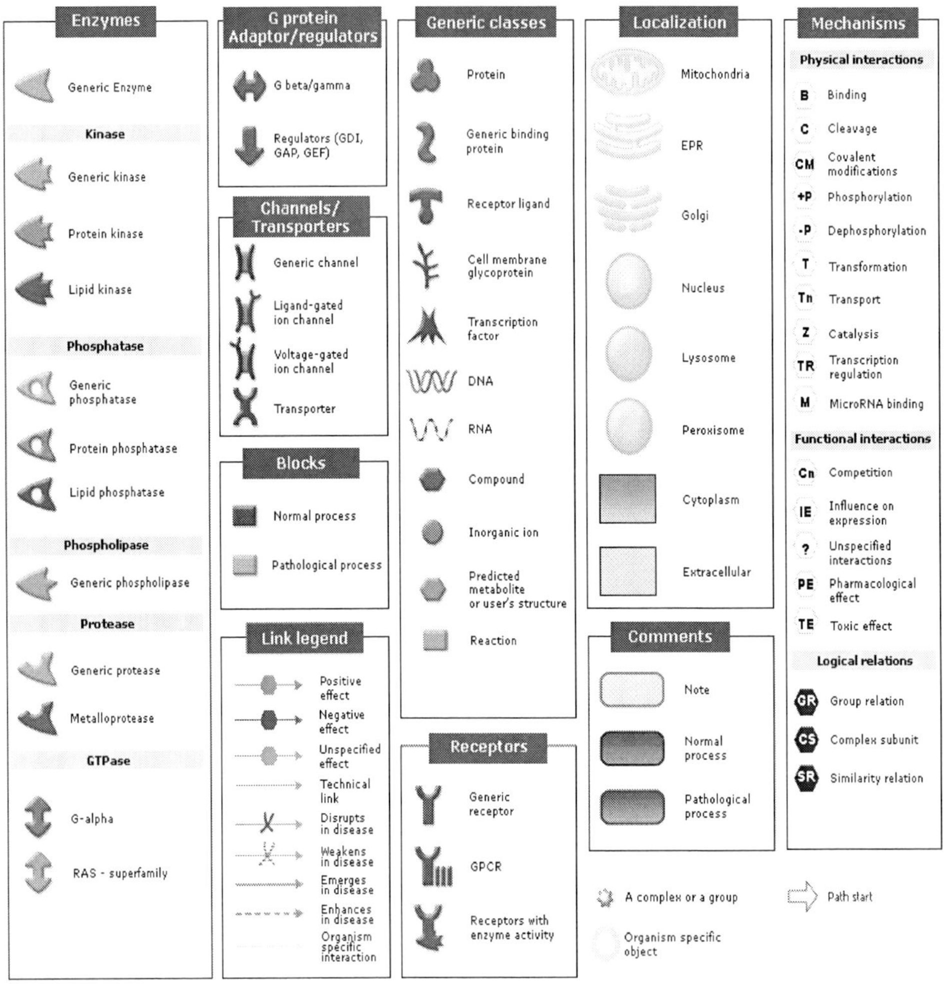

B

Figure 3. PPAR pathway – MetaCore.
The 249 unique genes/proteins were used as input for MetaCore pathway analysis.
A. Three PPAR isotypes PPARA (alpha), PPARD (delta/beta) and PPARG (gamma) complexed with retinoid X receptor RXRA (alpha) were involved in normal and pathological processes.
B. The explanation of the symbols used in GeneGo pathway.
This figure is adapted with permission from Singh and Li (2011) [7].

The pathways and networks of top 34 genes/proteins were found to be very similar to those of 249 unique genes/proteins interacting with phthalates DEHP/MEHP and DBP/BBP/MBP. Thus, these top 34 genes/proteins may serve as molecular biomarkers for assaying the toxicity of phthalates. Advances in the measurement of molecular biomarkers hold much promise in facilitating the study design and improving the epidemiological data. In view of limited human experimental evidence, human embryonic stem cells may be developed using systems biology with the top 34 genes/proteins as molecular biomarkers to assay the adverse health effects of phthalate toxicity [34, 35]. Human embryonic stem cells may also be differentiated into beating cardiomyoytes as new tool for testing cardiotoxicity [36]. Gene expression profiles of these differentiated cardiomyocytes have been found to be reminiscent of developing contractile cells in the normal human heart.

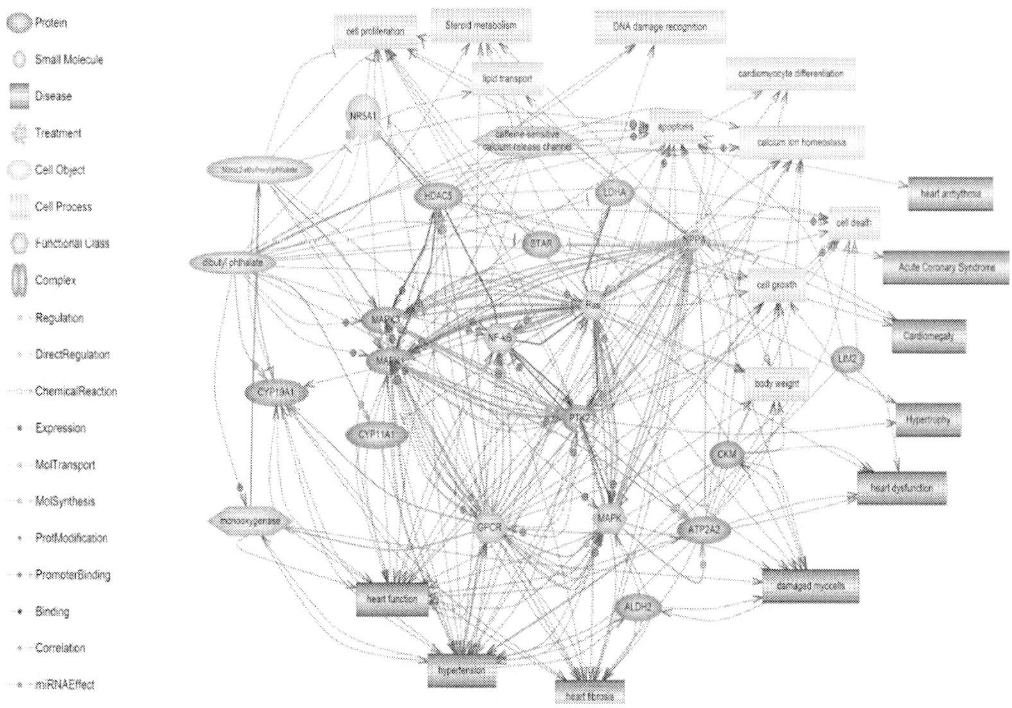

A

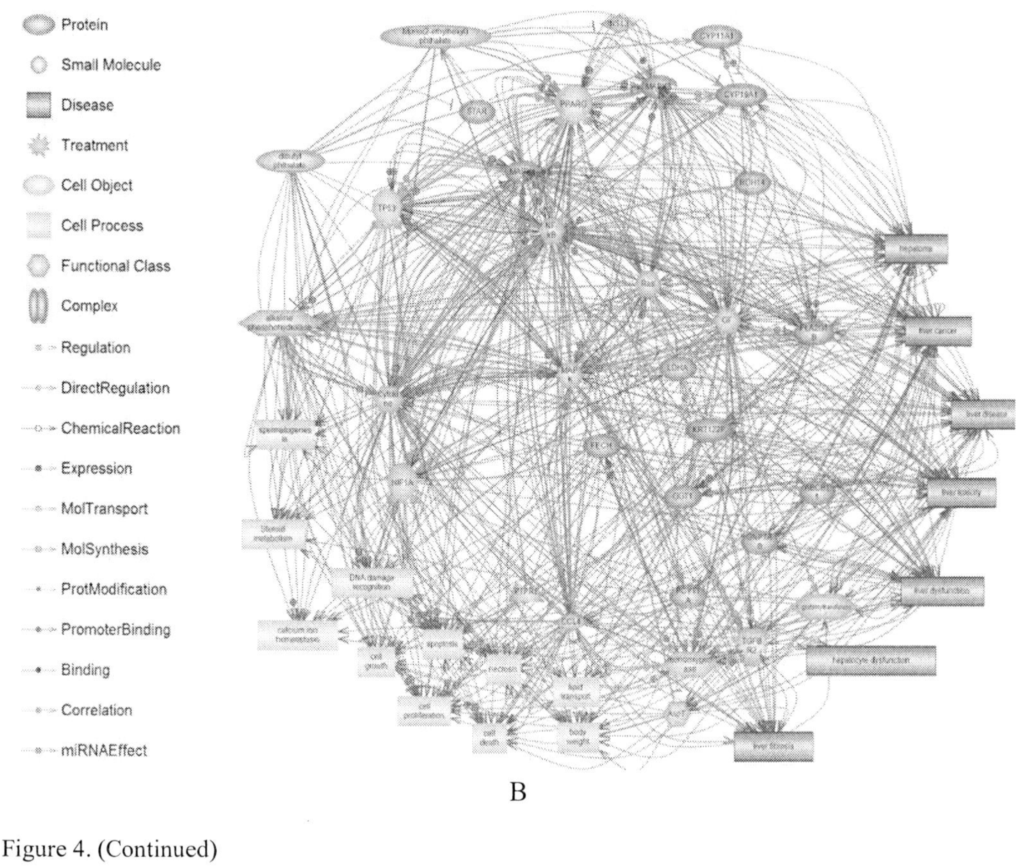

B

Figure 4. (Continued)

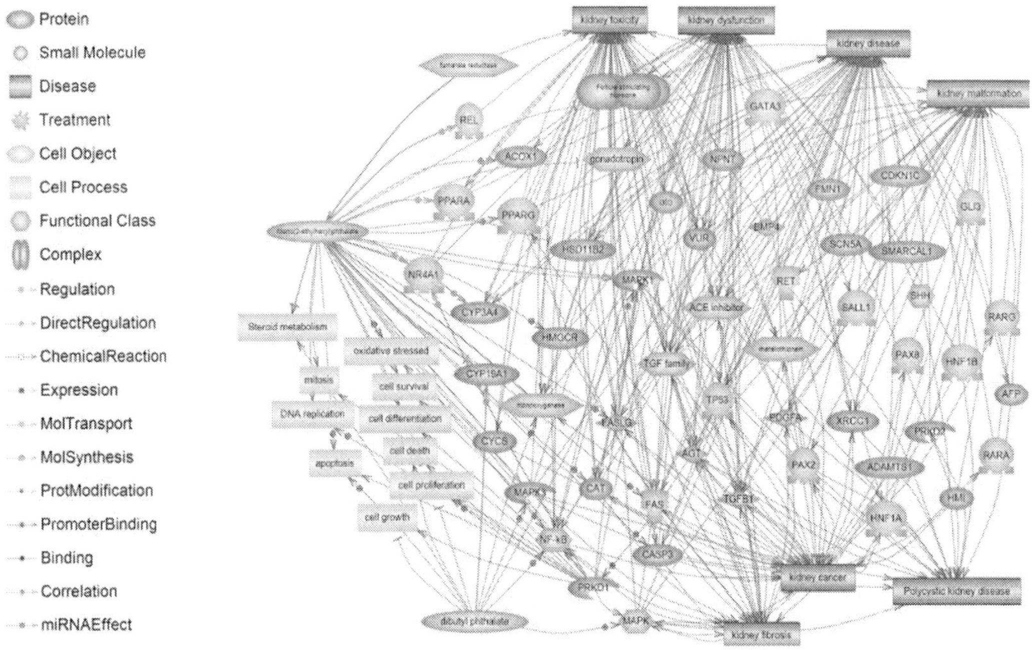

C

Figure 4. The pathways and networks of top three toxicities of MEHP and DBP analyzed by Pathway Studio.
The 249 unique genes/proteins were used as input for pathway and network analyses by Pathway Studio.
A, Cardiotoxicities;
B, Heptotoxicites;
C, Nephrotoxicities.
This figure is adapted with permission from Singh and Li (2011) [7].

Recently, DEHP was found to be illegally used as substitute for more expensive palm oil in food and drink in Taiwan for more than 20 years. The epidemiological studies on the 23 million Taiwanese, especially, the most sensitive populations such as children and women of childbearing age, are extremely valuable in assessing the adverse health effects of DEHP exposure, and an investigation should be carried out as soon as possible in the near future in order to provide the affected individuals some medical cares.

Conclusion

A total of 445 interactions curated in CTD were obtained between the five most frequently curated phthalates (DEHP/MEHP and DBP/BBP/MBP) and 249 unique genes/proteins. The pathways and networks of the top 34 genes/proteins were found to be very similar to those of the 249 unique genes/proteins. Thus, the top 34 genes/proteins may serve as molecular biomarkers of phthalate toxicities. The top three phthalate toxicity categories were found to be cardiotoxicity, hepatotoxicity and nephrotoxicity. Further studies are warranted to demonstrate the significance of these associations of phthalates with respect to the newly discovered cardio and rental toxicities. The 1,437 potential human diseases were inferred by the phthalate-gene/protein-disease relationships, and the top 20 diseases included cardiovascular diseases, liver diseases, urologic diseases, endocrine diseases, and genital diseases.

ACKNOWLEDGMENTS

We thank Dr. Sung-Liang Yu for helpful discussion and Miss Zong-Yun Tsai for technical assistance. This investigation was supported in part by Kaohsiung Medical University Research Foundation (KMUER004) to S.S.-L. Li.

REFERENCES

[1] Heudorf U, Mersch-Sundermann V, Angerer Jg. Phthalates: Toxicology and exposure. *International Journal of Hygiene and Environmental Health.* 2007;210(5):623-634.

[2] Wormuth M, Scheringer M, Vollenweider M, Hungerbühler K. What Are the Sources of Exposure to Eight Frequently Used Phthalic Acid Esters in Europeans? *Risk Analysis.* 2006;26(3):803-824.

[3] Thomas JA, Thomas MJ, Gangolli SD. Biological Effects of DI-(2-Ethylhexyl) Phthalate and Other Phthalic Acid Esters. *Critical Reviews in Toxicology.* 1984;13(4):283 - 317.

[4] Schwartz DA, Korach KS. Emerging Research on Endocrine Disruptors. *Environ Health Perspect.* 2007;115(1).

[5] Shelby MD. NTP-CERHR Monograph on the Potential Human Reproductive and Developmental Effects of Di(2-Ethylhexyl) Phthalate (DEHP). *NIH Publication.* 2006;No. 06-4476.

[6] Davis AP, Murphy CG, Saraceni-Richards CA, Rosenstein MC, Wiegers TC, Mattingly CJ. Comparative Toxicogenomics Database: a knowledgebase and discovery tool for chemical-gene-disease networks. *Nucl. Acids Res.* 2009;37:D786-792.

[7] Singh S, Li SS-L. Phthalates: Toxicogenomics and inferred human diseases. *Genomics.* 2011;97(3):148-157.

[8] Desvergne B, Feige JN, Casals-Casas C. PPAR-mediated activity of phthalates: A link to the obesity epidemic? *Molecular and Cellular Endocrinology.* 2009;304(1-2):43-48.

[9] Hatch E, Nelson J, Qureshi MM, *et al.* Association of urinary phthalate metabolite concentrations with body mass index and waist circumference: a cross-sectional study of NHANES data, 1999-2002. *Environmental Health.* 2008;7(1):27.

[10] Stahlhut RW, van Wijngaarden E, Dye TD, Cook S, Swan SH. Concentrations of urinary phthalate metabolites are associated with increased waist circumference and insulin resistance in adult U.S. males. *Environ Health Perspect.* 2007;115(6):876-882.

[11] Posnack NG, Lee NH, Brown R, Sarvazyan N. Gene expression profiling of DEHP-treated cardiomyocytes reveals potential causes of phthalate arrhythmogenicity. *Toxicology.* 2011;279(1-3):54-64.

[12] van Dartel DAM, Pennings JLA, Hendriksen PJM, van Schooten FJ, Piersma AH. Early gene expression changes during embryonic stem cell differentiation into cardiomyocytes and their modulation by monobutyl phthalate. *Reproductive Toxicology.* 2009;27(2):93-102.

[13] Doull J, Cattley R, Elcombe C, *et al.* A Cancer Risk Assessment of Di(2-ethylhexyl)phthalate: Application of the New U.S. EPA Risk Assessment Guidelines. *Regulatory Toxicology and Pharmacology.* 1999;29(3):327-357.

[14] Ito Y, Yamanoshita O, Asaeda N, *et al.* Di(2-ethylhexyl)phthalate Induces Hepatic Tumorigenesis through a Peroxisome Proliferator-activated Receptor α-independent Pathway. *Journal of Occupational Health.* 2007;49(3):172-182.

[15] Eveillard A, Mselli-Lakhal L, Mogha A, *et al.* Di-(2-ethylhexyl)-phthalate (DEHP) activates the constitutive androstane receptor (CAR): A novel signalling pathway sensitive to phthalates. *Biochemical Pharmacology.* 2009;77(11):1735-1746.

[16] Takashima K, Ito Y, Gonzalez FJ, Nakajima T. Different Mechanisms of DEHP-induced Hepatocellular Adenoma Tumorigenesis in Wild-type and Pparα-null Mice. *Journal of Occupational Health.* 2008;50(2):169-180.

[17] Rothenbacher K-P, Kimmel R, Hildenbrand S, Schmahl FW, Dartsch PC. Nephrotoxic effects of di-(2-ethylhexyl)phthalate (DEHP) hydrolysis products on cultured kidney epithelial cells. *Human and Experimental Toxicology.* 1998;17(6):336-342.

[18] Lahousse SA, Wallace DG, Liu D, Gaido KW, Johnson KJ. Testicular Gene Expression Profiling following Prepubertal Rat Mono-(2-ethylhexyl) Phthalate Exposure Suggests a Common Initial Genetic Response at Fetal and Prepubertal Ages. *Toxicol. Sci.* 2006;93(2):369-381.

[19] Saillenfait A-M, Sabaté J-P, Gallissot F. Effects of in utero exposure to di-n-hexyl phthalate on the reproductive development of the male rat. *Reproductive Toxicology.* 2009;28(4):468-476.

[20] Liu K, Lehmann KP, Sar M, Young SS, Gaido KW. Gene Expression Profiling Following In Utero Exposure to Phthalate Esters Reveals New Gene Targets in the Etiology of Testicular Dysgenesis. *Biology of Reproduction.* 2005;73(1):180-192.

[21] Swan SH, Main KM, Liu F, *et al.* Decrease in anogenital distance among male infants with prenatal phthalate exposure. *Environ Health Perspect.* 2005;113(8):1056-1061.

[22] Thompson CJ, Ross SM, Hensley J, *et al.* Differential Steroidogenic Gene Expression in the Fetal Adrenal Gland Versus the Testis and Rapid and Dynamic Response of the Fetal Testis to Di(n-butyl) Phthalate. *Biol Reprod.* 2005;73(5):908-917.

[23] McKee RH, Pavkov KL, Trimmer GW, Keller LH, Stump DG. An assessment of the potential developmental and reproductive toxicity of di-isoheptyl phthalate in rodents. *Reproductive Toxicology.* 2006;21(3):241-252.

[24] Martino-Andrade AJ, Chahoud I. Reproductive toxicity of phthalate esters. *Molecular Nutrition & Food Research.* 2010;54(1):148-157.

[25] Koch HM, Drexler H, Angerer J. An estimation of the daily intake of di(2-ethylhexyl)phthalate (DEHP) and other phthalates in the general population. *International Journal of Hygiene and Environmental Health.* 2003;206(2):77-83.

[26] Barr DB, Silva MJ, Kato K, *et al.* Assessing human exposure to phthalates using monoesters and their oxidized metabolites as biomarkers. *Environ Health Perspect.* 2003;111(9):1148-1151.

[27] Andrade AJM, Grande SW, Talsness CE, *et al.* A dose-response study following in utero and lactational exposure to di-(2-ethylhexyl) phthalate (DEHP): Effects on androgenic status, developmental landmarks and testicular histology in male offspring rats. *Toxicology.* 2006;225(1):64-74.

[28] Eveillard A, Lasserre F, de Tayrac M, *et al.* Identification of potential mechanisms of toxicity after di-(2-ethylhexyl)-phthalate (DEHP) adult exposure in the liver using a systems biology approach. *Toxicology and Applied Pharmacology.* 2009;236(3):282-292.

[29] Hauser R, Meeker JD, Duty S, Silva MJ, Calafat AM. Altered Semen Quality in Relation to Urinary Concentrations of Phthalate Monoester and Oxidative Metabolites. *Epidemiology.* 2006;17(6):682-691.

[30] Botelho G, Golin M, Bufalo A, Morais R, Dalsenter P, Martino-Andrade A. Reproductive Effects of Di(2-ethylhexyl)phthalate in Immature Male Rats and Its Relation to Cholesterol, Testosterone, and Thyroxin Levels. *Archives of Environmental Contamination and Toxicology.* 2009;57(4):777-784.

[31] Ghisari M, Bonefeld-Jorgensen EC. Effects of plasticizers and their mixtures on estrogen receptor and thyroid hormone functions. *Toxicology Letters.* 2009;189(1):67-77.

[32] Swan SH. Environmental phthalate exposure in relation to reproductive outcomes and other health endpoints in humans. *Environmental Research.* 2008;108(2):177-184.

[33] Feige JN, Gerber A, Casals-Casas C, *et al.* The pollutant diethylhexyl phthalate regulates hepatic energy metabolism via species-specific PPARalpha-dependent mechanisms. *Environmental Health Perspectives.* 2010;118(2):234-241.

[34] Li SS-L, Liu Y-H, Tseng C-N, Chung T-L, Lee T-Y, Singh S. Characterization and Gene Expression Profiling of Five New Human Embryonic Stem Cell Lines Derived in Taiwan. *Stem Cells and Development.* 2006;15(4):532-555.

[35] Meeker JD, Sathyanarayana S, Swan SH. Phthalates and other additives in plastics: human exposure and associated health outcomes. *Philosophical Transactions of the Royal Society B: Biological Sciences.* 2009;364(1526):2097-2113.

[36] Braam SR, Tertoolen L, van de Stolpe A, Meyer T, Passier R, Mummery CL. Prediction of drug-induced cardiotoxicity using human embryonic stem cell-derived cardiomyocytes. *Stem Cell Research.* 2010;4(2):107-116.

INDEX

B

C

E

F

G

J

K

L

M

N

O

P

Q

R

S

T

U

V

W

Y

Z